ENCYCLOPEDIA OF POLYMER SCIENCE AND TECHNOLOGY

Volume - I

ENCYCLOPEDIA OF POLYMER SCIENCE AND TECHNOLOGY
Volume - 1

by

Navid Naderpour
Bandar Imam Petrochemical Complex BIPC
Tarbiat Modares University of Tehran
Engineering Faculty-Polymer Department
Research & Development center of BIPC
Islamic Republic of Iran

Ebrahim Vasheghani Farahani
Tarbiat Modares University of Tehran
Engineering Faculty-Polymer Department
Islamic Republic of Iran

Adel Nejad Salim
National Petrochemical Company NPC
Islamic Republic of Iran

Reza Amiri
Bandar Imam Petrochemical Complex - BIPC
Islamic Republic of Iran

Simin Eydivand
Bandar Imam Petrochemical Complex BIPC
Research & Development center of BIPC
Islamic Republic of Iran

2009

SBS Publishers & Distributors Pvt. Ltd.
New Delhi

ISBN SET - 9788189741921
ISBN V-1 - 9788189741938

First Published in India in 2009

Published by:
SBS PUBLISHERS & DISTRIBUTORS PVT. LTD.
2/9, Ground Floor, Ansari Road, Darya Ganj,
New Delhi - 110002, INDIA
Tel: 23289119, 41563911
Email: mail@sbspublishers.com

Printed in India by Syndicate Binders, Noida.

Special Thanks to

Mr. KIYANOSH SHAMSEAMIRI - Manager, Basparan Complex
Mr. FATHE - Education Manager, NPC
Mr. FAKKARI - Head, Education Center, BIPC
And Specially Dr Nader Naderpour
for encouragement, help and advice.

We would also like to thank Dr HADADI ASL, Dr NASR JAFARI and Dr GHAFELEBASHI of Petrochemical Research and Technology Co. (NPC-RT).

We also thank
Dr M.NAVIDFAMILI,
Dr M.SEMSARZADEH,
& Dr M KOKABI
of Polymer Department of TARBIAT MODARES University

Mr. AKBARPOUR,
M.GHOLAMZADEH,
& D.MOHAMADI
of Basparan Engineering Complex

Mr. H.GODARZBOOD,
M.JOHARI,
M.HAMEDIAN,
A.MOAVI
& R.ADIBAN,
Researchers of R & D Center, BIPC
for providing us the visuals and figures for this Encyclopedia.

And finally we thank many people who have helped, often without knowing they were doing so, for exhortation and consolation throughout the years it has taken to create this work.

Preface

This publication entitled "Encyclopedia of Polymer Science and Technology" provides the readers with key information on the said subject in an affordable three volumes condensed format. This user-friendly reference publication offers quick access to all areas of polymer science and technology. It presents the state of the art in all areas of polymer science, technology and engineering, covering new imaging and analytical techniques and new methods of controlled polymer architecture. It also talks about related information about polymers, plastics, fibers, biomaterials and polymerization processes.

It follows a distinctly new approach while discussing the interdisciplinary nature of polymer science, technology and engineering. Modern polymer science, technology and engineering are firmly rooted both in the chemistry of macromolecules and their physical chemistry. Besides, this publication provides readers with information on all the important uses of synthetic polymers. This publication fulfills the much desired need for such compact volumes on the said subject. The elaborate bibliography/reference to all applied aspects makes this book an indispensable support for both students and professionals.

This "Encyclopedia of Polymer Science and Technology" gives an introduction to polymer science and technology in addition to a basic understanding of polymers and polymer nomenclature and polymer characterization. Natural polymers, synthetic polymers and their sources are discussed briefly. Other themes which are covered include the phenomena of observing and measuring polymer/colloids properties; polymer chemistry and chemical properties of polymers; chemistry of high polymers; polymer physics and physical properties of polymers; polymer structure, tacticity and other properties; and polymer measurement devices. This publication also addresses processes like methods of polymer manufacturing; methods of polymer synthesis and modification; design and synthesis of polymers with controlled structures; products and stuffs made from polymers; polymer processing, testing and analysis; polymer blends and composites; polymer fibre composite materials; design and manufacture of composites; biopolymers and biomaterials; smart polymers and polymeric materials; and polymers strength enhancement and degradation. Other aspects which are highlighted include computational polymer science and technology; theoretical polymer science and technology; applied polymer science and technology; polymer rheology; polymer crystallization; polymer material science and engineering; biomedical applications of polymers; macromolecules of interfaces and films; electronic and photonic molecular materials and devices; advanced polymer research; and polymer fibre applications.

The content of this publication on polymer science and technology is aimed for qualified and unqualified graduates who possess knowledge on polymer chemistry, polymer physics, polymer technology, polymer engineering and their applications. It also provides such students a skill on practical works related to polymer chemistry and processing. The objectives of this publication are to provide a broad range of knowledge in polymer science and technology as well as the skill/ability to apply this knowledge for industrial processes and modification of polymer products. This will also enhance opportunity of the graduate students for further education in polymer science and technology or related fields.

Contents

1

Introduction to Polymer Science and Technology

1.1 Polymer Science

Polymers are materials composed of long molecular chains that are well-accepted for a wide variety of applications. This unit explores these materials in terms of their chemical composition, associated properties and processes of manufacture from petrochemicals. The unit also shows a range of products in which polymers are used and explains why they are chosen in preference to many conventional materials.

Learning Outcomes

After studying this unit you should be able to:

- ❖ Isolate the key design features of a product which relate directly to the material(s) used in its construction;
- ❖ Indicate how the properties of polymeric materials can be exploited by a product designer;
- ❖ Describe the role of rubber-toughening in improving the mechanical properties of polymers;
- ❖ Identify the repeat units of particular polymers and specify the isomeric structures which can exist for those repeat units;
- ❖ Estimate the number—and weight-average molecular masses of polymer samples given the degree of polymerization and mass fraction of chains present;
- ❖ Calculate the molecular mass distribution for chain and step growth polymerizations from the concentrations of reactants and degree of conversion of monomer;

- Assess the effect of reactivity ratio on copolymer structures;
- Relate the crystal melting point T_m and the glass transition temperature T_g to molecular parameters, particularly chain length and composition;
- Interpret the mechanical properties of polymers using a viscoelastic master curve or dynamic mechanical information; and
- Make a preliminary selection of an appropriate polymer for a particular product specification.

Polymers Materials

The growth of polymer: Polymers, or materials composed of long molecular chains, are now well-accepted for a wide variety of applications, both structural and non-structural, and for mass-manufactured as well as one-off speciality.

Polymer Types

Traditionally, the industry has produced two main types of synthetic polymer—plastics and rubbers

Product Design and Manufacture

So what are the reasons for the continued growth in the use of polymers. It cannot be raw material cost, since the source of synthetic polymers is crude oil or natural gas, prices.

Molecular Engineering

It was the pioneering scientific work of Hermann Staudinger in the early part of the twentieth century which led to an understanding of the polymer state at an atomic and molecular level. Until then, plastics.

Chain Repeat Units

The repeat units of a range of polymers together with the monomer units from which they are derived. The simplest repeat unit is that for polyethylene, and consists of two carbon atoms.

Chain Configuration

The structure of repeat units is fixed by the chemical bonds between adjacent atoms. The shape or shapes thus created is known as the configuration, and for chains will be the chain configuration.

Chain Conformations

The repeat units or chains shown in the previous figures are all static representations of real chains and they are therefore of limited use. The key idea we need to explore real chains, and their influence.

Structure-Property Relationships

Given the large number of possible configurations in polymers, what guides to likely properties are available? We have already seen some of the effects on properties of changing tacticity.

Molecular Mass Distribution

In the peak (Mp) is at about 40 000. Number average () and weight average () molecular masses fall below and above the peak value. Determined using high temperature GPC.

Commercial Polymers

The increasing control of polymer structure by fine-tuned catalysis of polymerization opened up an enormous area for commercial exploitation, and new polymers are still being produced in this way.

Manufacture of Monomers

Primary sources of synthetic polymers: The most important primary sources of synthetic polymers are crude oil, natural gas and, to a minor extent, coal. Because all are primarily fuels rather than sources of materials, the manufacture of polymers.

Petrochemical Processing

Following distillation of petroleum into the major fractions (gasoline C5 up to 95 °C, naphtha 75-175 °C, kerosine 175-225 °C), the naphtha cut is subjected to cracking to yield smaller, double bonded.

Petrochemical Intermediates and Monomers

About 80 per cent of all petrochemicals end up in polymers, the most important building blocks being ethylene, propylene, butadiene and benzene. The first three can be polymerized directly but an important.

The Petrochemical Industry

The four-fold increase in the price of oil in 1973-4, together with associated political events, proved a powerful stimulus in the development and exploitation of North Sea crude oil. Increasing the price.

Polymerization

Understanding the polymerization process: Converting monomer to long chain polymer is the final step in the polymer manufacturing sequence. Polymerization is usually highly favourable in thermodynamic terms, mainly on energetic grounds because—

1. Chain and step growth,
2. Chain growth polymerization,
3. Step growth polymerization,
4. Copolymerization, and
5. Polymer grades.

Physical Properties of Polymers

The behaviour of polymers: The manufacture of polymer products is controlled by two often conflicting demands: the quality of the finished article in terms of its response to its environment and the ease or difficulty of processing.

1. Viscoelasticity of polymers,
2. Viscoelasticity and master curves,
3. Dynamic mechanical properties,
4. Orientation in polymers, and
5. Crystallisation of polymers.

Design in Polymers

Polymeric materials offer substantial benefits over conventional materials in terms of their low density, relative freedom from corrosion, transparency or translucency, and a range of physical properties.

1. Manufacturing and process methods,
2. Materials selection,
3. Case history: the Topper boat, and
4. Market experience.

1.2 Macromolecular Science

Macromolecular science or Polymer science is the subfield of materials science concerned with polymers, primarily synthetic polymers such as plastics. The field of polymer science includes researchers in multiple disciplines including chemistry, physics, and engineering.

This science comprises three main sub-disciplines:

- Polymer chemistry or macromolecular chemistry, concerned with the chemical synthesis and chemical properties of polymers.
- Polymer physics, concerned with the bulk properties of polymer materials and engineering applications.
- Polymer characterization is concerned with the analysis of chemical structure and morphology and the determination of physical properties in relation to compositional and structural parameters.

History of Polymer Science

Henri Braconnot's work in the 1830s is perhaps the first modern example of polymer science. Braconnot, along with Christian Schönbein and others, developed derivatives of the natural polymer cellulose producing new, semi-synthetic materials, such as celluloid and cellulose acetate. The term "polymer" was coined in 1833 by Jöns Jakob Berzelius, though Berzelius did little that would be considered polymer science in the modern sense. In the 1840s, Friedrich Ludersdorf and Nathaniel Hayward independently discovered that adding sulfur to raw natural rubber (polyisoprene) helped prevent the material from becoming sticky. In 1844 Charles Goodyear received a U.S. patent for vulcanizing rubber with sulfur and heat. Thomas Hancock had received a patent for the same process in the UK the year before. Vulcanized rubber represents the first commercially successful product of polymer research. In 1884 Hilaire de Chardonnet started the first artificial fiber plant based on regenerated cellulose, or viscose rayon, as a substitute for silk, but it was very flammable. In 1907 Leo Baekeland invented the first synthetic polymer, a thermosetting phenol-formaldehyde resin called Bakelite.

Despite significant advances in polymer synthesis, the molecular nature of the polymer was not understood until the work of Hermann Staudinger in 1922. Prior to Staudinger's work, polymers were understood in terms of the association theory or aggregate theory which originated with Thomas Graham in 1861. Graham proposed that cellulose and other polymers were "colloids", aggregates of molecules small molecular mass connected by an unknown intermolecular force. Hermann Staudinger was the first to propose that polymers consisted of long chains of atoms held together by covalent bonds. It took over a decade for Staudinger's work to gain wide acceptance in the scientific community, work for which he was awarded the Nobel Prize in 1953.

The World War II era marked the emergence of a strong commercial polymer industry. The limited or restricted supply of natural materials such as silk and latex necessitated the increased production of synthetic substitutes, such as rayon and neoprene. In the intervening years, the development of advanced polymers such as Kevlar and Teflon have continued to fuel a strong and growing polymer industry.

The growth in industrial applications was mirrored by the establishment of strong academic programs and research institute. In 1946, Herman Mark established the Polymer Research Institute at Brooklyn Polytechnic, the first research facility in the United States dedicated to polymer research. Mark is also recognized as a pioneer in establishing curriculum and pedagogy for the field of polymer science. In 1950, the POLY division of the American Chemical Society was formed, and has since grown to the second-largest division in this association with nearly 8,000 members.

Nobel Prizes Related to Polymer Science

2005 (Chemistry): Robert Grubbs, Richard Schrock, Yves Chauvin for olefin metathesis

2000 (Chemistry): Alan G. MacDiarmid, Alan J. Heeger, and Hideki Shirakawa for work on electroactive polymers contributing to the advent of molecular electronics

1991 (Physics): Pierre-Gilles de Gennes for developing a generalized theory of phase transitions with particular applications to describing ordering and phase transitions in polymers.

1974 (Chemistry): Paul J. Flory for contributions to theoretical polymer chemistry.

1963 (Chemistry): Giulio Natta and for contributions in polymer synthesis. (Ziegler-Natta catalysis).

1953 (Chemistry): Hermann Staudinger for contributions to the understanding of macromolecular chemistry.

1.3 Materials Science

Materials science or materials engineering is an interdisciplinary field involving the properties of matter and its applications to various areas of science and engineering. This science investigates the relationship between the structure of materials at atomic or molecular scale and their macroscopic properties. It includes elements of applied physics and chemistry, as well as chemical, mechanical, civil and electrical engineering. With significant media attention to nanoscience and nanotechnology in recent years, materials science has been propelled to the forefront at many universities. It is also an important part of forensic

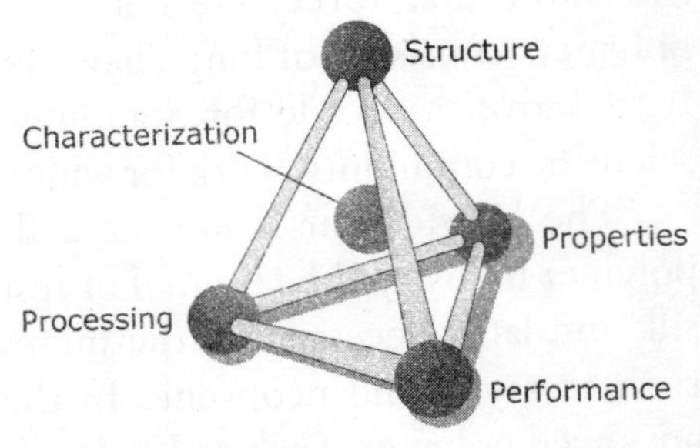

Fig. 1.1: The Materials Science Tetrahedron

engineering and forensic materials engineering, the study of failed products and components.

History

The material of choice of a given era is often its defining point; the Stone Age, Bronze Age, and Steel Age are examples of this. Materials science is one of the oldest forms of engineering and applied science, deriving from the manufacture of ceramics. Modern materials science evolved directly from metallurgy, which itself evolved from mining. A major breakthrough in the understanding of materials occurred in the late 19th century, when Willard Gibbs demonstrated that thermodynamic properties relating to atomic structure in various phases are related to the physical properties of a material. Important elements of modern materials science are a product of the space race: the understanding and engineering of the metallic alloys, and silica and carbon materials, used in the construction of space vehicles enabling the exploration of space. Materials science has driven, and been driven by, the development of revolutionary technologies such as plastics, semiconductors, and biomaterials.

Before the 1960s (and in some cases decades after), many *materials science* departments were named metallurgy departments, from a 19th and early 20th century emphasis on metals. The field has since broadened to include every class of materials, including: ceramics, polymers, semiconductors, magnetic materials, medical implant materials and biological materials.

Fundamentals

In materials science, rather than haphazardly looking for and discovering materials and exploiting their properties, one instead aims to understand materials fundamentally so that new materials with the desired properties can be created.

The basis of all materials science involves relating the desired properties and relative performance of a material in a certain application to the structure of the atoms and phases in that material through characterization. The major determinants of the structure of a material and thus of its properties are its constituent chemical elements and the way in which it has been processed into its final form. These, taken together and related through the laws of thermodynamics, govern a material's microstructure, and thus its properties.

An old adage in materials science says: "materials are like people; it is the defects that make them interesting". The manufacture of a perfect crystal of a material is currently physically impossible. Instead materials scientists manipulate the defects in crystalline materials such as precipitates, grain boundaries (Hall-Petch relationship), interstitial atoms, vacancies or substitutional atoms, to create materials with the desired properties.

Not all materials have a regular crystal structure. Polymers display varying degrees of crystallinity, and many are completely non-crystalline. Glasses, some ceramics, and many natural materials are amorphous, not possessing any long-range order in their atomic arrangements. The study of polymers combines elements of chemical and statistical thermodynamics to give thermodynamic, as well as mechanical, descriptions of physical properties.

In addition to industrial interest, materials science has gradually developed into a field which provides tests for condensed matter or solid state theories. New physics emerge because of the diverse new material properties which need to be explained.

Materials in Industry

Radical materials advances can drive the creation of new products or even new industries, but stable industries also employ materials scientists to make incremental improvements and troubleshoot issues with currently used materials. Industrial applications of materials science include materials design, cost-benefit tradeoffs in industrial production of materials, processing techniques (casting, rolling, welding, ion implantation, crystal growth, thin-film deposition, sintering, glassblowing, etc.), and analytical techniques (characterization techniques such as electron microscopy, x-ray diffraction, calorimetry, nuclear microscopy (HEFIB), Rutherford backscattering, neutron diffraction, etc.).

Besides material characterisation, the material scientist/engineer also deals with the extraction of materials and their conversion into useful forms. Thus ingot casting, foundry techniques, blast furnace extraction, and electrolytic extraction are all part of the required knowledge of a metallurgist/engineer. Often the presence, absence or variation of minute quantities of secondary elements and compounds in a bulk material will have a great impact on the final properties of the materials produced, for instance, steels are classified based on 1/10th and 1/100 weight percentages of the carbon and other alloying elements they contain. Thus, the extraction and purification techniques employed in the extraction of iron in the blast furnace will have an impact of the quality of steel that may be produced.

The overlap between physics and materials science has led to the offshoot field of *materials physics*, which is concerned with the physical properties of materials. The approach is generally more macroscopic and applied than in condensed matter physics. See important publications in materials physics for more details on this field of study.

The study of metal alloys is a significant part of materials science. Of all the metallic alloys in use today, the alloys of iron (steel, stainless steel, cast iron, tool steel, alloy steels) make up the largest proportion both by quantity and commercial value. Iron alloyed with various proportions of carbon gives low, mid and high carbon steels. For the steels, the hardness and tensile strength of the steel is directly related to the amount of carbon present, with increasing carbon levels also leading to lower ductility and toughness. The addition of silicon and graphitization will produce cast irons (although some cast irons are made precisely with no graphitization). The addition of chromium, nickel and molybdenum to carbon steels (more than 10%) gives us stainless steels.

Other significant metallic alloys are those of aluminium, titanium, copper and magnesium. Copper alloys have been known for a long time (since the Bronze Age), while the alloys of the other three metals have been relatively recently developed. Due to the chemical reactivity of these metals, the electrolytic extraction processes required were only developed relatively recently. The alloys of aluminium, titanium and magnesium are also known and valued for their high strength-to-weight ratios and, in the case of magnesium, their ability to provide electromagnetic shielding. These materials are ideal for situations where high strength-to-weight ratios are more important than bulk cost, such as in the aerospace industry and certain automotive engineering applications.

Other than metals, polymers and ceramics are also an important part of materials science. Polymers are the raw materials (the resins) used to make what we commonly call plastics. Plastics are really the final product, created after one or more polymers or additives have been added to a resin during processing, which is then shaped into a final form. Polymers which have been around, and which are in current widespread use, include polyethylene, polypropylene, PVC, polystyrene, nylons, polyesters, acrylics, polyurethanes, and polycarbonates. Plastics are generally classified as "commodity", "specialty" and "engineering" plastics.

PVC (polyvinyl-chloride) is widely used, inexpensive, and annual production quantities are large. It lends itself to an incredible array of applications, from artificial leather to electrical insulation and cabling, packaging and containers. Its fabrication and processing are simple and well-established. The versatility of PVC is due to the wide range of plasticisers and other additives that it accepts. The term "additives" in polymer science refers to the chemicals and compounds added to the polymer base to modify its material properties.

Polycarbonate would be normally considered an engineering plastic (other examples include PEEK, ABS). Engineering plastics are valued for their superior strengths and other special material properties. They are usually not used for disposable applications, unlike commodity plastics.

Specialty plastics are materials with unique characteristics, such as ultra-high strength, electrical conductivity, electro-fluorescence, high thermal stability, etc.

It should be noted here that the dividing line between the various types of plastics is not based on material but rather on their properties and applications. For instance, polyethylene (PE) is a cheap, low friction polymer commonly used to make disposable shopping bags and trash bags, and is considered a commodity plastic, whereas Medium-Density Polyethylene MDPE is used for underground gas and water pipes, and another variety called Ultra-high Molecular Weight Polyethylene UHMWPE is an engineering plastic which is used extensively as the glide rails for industrial equipment and the low-friction socket in implanted hip joints.

Another application of material science in industry is the making of composite materials. Composite materials are structured materials composed of two or more macroscopic phases. An example would be steel-reinforced concrete; another can be seen in the "plastic" casings of television sets, cell-phones and so on. These plastic casings are usually a composite material made up of a thermoplastic matrix such as acrylonitrile-butadiene-styrene (ABS) in which calcium carbonate chalk, talc, glass fibres or carbon fibres have been added for added strength, bulk, or electro-static dispersion. These additions may be referred to as reinforcing fibres, or dispersants, depending on their purpose.

Classes of Materials (by Bond Types)

Materials science encompasses various classes of materials, each of which may constitute a separate field. Materials are sometimes classified by the type of bonding present between the atoms:

1. Ionic crystals
2. Covalent crystals
3. Metals
4. Intermetallics
5. Semiconductors
6. Polymers
7. Composite materials
8. Vitreous materials

Sub-fields of Materials Science

- *Nanotechnology*—Rigorously, the study of materials where the effects of quantum confinement, the Gibbs-Thomson effect, or any other effect only present at the nanoscale is the defining property of the material; but more commonly, it is the creation and study of materials whose defining structural properties are anywhere from less than a nanometer to one hundred nanometers in scale, such as molecularly engineered materials.

- ***Microtechnology***—Study of materials and processes and their interaction, allowing microfabrication of structures of micrometric dimensions, such as MicroElectroMechanical Systems (MEMS).
- ***Crystallography***—The study of how atoms in a solid fill space, the defects associated with crystal structures such as grain boundaries and dislocations, and the characterization of these structures and their relation to physical properties.
- ***Materials Characterization***—Such as diffraction with x-rays, electrons, or neutrons, and various forms of spectroscopy and chemical analysis such as Raman spectroscopy, energy-dispersive spectroscopy (EDS), chromatography, thermal analysis, electron microscope analysis, etc., in order to understand and define the properties of materials.
- ***Metallurgy***—The study of metals and their alloys, including their extraction, microstructure and processing.
- ***Biomaterials***—Materials that are derived from and/or used with biological systems.
- ***Electronic and magnetic materials***—Materials such as semiconductors used to create integrated circuits, storage media, sensors, and other devices.
- **Tribology**—The study of the wear of materials due to friction and other factors.
- ***Surface science/Catalysis***—Interactions and structures between solid-gas solid-liquid or solid-solid interfaces.
- ***Ceramography***—The study of the microstructures of high-temperature materials and refractories, including structural ceramics such as RCC, polycrystalline silicon carbide and transformation toughened ceramics Some practitioners often consider rheology a sub-field of materials science, because it can cover any material that flows. However, modern rheology typically deals with non-Newtonian fluid dynamics, so it is often considered a sub-field of continuum mechanics. Glass Science—any non-crystalline material including inorganic glasses, vitreous metals and non-oxide glasses.
- ***Forensic engineering***—The study of how products fail, and the vital role of the materials of construction.
- ***Forensic materials engineering***—The study of material failure, and the light it sheds on how engineers specify materials in their product.

Topics that Form the Basis of Materials Science

- Thermodynamics, statistical mechanics, kinetics and physical chemistry, for phase stability, transformations (physical and chemical) and diagrams.

- Crystallography and chemical bonding, for understanding how atoms in a material are arranged.
- Mechanics, to understand the mechanical properties of materials and their structural applications.
- Solid-state physics and quantum mechanics, for the understanding of the electronic, thermal, magnetic, chemical, structural and optical properties of materials.
- Diffraction and wave mechanics, for the characterization of materials.
- Chemistry and polymer science, for the understanding of plastics, colloids, ceramics, liquid crystals, solid state chemistry, and polymers.
- Biology, for the integration of materials into biological systems.
- Continuum mechanics and statistics, for the study of fluid flows and ensemble systems.
- Mechanics of materials, for the study of the relation between the mechanical behavior of materials and their microstructures.

Important Journals

- Chemistry of Materials
- Nature Materials
- Acta Materialia
- JOM
- Advanced Materials
- Computational materials science
- Advanced Functional Materials
- Journal of Materials Chemistry
- Journal of Materials Online—Open Access
- Metallurgical and Materials Transactions
- Journal of Materials Research
- Journal of Materials Science
- Federation of European Materials Science Societies Newsletter
- AMMTIAC eNews/Quarterly Advanced materials, manufacturing, and testing. (Free subscription)

Polymers and small organic molecules each play a key role in the design and development of novel, high performance materials. Polymers are composed of molecules with large molecular mass and have repeating structural units. They find wide use in almost all consumer products including plastic parts, fishing line, gaskets, tires, automobile bumpers, adhesives, film supports, abrasion-resistant coatings, lenses and optical coatings, and photoresists for semiconductor manufacture. Digital printing technologies use a variety of polymers as critical components, including:

- Inkjet (inks and media),
- Electrophotography (toners), and
- Thermal printing.

New polymer technologies are used across an increasingly broad spectrum of products.

Scientific work in this area includes polymer synthesis to produce materials with novel properties, as well as polymer physics and rheology to better understand the materials' properties and interactions.

For example, polymer physics helps determine the behavior of ink leaving an inkjet printhead. Changing the morphology of certain polymers, can produce new nanostructured materials (after microphase separation). Our researchers also work to better understand the role played by polymers at interfaces, such as stabilizing nanoparticle dispersions in liquids, or improving adhesion between two layers.

Organic molecules including dyes, pigments and dielectric materials can be specifically tailored to provide the physical and chemical properties required for a variety of imaging applications. A strong understanding of structure/ activity relationships guides the design and synthesis of small organic molecules, enabling the invention of media possessing improved color gamut, robustness and image quality.

Our research teams are working to develop materials that can be used for microelectronics, optical systems, adhesives, dispersants or interfacial control, binders, controlled surface functionalization, sensors, and polymer-based nanostructures. Synthetic approaches to produce novel architectures coupled with a thorough understanding of the materials properties will be required to control and optimize interactions in complex systems.

1.4 Careers in Polymer Science and Engineering

What is Polymer Science?

Polymer Science is a typical multi-disciplinary subject, containing elements of organic chemistry (polymer synthesis), analytical chemistry (structural characterisation), physical chemistry (conformation, interactions in solution), physics (glass transition, crystallisation) and mechanics (rheology, time-dependent mechanical properties). In certain areas of polymer science topics such as surface chemistry, biochemistry and medicine are highly significant. The engineering aspects of polymer science are important from the application point of view, for example, materials selection, processing of polymers, strength and fatigue properties, and degradation. Finally, the impact of polymers on our environment is an increasingly important subject. In this area biodegradation of polymers and the development of new materials based on biopolymers are topics of current interest.

Who Needs Polymer Science?

Approx. 80 per cent of all chemical engineers professionally come into contact with polymers and polymer science in their daily activities. Examples are materials for chemical equipment, binders in paints and adhesives, functional polymeric chemicals in the process industry, rheology additives in the food industry, packaging materials, components in the automotive industry, controlled release pharmaceuticals, medical equipment, and polymer production and polymer processing. Knowledge of polymer science is consequently very useful within many different areas of trade and industry and not only in the polymer producing and processing industry.

How to Study Polymer Science and Engineering?

In the Chemical Engineering curriculum at Lund Institute of Technology a number of courses are offered by the Polymer Science & Engineering Department. In the third year an introductory course on Polymer Technology is given for all students in the program. In the fourth year a number of courses are offered, which are focused on fundamental organic and physical polymer chemistry, polymer physics, and processing and applications of polymers. These courses contain both lectures and laboratory assignments. Contacts with industry play an important role, with lectures, excursions, and projects. In the final year, the MSc project can be carried out either at the department with direct supervision from one of the researchers, or in collaboration with industry. The PSE department has a wide network of industrial contacts in Sweden. It is also possible to run the MSc project abroad.

What about Jobs?

During the last decade the number of jobs available for PSE graduates has been larger than the number of students. Positions have been offered in R&D, production, sales support, teaching, etc. Those students who are interested in academic research have good opportunities at the PSE department for continued studies to a Licentiate or a PhD degree in Polymer Science and Engineering. Many of the research projects are run in collaboration with industry, which gives the students good opportunities for R&D jobs in industry after graduation.

1.5 Polymer Definitions

Polymers, macromolecules, high polymers, and giant molecules are high-molecular-weight materials composed of repeating subunits. These materials may be organic, inorganic, or organometallic, and synthetic or natural in origin. Polymers are essential materials for almost every industry as adhesives, building materials, paper, cloths, fibers, coatings, plastics, ceramics, concretes, liquid crystals, photoresists, and coatings. They are also major components in soils and plant and animal life. They are important in nutrition, engineering, biology, medicine, computers, space exploration, health, and the environment.

Natural inorganic polymers include diamonds, graphite, sand, asbestos, agates, chert, feldspars, mica, quartz, and talc. Natural organic polymers include polysaccharides (or polycarbohydrates) such as starch and cellulose, nucleic acids, and proteins. Synthetic inorganic polymers include boron nitride, concrete, many high-temperature superconductors, and a number of glasses. Siloxanes or polysiloxanes represent synthetic organometallic polymers.

Synthetic polymers used for structural components weigh considerably less than metals, helping to reduce the consumption of fuel in vehicles and aircraft. They even outperform most metals when measured on a strength-per-weight basis. Polymers have been developed which can also be used for engineering purposes such as gears, bearings, and structural members.

Nomenclature

Many polymers have both a common name and a structure-based name specified by the International Union of Pure and Applied Chemistry (IUPAC). Some polymers are commonly known by their acronyms. Some companies use trade names to identify the specific polymeric products they manufacture. For example, Fortrel® polyester is a poly(ethylene terephthalate) (PET) fiber. Polymers are often generically named, such as rayon, polyester, and nylon.

Composition

Polymer structures can be represented by similar or identical repeat units. These are derived from smaller molecules, called monomers, which react to form the polymer. Propylene monomer and the repeat unit it forms in polypropylene are shown on next page.

With the exception of its end groups, polypropylene is composed entirely of this repeat unit. The number of units (n) in a polymer chain is called the degree of polymerization (DP). Other polymers, such as proteins, can be described in terms of the approximate repeat unit where the nature of R (a substituted atom or group of atoms) varies.

$$CH_2{-}CH(CH_3) \longrightarrow \left[CH_2{-}CH(CH_3) \right]_n \qquad \left(NH{-}C({=}O){-}CH(R) \right)_n$$

Monomer **Polymer repeat unit** **Protein repeat unit**

Primary Structure

The sequence of repeat units within a polymer is called its primary structure. Unsymmetrical reactants, such as substituted vinyl monomers, react almost exclusively to give a "head-to-tail" product, in which the R substituents occur on alternate carbon atoms. A variety of head-to-head structures are also possible.

Each R-substituted carbon atom is a chiral center (an atom in a molecule attached to four different groups) with different geometries possible. Arrangements where the substitutes on the chiral carbon are random are referred to as atactic structures. Arrangements where the geometry about the chiral carbon alternates are said to be syndiotactic. Structures where the geometry about the chiral atom has the same geometry are said to be isotactic or stereoregular.

Stereoregular polymers are produced using special stereoregulating catalyst systems. A series of soluble catalysts have been developed that yield products with high stereoregularity and low chain-size disparity. As expected, polymers with regular structures—that is, isotactic and syndiotactic structures—tend to be more crystalline and stronger.

Polymers can be linear or branched with varying amounts and lengths of branching. Most polymers contain some branching.

Copolymers are derived from two different monomers, which may be represented as A and B. There exists a large variety of possible structures and, with each structure, specific properties. These varieties include alternating, random, block, and graft see (illustration).

$$CH_2{-}CH(CH_3) \longrightarrow \left[CH_2{-}CH(CH_3) \right]_n \qquad \left(NH{-}C({=}O){-}CH(R) \right)_n$$

Monomer **Polymer repeat unit** **Protein repeat unit**

Copolymer Structures

(a) alternating, (b) random, (c) block, (d) graft.

Secondary Structure

This refers to the localized shape of the polymer, which is often the consequence of hydrogen bonding. Most flexible to semiflexible linear polymer chains tend toward two structures—helical and pleated sheet/skirtlike. The pleated skirt arrangement is most prevalent for polar materials where hydrogen bonding can occur. In nature, protein tissue is often of a pleated skirt arrangement. For both polar and nonpolar polymer chains, there is a tendency toward helical formation with the inner core having "like" secondary bonding forces.

Tertiary Structure

This refers to the overall shape of a polymer, such as in polypeptide folding. Globular proteins approximate rough spheres because of a complex combination of environmental and molecular constraints, and bonding opportunities. Many natural and synthetic polymers have "superstructures," such as the globular proteins and aggregates of polymer chains, forming bundles and groupings.

Quaternary Structure

This refers to the arrangement in space of two or more polymer subunits, often a grouping of tertiary structures. For example, hemoglobin (quaternary structure) is essentially the combination of four myoglobin (tertiary structure) units. Many crystalline synthetic polymers form spherulites.

Synthesis

For polymerization to occur, monomers must have at least two reaction points or functional groups. There are two main reaction routes to synthetic polymer formation—addition and condensation. In chain-type kinetics, initiation starts a series of monomer additions that result in the reaction mixture consisting mostly of unreacted monomer and polymer. Vinyl polymers, derived from vinyl monomers and containing only carbon in their backbone, are formed in this way. Examples of vinyl polymers include polystyrene, polyethylene, polybutadiene, polypropylene (see structure), and poly(vinyl chloride).

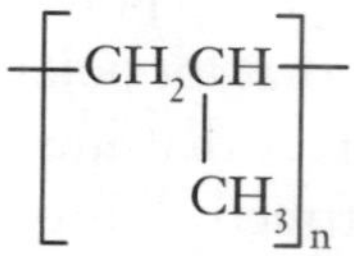

$$\left[-CH_2\underset{\displaystyle CH_3}{\underset{|}{C}H}- \right]_n$$

The second main route is a step-wise polymerization. Polymerization occurs in a step-wise fashion so that the average chain size within the reaction mixture may have an overall degree of polymerization of 2, then 5, then 10, and so on, until the entire mixture contains largely polymer with little or no monomer left. Polymers typically produced using the step-wise process are called condensation polymers, and include polyamides, polycarbonates, polyesters, and polyurethanes (*see* structures).

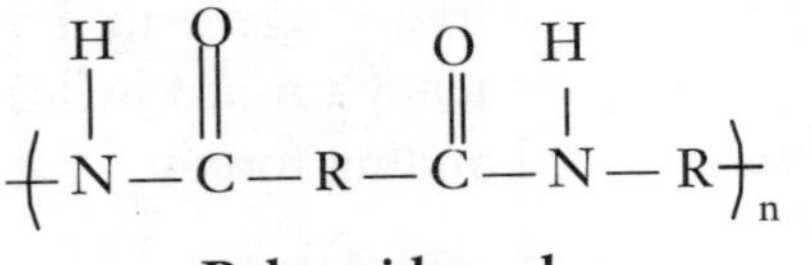

$$\left[-\underset{\text{H}}{\text{N}}-\overset{\text{O}}{\overset{\|}{\text{C}}}-\text{R}-\overset{\text{O}}{\overset{\|}{\text{C}}}-\overset{\text{H}}{\overset{|}{\text{N}}}-\text{R}-\right]_n$$

Polyamide, nylon

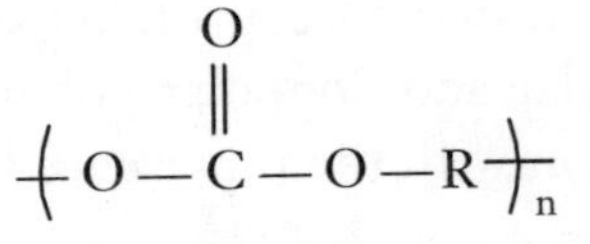

$$\left[-\text{O}-\overset{\text{O}}{\overset{\|}{\text{C}}}-\text{O}-\text{R}-\right]_n$$

Polycarbonate

$$\left[-\text{O}-\text{R}-\text{O}-\overset{\text{O}}{\overset{\|}{\text{C}}}-\overset{\text{H}}{\overset{|}{\text{N}}}-\text{R'}-\overset{\text{H}}{\overset{|}{\text{N}}}-\overset{\text{O}}{\overset{\|}{\text{C}}}-\right]$$

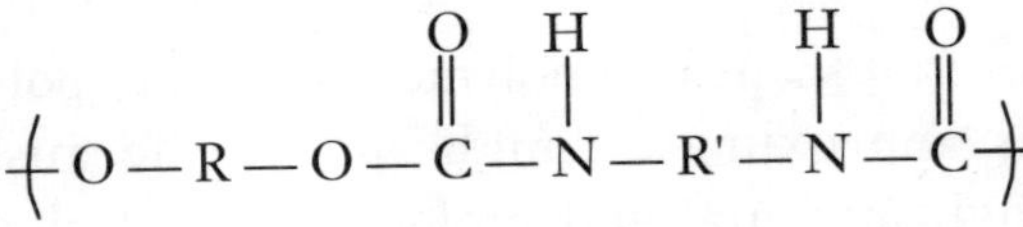

Polyurethane, PU

Condensation polymer chains are characterized as having a noncarbon atom in their backbone. For polyamides the noncarbon is nitrogen (N), while for polycarbonates it is oxygen (O). Condensation polymers are synthesized using melt (the reactants are heated causing them to melt), solution (the reactants are dissolved), and interfacial (the reactants are dissolved in immiscible solvents) techniques.

Molecular Properties

These are used to help determine the structure and behavior of the polymer. The molecular weight of a particular polymer chain is the product of the number of units times the molecular weight of the repeating unit. Two statistical averages describe polymers, the number-average molecular weight and the weight-average molecular weight.

Size is the most important property of polymers allowing for storage of information (nucleic acids and proteins). Polymeric materials remember any action that distorts or moves polymer chains or segments (such as bending, stretching, and melting). Size also accounts for an accumulation of the interchain and intrachain secondary attractive forces called van der Waals forces. For nonpolar polymers, such as polyethylene, the attractive forces for each repeating unit are less than that for polar polymers. Polyvinyl chloride, a polar polymer, has attractive forces that include both dispersion and dipole-dipole forces so that the total attractive forces are proportionally larger than those for polyethylene. Polymers with hydrogen bonding (such as proteins, polysaccharides, nucleic acids, and nylons) have attractive forces that are even greater. Hydrogen bonding is so strong in cellulose that cellulose is not soluble in water until the inter—and intrachain hydrogen bonds are broken.

Polymers often have a combination of ordered regions, called crystalline regions, and disordered or amorphous regions. Crystalline regions are more rigid, contributing to strength and resistance to external forces. The amorphous regions contribute to polymers' flexibility. Most commercial polymers have a balance between amorphous and crystalline regions, allowing a balance between flexibility and strength.

Polymers are viscoelastic materials. Ductile polymers, such as polyethylene and polypropylene, "give" or "yield," and at high elongations some strengthening and orientation occur. A brittle polymer, such as polystryene, does not give much and breaks at a low elongation. A fiber, a polymer material that is much longer than it is wide, exhibits high strength, high stiffness, and little elongation.

Materials

Fibers are polymer materials that are strong in one direction, and they are much longer (>100 times) than they are wide. Elastomers (or rubbers) are polymeric materials that can be distorted through the application of force, and when the force is removed, the material returns to its original shape. Plastics are materials that have properties between fibers and elastomers—they are hard and flexible. Coatings and adhesives are generally derived from polymers that are members of other groupings (for example, polysiloxanes are elastomers, but also are used as adhesives). Industrially important adhesives and coatings include laminates, sealants and caulks, composites, films, polyblends, liquid crystals, ceramics, cements, and smart materials.

Additives

Processed polymeric materials are generally a combination of the polymer and the materials that are added to modify its properties, assist in processing, and introduce new properties. Additives can be solids, liquids, or gases. Typical additives are plasticizers, antioxidants, colorants, fillers, and reinforcements.

Recycling

Many polymers are thermoplastics, that is, they can be reshaped through application of heat and pressure and used in the production of other thermoplastic materials. The recycling of thermosets, polymers that do not melt but degrade prior to softening, is more difficult. These materials are often ground into a fine powder, are blended with additives (often adhesives or binders), and then are reformed.

Polymer Overview

While the term polymer in popular usage suggests "plastic", polymers comprise a large class of natural and synthetic materials with a variety of properties and purposes. Natural polymer materials such as shellac and amber have been in use for centuries. Biopolymers such as proteins (for example hair, skin and part of the bone structure) and nucleic acids play crucial roles in biological processes. A variety of other natural polymers exist, such as cellulose, which is the main constituent of wood and paper.

Historical Development

The term polymer was coined in 1833 by Jöns Jakob Berzelius. Around the same time Henri Braconnot did pioneering work in derivative cellulose compounds, perhaps the earliest important work in polymer science. The development of vulcanization later in the nineteenth century improved the durability of the natural polymer rubber, signifying the first popularized semi-synthetic polymer. The first wholly synthetic polymer, Bakelite, was introduced in 1909.

Despite significant advances in synthesis and characterization of polymers, a proper understanding of polymer molecular structure did not come until the 1920s. Before that, scientists believed that polymers were clusters of small molecules (called colloids), without definite molecular weights, held together by an unknown force, a concept known as association theory. In 1922, Hermann Staudinger proposed that polymers consisted of long chains of atoms held together by covalent bonds, an idea which did not gain wide acceptance for over a decade, and for which Staudinger was ultimately awarded the Nobel Prize. An important contribution to synthetic polymer science was given by the Italian chemist Giulio Natta and Karl Ziegler who won the Nobel Prize in Chemistry in 1963 for the development of the Ziegler-Natta catalyst. In the intervening century, synthetic polymer materials such as Nylon, polyethylene, Teflon, and silicone have formed the basis for a burgeoning polymer industry.

Synthetic polymers today find application in nearly every industry and area of life. Polymers are widely used as adhesives and lubricants, as well as structural components for products ranging from childrens' toys to aircraft. Polymers such as poly(methyl methacrylate) find application as photoresist materials used in semiconductor manufacturing and low-k dielectrics for use in high-performance microprocessors. Future applications include flexible polymer-based substrates for electronic displays and improved time-released and targeted drug delivery.

Most polymer research may be categorized as polymer science, a sub-discipline of materials science which includes researchers in chemistry (especially organic chemistry), physics, and engineering. Polymer science may be roughly divided into two subdisciplines:

- Polymer chemistry or macromolecular chemistry, concerned with the chemical synthesis and chemical properties of polymers.
- Polymer physics, concerned with the bulk properties of polymer materials and engineering applications.

The field of polymer science is generally concerned with synthetic polymers, such as plastics, or chemical treatment and modification of natural polymers.

The study of biological polymers, their structure, function, and method of synthesis is generally the purview of biology, biochemistry, and biophysics. These disciplines share some of the terminology familiar to polymer science, especially when describing the synthesis of biopolymers such as DNA or polysaccharides. However, usage differences persist, such as the practice of using the term macromolecule to describe large non-polymer molecules and complexes of multiple molecular components, such as hemoglobin. Substances with distinct biological function are rarely described in the terminology of polymer science. For example, a protein is rarely referred to as a copolymer.

Polymer Synthesis

Polymers are synthesized by three primary methods: organic synthesis in a laboratory or factory, biological synthesis in living cells and organisms, or by chemical modification of naturally occurring polymers.

Organic Synthesis

In 1907, Leo Baekeland created the first completely synthetic polymer, Bakelite, by reacting phenol and formaldehyde at precisely controlled temperature and pressure. Subsequent work by Wallace Carothers in the 1920s demonstrated that polymers could be synthesized rationally from their constituent monomers. The intervening years have shown significant developments in rational polymer synthesis. Most commercially important polymers today are entirely synthetic and produced in high volume, on appropriately scaled organic synthetic techniques.

Laboratory synthetic methods are generally divided into two categories, condensation polymerization and addition polymerization. However, some newer methods such as plasma polymerization do not fit neatly into either category. Synthetic polymerization reactions may be carried out with or without a catalyst. Efforts towards rational synthesis of biopolymers via laboratory synthetic methods, especially artificial synthesis of proteins, is an area of intense research.

Biological Synthesis

Natural polymers and biopolymers formed in living cells may be synthesized by enzyme-mediated processes, such as the formation of DNA catalyzed by DNA polymerase. The synthesis of proteins involves multiple enzyme-mediated processes to transcribe genetic information from the DNA and subsequently translate that information to synthesize the specified protein. The protein may be modified further following translation in order to provide appropriate structure and function.

Modification of Natural Polymers

Many commercially important polymers are synthesized by chemical modification of naturally occurring polymers. Prominent examples include the reaction of nitric acid and cellulose to form nitrocellulose and the formation of vulcanized rubber by heating natural rubber in the presence of sulfur.

Polymer Structure and Properties

Types of polymer 'properties' can be broadly divided into several categories based upon scale. At the nano-micro scale are properties that directly describe the chain itself. These can be thought of as polymer structure. At an intermediate mesoscopic level are properties that describe the morphology of the polymer matrix in space. At the macroscopic level are properties that describe the bulk behavior of the polymer.

Structure

The structural properties of a polymer relate to the physical arrangement of monomers along the backbone of the chain. Structure has a strong influence on the other properties of a polymer. For example, a linear chain polymer may be soluble or insoluble in water depending on whether it is composed of polar monomers (such as ethylene oxide) or nonpolar monomers (such as styrene). On the other hand, two samples of natural rubber may exhibit different durability even though their molecules comprise the same monomers. Polymer scientists have developed terminology to precisely describe both the nature of the monomers as well as their relative arrangement:

Monomer Identity

The identity of the monomers comprising the polymer is generally the first and most important attribute of a polymer. Polymer nomenclature is generally based upon the type of monomers comprising the polymer. Polymers that contain only a single type of monomer are known as homopolymers, while polymers containing a mixture of monomers are known as copolymers. Poly(styrene), for example, is composed only of styrene monomers, and is therefore is classifed as a homopolymer. Ethylene-vinyl acetate, on the other hand, contains more than one variety of monomer and is thus a copolymer. Some biological polymers are composed of a variety of different but structurally related monomers, such as polynucleotides composed of nucleotide subunits.

A polymer molecule containing ionizable subunits is known as a polyelectrolyte. An ionomer is a subclass of polyelectrolyte with a low fraction of ionizable subunits.

Chain Linearity

The simplest form of polymer molecule is a straight chain or linear polymer, composed of a single main chain. The flexibility of an unbranched chain polymer is characterized by its persistence length. A branched polymer molecule is composed of a main chain with one or more substituent side chains or branches. Special types of branched polymers include star polymers, comb polymers, and brush polymers. If the polymer contains a side chain that has a different composition or configuration than the main chain the polymer is called a graft or grafted polymer. A cross-link suggests a branch point from which four or more distinct chains emanate. A polymer molecule with a high degree of crosslinking is referred to as a polymer network. Sufficiently high crosslink concentrations may lead to the formation of an 'infinite network', also known as a 'gel', in which networks of chains are of unlimited extend—there is essentially all chains have linked into one molecule.

Chain Size

Polymer bulk properties may be strongly dependent on the size of the polymer chain. Like any molecule, a polymer molecule's size may be described in terms of molecular weight or mass. In polymers, however, the molecular mass may be expressed in terms of degree of polymerization, essentially the number of monomer units which comprise the polymer. For synthetic polymers, the molecular weight is expressed statistically to describe the distribution of molecular weights in the sample. This is because of the fact that almost all industrial processes produce a distribution of polymer chain sizes. Examples of such statistics include the number average molecular weight and weight average molecular weight. The ratio of these two values is the polydispersity index, commonly used to express the "width" of the molecular weight.

The space occupied by a polymer molecule is generally expressed in terms of radius of gyration or excluded volume.

Monomer Arrangement in Copolymers

Monomers within a copolymer may be organized along the backbone in a variety of ways.

- Alternating copolymers possess regularly alternating monomer residues
- Periodic copolymers have monomer residue types arranged in a repeating sequence
- Random copolymers have a random sequence of monomer residue types
- Statistical copolymers have monomer residues arranged according to a known statistical rule
- Block copolymers have two or more homopolymer subunits linked by covalent bonds. Block copolymers with two or three distinct blocks are called diblock copolymers and triblock copolymers, respectively.

Tacticity in Polymers with Chiral Centers

This property describes the relative stereochemistry of chiral centers in neighboring structural units within a macromolecule. There are three types: isotactic, atactic, and syndiotactic.

Morphological Properties

Crystallinity

When applied to polymers, the term crystalline has a somewhat ambiguous usage. In some cases, the term crystalline finds identical usage to that used in conventional crystallography. For example, the structure of a crystalline protein or polynucleotide, such as a sample prepared for x-ray crystallography, may be defined in terms of a conventional unit cell composed of one or more polymer molecules with cell dimensions of hundreds of angstroms or more.

A synthetic polymer may be described as crystalline if it contains regions of three-dimensional ordering on atomic (rather than macromolecular) length scales, usually arising from intramolecular folding and/or stacking of adjacent chains. Synthetic polymers may consist of both crystalline and amorphous regions; the *degree of crystallinity* may be expressed in terms of a weight fraction or volume fraction of crystalline material. Few synthetic polymers are entirely crystalline.

Bulk Properties

The bulk properties of a polymer are those most often of end-use interest. These are the properties that dictate how the polymer actually behaves on a macroscopic scale.

Tensile Strength

The tensile strength of a material quantifies how much stress the material will endure before failing. This is very important in applications that rely upon polymer's physical strength or durability. For example, a rubber band with a higher tensile strength will hold a greater weight before snapping. In general tensile strength increases with polymer chain length.

Young's Modulus of Elasticity

This parameter quantifies the elasticity of the polymer. It is defined, for small strains, as the ratio of rate of change of stress to strain. Like tensile strength this is highly relevant in polymer applications involving the physical properties of polymers, such as rubber bands.

Transport Properties

Transport properties such as diffusivity relate to how rapidly molecules move through the polymer matrix. These are very important in many applications of polymers for films and membranes.

Pure Component Phase Behavior

Melting Point

The term "melting point" when applied to polymers suggests not a solid-liquid phase transition but a transition from a crystalline or semi-crystalline phase to a solid amorphous phase. Though abbreviated as simply "T_m", the property in question is more properly called the "crystalline melting temperature". Among synthetic polymers, crystalline melting is only discussed with regards to thermoplastics, as polymers will decompose at high temperatures rather than melt.

Boiling Point

The boiling point of a polymer substance is never defined due to the fact that polymers will decompose before reaching theoretical boiling temperatures.

Glass Transition Temperature (T_g)

A parameter of particular interest in synthetic polymer manufacturing is the glass transition temperature (T_g), which describes the temperature at which amorphous polymers undergo a second order phase transition from a rubbery, viscous amorphous solid to a brittle, glassy amorphous solid. The glass transition temperature may be engineered by altering the degree of branching or cross-linking in the polymer or by the addition of plasticizer.

Polymer Solution Behavior

In general, polymeric mixtures are far less miscible than mixtures of small molecule materials. This effect is a result of the fact that the driving force for mixing is usually entropics, not energetics. In other words, miscible materials usually form a solution not because their interaction with each other is more favorable than their self-interaction but because of an increase in entropy and hence free energy associated with increasing the amount of volume available to each component. This increase in entropy scales with the number of particles (or moles) being mixed. Since polymeric molecules are much larger and hence generally have much higher specific volumes than small molecules, the number of molecules involved in a polymeric mixture are far less than the number in a small molecule mixture of equal volume. The energetics of mixing, on the other hand, are comparable on a per volume basis for polymeric and small molecule mixtures. This tends to increase the free energy of mixing for polymer solutions and thus make solvation less favorable. Thus, concentrated solutions of polymers are far rarer than those of small molecules.

In dilute solution, the properties of the polymer are characterized by the interaction between the solvent and the polymer. In a *good solvent*, the polymer appears swollen and occupies a large volume. In this scenario, intermolecular forces between the solvent and monomer subunits dominate over intramolecular interactions. In a *bad solvent* or *poor solvent*, intramolecular forces dominate and the chain contracts. In the *theta solvent*, or the state of the polymer solution where the value of the second virial coefficient becomes 0, the intermolecular polymer-solvent repulsion balances exactly the intramolecular monomer-monomer attraction. Under the theta condition (also called the Flory condition) the polymer behaves like an ideal random coil.

Polymer Structure/Property Relationships

Polymer bulk properties are strongly dependent upon their structure and mesoscopic behavior. A number of qualitative relationships between structure and properties are known.

Chain Length

Increasing chain length tends to decrease chain mobility, increase strength and toughness, and increase the glass transition temperature (Tg). This is a result of the increase in chain interactions such as Van der Waals attractions and entanglements that come with increased chain length. These interactions tend to fix the individual chains more strongly in position and resist deformations and matrix breakup, both at higher stresses and higher temperatures.

Branching

Branching of polymer chains also affect the bulk properties of polymers. Long chain branches may increase polymer strength, toughness, and Tg due to an increase in the number of entanglements per chain. Random length and atactic short chains, on the other hand, may reduce polymer strength due to disruption of organization. Short side chains may likewise reduce crystallinity due to disruption of the crystal structure. Reduced crystallinity may also be associated with increased transparency due to light scattering by small crystalline regions. A good example of this effect is related to the range of physical attributes of polyethylene. High density polyethylene (HDPE) has a very low degree of branching, is quite stiff, and is used in applications such as milk jugs. Low density polyethylene (LDPE), on the other hand, has significant numbers of short branches, is quite flexible, and is used in applications such as plastic films. The branching index of the polymer is a parameter that characterizes the effect of long-chain branches on the size of a branched macromolecule in solution.

Chemical Cross-Linking

Cross linking tends to increase T_g and increase strength and toughness. Cross linking consists of the formation of chemical bonds between chains. Among other applications, this process is used to strengthen rubbers in a process known as Vulcanization, which is based on cross linking by sulfur. Car tires, for example, are highly cross linked in order to reduce the leaking of air out of the tire and to toughen the tires durability. Eraser rubber, on the other hand, is not cross linked to allow flaking of the rubber and prevent damage to the paper.

Inclusion of Plasticizers

Inclusion of Plasticizers tends to lower Tg and increase polymer flexibility. Plasticizers are generally small molecules that are chemically similar to the polymer and create gaps between polymer chains for greater mobility and reduced interchain interactions. A good example of the action of plasticizers is related to polyvinylchlorides or PVCs. A uPVC or unplastiscized polyvinylchloride is used for things such as pipes. A pipe has no plasticizers in it because it needs to remain strong and heat resistant. Plasticized PVC is used for clothing for a flexible quality. Plasticizers are also put in some types of cling film to make the polymer more flexible.

Degree of Crystallinity

Increasing degree of crystallinity tends to make a polymer more rigid. It can also lead to greater brittlness. Polymers with degree of crystallinity approaching zero or one will tend to be transparent, while polymers with intermediate degrees of crystallinity will tend to be opaque due to light scattering by crystalline/glassy regions.

Standardized Polymer Nomenclature

There are multiple conventions for naming polymer substances. Many commonly used polymers, such as those found in consumer products, are referred to by a *common* or *trivial* name. The trivial name is assigned based on historical precedent or popular usage rather than a standardized naming convention. Both the American Chemical Society and IUPAC have proposed standardized naming conventions; the ACS and IUPAC conventions are similar but not identical. Examples of the difference between the various naming conventions are given in the table below:

Common Name	*ACS Name*	*IUPAC Name*
Poly(ethylene oxide) or (PEO)	poly(oxyethylene)	poly(oxyethylene)
Poly(ethylene terephthalate) or (PET)	poly(oxy-1,2-ethanediyloxycarbonyl -1,4-phenylenecarbonyl)	poly(oxyethyleneoxyterephth = aloyl)
Nylon	poly[imino(1-oxo-1,6-hexanediyl)]	poly[imino(1-oxohexane-1,6-diyl)]

In both standardized conventions the polymers names are intended to reflect the monomer(s) from which they are synthesized rather than the precise nature of the repeating subunit. For example, the polymer synthesized from the simple alkene ethene is called polyethylene, retaining the -ene suffix even though the double bond is removed during the polymerization process:

$$\left(H_2C{=}CH_2 \right)_n \xrightarrow[\text{polymerisation}]{\text{radical addition}} \left(-CH_2-CH_2- \right)_n$$

The Polymerisation of ethene into poly(ethene)

$$\left(-CH_2-CH_2- \right)_n$$

Chemical Properties of Polymers

The attractive forces between polymer chains play a large part in determining a polymer's properties. Because polymer chains are so long, these interchain forces are amplified far beyond the attractions between conventional molecules. Different side groups on the polymer can lend the polymer to ionic bonding or hydrogen bonding between its own chains. These stronger forces typically result in higher tensile strength and melting points.

The intermolecular forces in polymers can be affected by dipoles in the monomer units. Polymers containing amide or carbonyl groups can form hydrogen bonds between adjacent chains; the partially positively charged hydrogen atoms in N-H groups of one chain are strongly attracted to the partially negatively charged oxygen atoms in C = O groups on another. These strong hydrogen bonds, for example, result in the high tensile strength and melting point of polymers containg urethane or urea linkages. Polyesters have dipole-dipole bonding between the oxygen atoms in C = O groups and the hydrogen atoms in H-C groups. Dipole bonding is not as strong as hydrogen bonding, so a polyester's melting point and strength are lower than Kevlar's (Twaron), but polyesters have greater flexibility.

Ethene, however, has no permanent dipole. The attractive forces between polyethylene chains arise from weak van der Waals forces. Molecules can be thought of as being surrounded by a cloud of negative electrons. As two polymer chains approach, their electron clouds repel one another. This has the effect of lowering the electron density on one side of a polymer chain, creating a slight positive dipole on this side. This charge is enough to actually attract the second polymer chain. Van der Waals forces are quite weak, however, so polyethene can have a lower melting temperature compared to other polymers.

Polymer Characterization

The characterization of a polymer requires several parameters which need to be specified. This is because a polymer actually consists of a statistical distribution of chains of varying lengths, and each chain consists of monomer residues which affect its properties.

A variety of lab techniques are used to determine the properties of polymers. Techniques such as wide angle X-ray scattering, small angle X-ray scattering, and small angle neutron scattering are used to determine the crystalline structure of polymers. Gel permeation chromatography is used to determine the number average molecular weight, weight average molecular weight, and polydispersity. FTIR, Raman and NMR can be used to determine composition. Thermal properties such as the glass transition temperature and melting point can be determined by differential scanning calorimetry and dynamic mechanical analysis. Pyrolysis followed by analysis of the fragments is one more technique for determining the possible structure of the polymer.

Polymer Degradation

Polymer degradation is a change in the properties—tensile strength, colour, shape, etc—of a polymer or polymer based product under the influence of one or more environmental factors such as heat, light or chemicals. It is often due to the hydrolysis of the bonds connecting the polymer chain, which in turn leads to a decrease in the molecular mass of the polymer. These changes may be undesirable, such as changes during use, or desirable, as in biodegradation or deliberately lowering the molecular mass of a polymer. Such changes occur primarily because of the effect of these factors on the chemical composition of the polymer.

The degradation of polymers to form smaller moleculars may proceed by *random scission* or *specific scission*. The degradation of polyethylene occurs by *random scission*—that is by a random breakage of the linkages (bonds) that hold the atoms of the polymer together. When heated above 450 Celsius it degrades to form a mixture of hydrocarbons. Other polymers—like polyalphamethylstyrene—undergo 'specific' chain scission with breakage occurring only at the ends. They literally unzip or depolymerize to become the constituent monomer.

In a finished product such a change is to be prevented or delayed. However the degradation process can be useful from the view points of understanding the structure of a polymer or recycling/reusing the polymer waste to prevent or reduce environmental pollution. Polylactic acid and Polyglycolic acid, for example, are two polymers that are useful for their ability to degrade under aqueous conditions. A copolymer of these polymers is used for biomedical applications such as hydrolysable stitches that degrade over time after they are applied to a wound. These materials can also be used for plastics that will degrade over time after they are used and will therefore not remain as litter.

Industry

Today there are primarily six commodity polymers in use, namely polyethylene, polypropylene, polyvinyl chloride, polyethylene terephthalate, polystyrene and polycarbonate. These make up nearly 98 per cent of all polymers and plastics encountered in daily life.

Each of these polymers has its own characteristic modes of degradation and resistances to heat, light and chemicals.

Cracking of Polymers

Cracking refers to thermally or otherwise degrading the polymer to recover either monomers or oligomers.

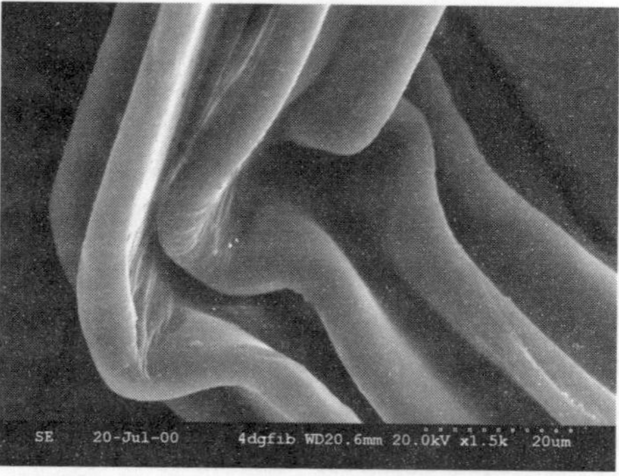

Fig. 1.2: Image of a bend in a polyester fiber with a high surface area, as seen at high magnification with a scanning electron microscope. Polyester is a category of thermoplastic polymers.

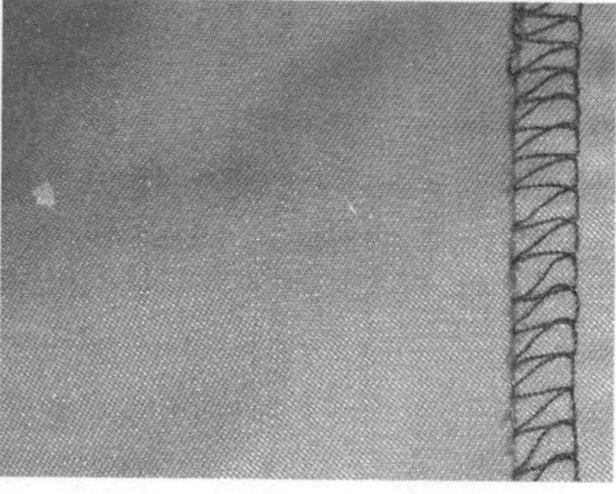

Fig. 1.3: Close-up of a polyester shirt.

Fig. 1.4: A Bakelite distributor rotor.

A *polymer* (from the Greek words *polys*, meaning "many," and *meros*, meaning "parts") is a chemical compound consisting of large molecules, each of which is a long chain made up of small structural units that are linked together by covalent chemical bonds. Each structural unit, called a *monomer* (Greek word *monos* means "alone" or "single"), is a small molecule of low-to-moderate molecular weight. Within a given polymer molecule, the monomers are usually identical or similar in structure. The chemical reaction by which monomers are linked together to form polymers is called *polymerization*.

Polymers form a large, diverse group of materials. Within each living organism, polymers (biopolymers) such as DNA, RNA, proteins, and polysaccharides perform specific functions that enable the organism to survive, grow, and reproduce. In addition, natural polymers—such as cotton, flax, jute, silk, and wool—have long been used for the production of clothing, rope, carpeting, felt, insulation, and upholstery. More recently, scientists have discovered how to produce new polymers with a wide range of properties, at relatively low cost. Their work has given birth to a proliferation of plastics, artificial fibers, and synthetic rubber. Consequently, synthetic polymers are being used for numerous products in homes, schools, offices, factories, recreational facilities, and means of transportation and communication. Thus, artificial polymers have become an integral part of our modern technological society.

On the downside, most artificial polymers are not biodegradable, and factories and incineration furnaces often release chemical pollutants. To help solve these problems, recycling programs have been instituted in many countries, and manufacturing plants and incinerators are now fitted with pollutant traps. In addition, biodegradable polymers are being sought out.

General Characteristics and Classification

Most polymers are organic—that is, their long chains have backbones of mostly carbon atoms. There are also some inorganic polymers, such as the silicones, which have a backbone of alternating silicon and oxygen atoms.

Polymer chains may or may not be cross-linked with one another. Thus the molecules of a polymer can have various topologies (shapes), such as linear (unbranched), branched, network (cross-linked 3-dimensional structure), comb, or star. The properties of a polymer depend on these shapes and on the structures of the monomers that make up the chains. For example, branched polymer chains cannot line up as close to one another as linear chains can. As a result, intermolecular bonds between branched chains are weaker, and such materials have lower densities, lower melting points, and lower tensile strength. Also, properties such as the solubility, flexibility, and strength of the polymer vary according to the types of monomers in the chains.

Polymers are typically classified as follows:

- *Thermoplastics*: A thermoplastic is a material that is deformable, melts to a liquid when heated, and freezes to a brittle, glassy state when cooled sufficiently. Most thermoplastics are polymers whose molecules have linear or branched structures. The molecules associate with one another through various interactions: weak van der Waals forces, as in the case of polyethylene and polypropylene; stronger dipole-dipole interactions; hydrogen bonding, as in the case of nylon; or the stacking of aromatic rings, as in the case of polystyrene.
- *Thermosets* (or *thermosetting plastics*): These are materials that are taken through a "curing" process with the addition of energy. The energy may be in the form of heat (generally above 200 °C), a chemical reaction, or irradiation. Thermoset materials are usually liquidy, powdery, or malleable prior to curing, and designed to be molded into their final form or used as adhesives. During the curing process, molecules of the starting material become cross-linked and take on a stronger form. Once cured, the thermoset cannot be remelted and remolded. Examples of thermosets are vulcanized rubber, Bakelite (used in electrical insulators), melamine (used in worktop surfaces), and epoxy resin (used as an adhesive).
- *Elastomers*: The term elastomer is applied to an "elastic polymer"—that is, a polymer that returns to its original shape when a load is removed. Elastomers are usually thermosets (that require curing), but some are thermoplastic. The long polymer chains become cross-linked during curing and account for the flexible nature of the material. The molecular form of elastomers has been likened to a "spaghetti and meatball" structure, where the meatballs signify cross-links between the flexible spaghetti strands (polymer chains). Most elastomers are rubbers, and the term *elastomer* is

often used interchangeably with the term *rubber.* Examples of thermoplastic elastomers are Hytrel and Santoprene.

- *Coordination polymers*: In a coordination polymer, many metal centers are interconnected through ligand bridges. Most of the common halides and oxides are coordination polymers. In a more conventional sense, the term coordination polymer is reserved for compounds where the metals are bridged by polyatomic ligands, such as cyanide and carboxylates. One of the most popular bridging ligands used in the synthesis of these polymers is a tricarboxylic acid called BTC (benzene-1,3,5-tricarboxylic acid). The polymers are metal salts of this acid. Another coordination polymer is Prussian Blue, which is based on Fe-CN-Fe linkages.
- *Biopolymers* (biological polymers): Biopolymers are a special class of polymers produced within living organisms. They include starch, proteins, peptides, DNA, and RNA. Their monomer units are sugars, amino acids (for proteins and peptides), and nucleotides (for DNA and RNA). Unlike synthetic (artificially produced) polymers, each biopolymer has a well-defined structure. Many biopolymers spontaneously fold into characteristic shapes that determine their biological functions.

Synthetic polymers are often named after the monomer from which they are made. For example, polyethene (also called polyethylene) is the name given to the polymer formed when thousands of ethene (ethylene) molecules are bonded together. The polyethene molecules are straight or branched chains of repeating $-CH_2-CH_2-$ units (with a $-CH_3$ at each terminus). The polymerization reaction can be written as follows.

$$\left(\mathrm{H_2C{=}CH_2} \right)_n \xrightarrow[\text{polymerization}]{\text{radical addition}} \sim\mathrm{CH_2{-}CH_2{-}CH_2{-}CH_2{-}CH_2{-}CH_2}\sim$$

The polymerization of ethene into poly(ethene).

The product may also be written as:

$$\left(\begin{array}{cc} \mathrm{H} & \mathrm{H} \\ | & | \\ -\mathrm{C} & -\ \mathrm{C}- \\ | & | \\ \mathrm{H} & \mathrm{H} \end{array} \right)_n$$

By contrast, biopolymers have been named apart from their monomeric constitution. For instance, proteins are polymers of amino acids. Typically, each protein chain is made up of hundreds of amino acid monomers, and the sequence of these monomers determines its shape and biological function.

Whereas polyethylene forms spontaneously under the right conditions, the synthesis of biopolymers such as proteins and nucleic acids requires the help of specialized biological machinery, including enzymes that catalyze the reactions. Unlike synthetic polymers, these biopolymers (other than carbohydrates) have exact sequences and lengths. Since the 1950s, catalysts have also revolutionized the development of synthetic polymers. By allowing more careful control over polymerization reactions, polymers with new properties—such as the ability to emit colored light—have been manufactured.

Copolymerization

Copolymerization involves the linking together of two or more different monomers, producing chains with varied properties. For example, a protein can be called a copolymer—one in which different amino acid monomers are linked together. Depending on the sequence of amino acids, the protein chains have different shapes and functions.

When ethene is copolymerized with small amounts of 1-hexene (or 4-methyl-1-pentene), the product is called linear low-density polyethene (LLDPE). The C_4 branches resulting from the hexene lower the density and prevent large crystalline regions from forming in the polymer, as they do in high-density polyethene (HDPE). This means that LLDPE can withstand strong tearing forces while maintaining flexibility.

The polymerization reaction may be carried out in a stepwise manner, to produce a structure with long sequences (or blocks) of one monomer alternating with long sequences of the other. The product is called a *block copolymer.*

In the case of some copolymers, called graft copolymers, entire chains of one kind (such as polystyrene) are made to grow out of the sides of chains of another kind (such as polybutadiene). The resultant product is less brittle and more impact-resistant. Thus, block and graft copolymers can combine the useful properties of both constituents and often behave as quasi-two-phase systems.

The formation of nylon is an example of step-growth polymerization, or condensation polymerization. The two types of monomers can have different R and R' groups, shown in the diagram below. The properties of the nylon can vary, depending on the R and R' groups in the monomers used.

$$\mathrm{HOOC{-}R{-}COOH} + \left[\mathrm{H_2N{-}R'{-}NH_2}\right]_n \longrightarrow *\!\left(\overset{\displaystyle O}{\overset{\|}{C}}-R-\overset{\displaystyle O}{\overset{\|}{C}}-\underset{\displaystyle H}{\underset{|}{N}}-R'-\underset{\displaystyle H}{\underset{|}{N}}\right)_n\!*$$

The first commercially successful, completely synthetic polymer was nylon 6,6, with four carbon atoms in the R group (adipic acid) and six carbon atoms in the R′ group (hexamethylene diamine). Each monomer actually contributes 6 carbon atoms (including the two carboxyl carbons of adipic acid)—hence the name nylon 6,6. In naming nylons, the number of carbons from the diamine is given first, and the number from the diacid, second. Kevlar is an aromatic nylon in which both R and R' are benzene rings.

Copolymers illustrate the point that the *repeating unit* in a polymer—such as a nylon, polyester, or polyurethane—is often made up of two (or more) monomers.

Physical Properties of Polymers

Polymer chains have markedly unique physical properties, as follows.

- *Molar mass distribution*: During a polymerization reaction, polymer chains terminate after varying degrees of chain lengthening. The reaction produces an ensemble of differing chain lengths of differing molecular masses, with a (Gaussian) distribution around an average value. The molar mass distribution in a polymer describes this distribution of molecular masses for different chain lengths. Biopolymers, however, have well-defined structures, and they therefore do not have a molar mass distribution.
- *Degree of polymerization*: This is the number of monomer units in an average polymer chain, at time t in a polymerization reaction. For most industrial purposes, synthetic polymer chains need to have thousands or tens of thousands of monomer units.
- Crystallinity, and thermal phase transitions:
  - (a) *Melting point* (T_m): Thermoplastic (non-cross-linked) polymers have a melting temperature above which their crystalline structure entirely disappears.
  - (b) *Glass transition temperature* (T_g): The glass transition temperature of a material is the temperature below which its molecules have little relative mobility. This temperature is usually applicable to glasses and plastics that have wholly or partially amorphous phases. Thermoplastic (non-cross-linked) polymers have a T_g value below which they become rigid and brittle, and can crack and shatter under stress. (The T_g value is lower than T_m.) Above T_g, the polymer becomes rubbery and capable of deformation without fracture. This is one of the properties that make many plastics useful. Such behavior, however, is not exhibited by cross-linked thermosetting plastics—once cured, they are set for life, never deforming or melting when heated.
- *Stereoregularity (or tacticity)*: This property describes the arrangement of functional groups on the backbone of carbon chains.

Chemical Properties of Polymers

The attractive forces between polymer chains play a large part in determining a polymer's properties. Given that polymer chains are so long, these interchain forces are amplified far beyond the attractions between conventional molecules. Also, longer chains are more *amorphous* (randomly oriented). Polymers can be visualized as tangled spaghetti chains—the more tangled the chains, the more difficult it is to pull any one strand out. These stronger forces typically result in high tensile strength and melting points.

The intermolecular forces in polymers are determined by dipoles in the monomer units. For example, polymers containing amide groups can form hydrogen bonds between adjacent chains. The somewhat positively charged hydrogen atoms in the N-H groups of one chain are strongly attracted to the somewhat negatively charged oxygen atoms in the C = O groups on another. Such strong hydrogen bonds are responsible for the high tensile strength and melting point of Kevlar.

In the case of polyesters, there is dipole-dipole bonding between the oxygen atoms in C = O groups and the hydrogen atoms in C-H groups. Dipole bonding is not as strong as hydrogen bonding, so the polyester's melting point and strength are lower than Kevlar's, but polyesters have greater flexibility.

If one considers polyethene, the monomer units (ethene) have no permanent dipole. Attractive forces between polyethene chains arise from weak van der Waals forces. Molecules can be thought of as being surrounded by a cloud of negative electrons. As two polymer chains approach, their electron clouds repel one another. This has the effect of lowering the electron density on one side of a polymer chain, creating a slight positive charge on this side. This charge is enough to attract the second polymer chain. Van der Waals forces are quite weak, however, so polyethene melts at low temperatures.

Applications

Applications of Synthetic Polymers

- *Acrylonitrile butadiene styrene* (*ABS*): This is a common thermoplastic, appropriate for making light but rigid products such as automotive body parts, protective head gear, golf club heads, and LEGO® toys.
- *Polyacrylates* (*acrylic*): Noted for their transparency and resistance to breakage, polyacrylates may be used as substitutes for window glass. A familiar product in this group is Plexiglas®.
- *Cellulose acetate*: It is used as a film base in photography, as a component in some adhesives, and as a synthetic fiber. The fiber form is used for dresses,

draperies, upholstery, diapers, cigarette filters and other filters, and fiber-tip pens.

- *Ionomers*: These are useful for golf ball covers, semipermeable membranes, dental cements, and fuel cells.
- *Liquid crystal polymers*: Uses for this group of polymers include electrical and electronic applications, automotive parts, and engineering parts.
- *Polyamides, such as nylon and Kevlar®*: Nylon fibers are used in clothing, parachutes, ropes, carpets, guitar and racket strings, and fishing nets. Kevlar® is used in applications ranging from bicycles to bulletproof jackets.
- *Polyesters, such as polyethylene terephthalate (PET) and polycarbonates*: Polyester fibers are used to make fabrics for personal clothing, bed sheets, bedspreads, curtains, and so forth. In addition, polyesters are used to make bottles, films, liquid crystal displays, holograms, filters, and electrical insulation. Thermosetting polyester resins are commonly used as casting materials, fiberglass laminating resins, and nonmetallic auto-body fillers. Polyesters are also widely used as a finish on high-quality wooden products like guitars, pianos, and vehicle or yacht interiors.
- *Polytetrafluoroethylene* (*Teflon®*): Among its many uses, it is suitable as an insulator in cables and connector assemblies and as a material for printed circuit boards (at microwave frequencies), bearings, bushings, and gears.
- *Polyethylene* (*polyethene, PE*): The polyethylenes are a widely used group of materials and are classified according to their molecular weight, density, and branching. For instance, ultra high molecular weight PE (UHMWPE) is used for can—and bottle-handling machine parts, moving parts on weaving machines, bearings, gears, artificial joints, and the newer bulletproof vests. High-density PE (HDPE) is used in making milk jugs, detergent bottles, margarine tubs, and garbage containers. Low-density PE (LDPE) is used for film wrap and plastic bags, as well as for some rigid containers.
- *Melamine resin*: Combined with formaldehyde, it produces a thermoset plastic that is used to make decorative wall panels, laminates, kitchen utensils, and plates. It is the main constituent of Formica® and Arborite®.
- *Epoxy resin*: It is used for many applications, including coatings, adhesives, and composite materials, such as those using carbon fiber and fiberglass reinforcements.
- *Polybutadiene* (BR): This synthetic rubber has a high resistance to wear and is used mainly for the manufacture of tires.
- *Polychloroprene* (*Neoprene*): This synthetic rubber has many applications, such as for wetsuits, electrical insulation, car fan belts, gaskets, hoses, corrosion-resistant coatings, and as padding in metal cases.

Applications of bBiopolymers

- *Cotton*: This soft fiber, which grows around the seeds of the cotton plant (*Gossypium* species), consists of nearly pure cellulose. It is most often spun into thread and used to make a soft, breathable textile, the most widely used natural fiber in clothing today.
- *Flax*: Flax fibers have been used for the production of linen for 5,000 years. The best grades are used for fabrics such as damasks, lace, and sheeting. Coarser grades are used for manufacturing twine and rope. Flax fiber is also a raw material for the high-quality paper used for banknotes.
- *Hemp*: Hemp fibers, obtained from the *Cannabis* species of plants, are used to make cordage and clothing.
- *Jute*: Jute fibers, composed of plant cellulose and lignin, are used to make coarse fabrics (called burlap or hessian cloth) and sacks (called gunny bags).
- *Kenaf*: Kenaf fibers, made by the kenaf plant *(Hibiscus cannabinus)*, are used for the manufacture of rope, twine, coarse cloth, and paper.
- *Silk*: This protein fiber, obtained from the cocoons of silkworm larvae, is woven into textiles.
- *Wool*: This protein fiber, derived mainly from the fur of sheep and goats, is used for making clothing, carpeting, felt, insulation, and upholstery. It is also used to absorb odors and noise in heavy machinery and stereo speakers.
- *Zein*: This protein, found in maize, is used in the manufacture of textile fibers, biodegradable plastics, printing inks, and adhesives. It is also used as a coating for candy, nuts, fruit, and encapsulated foods and drugs.

Natural Functions of Biopolymers

- *Proteins*. There are different types of proteins that are involved in numerous functions in each living cell. Examples include:
  - Catalysis of biochemical reactions, carried out by numerous enzymes
  - Transport and storage of small molecules and ions
  - Immune defense, such as by forming antibodies
  - Sending and receiving signals, such as by receptors on cell surfaces
  - Structural support, such as components of the skin, hair, and bone.
  - Coordinated motion, such as the components of muscles and molecular motors.
  - Control of cell growth, such as by factors that control the synthesis of messenger RNA and proteins.
- *RNA (ribonucleic acid)*. There are different types of RNA that carry out different functions. Examples include:

- *Messenger RNA (mRNA)*: Various mRNAs get their information from DNA and serve as templates for the synthesis of proteins.
- *Transfer RNA (tRNA)*: Specific tRNA molecules carry specific amino acids and transfer them to growing protein chains.
- *Ribosomal RNA (rRNA)*: rRNA molecules are part of cellular structures called *ribosomes*, which function as "workbenches" on which proteins are synthesized.
- *Ribozymes*: These are RNA molecules that can function as enzymes, that is, they can catalyze chemical reactions.
- *Small interfering RNA (siRNA)*: Among their various functions, siRNAs are involved in pathways by which they interfere with the expression of specific genes.

❖ *DNA (deoxyribonucleic acid)*: A constituent of the chromosomes (and organelles such as mitochondria and chloroplasts) of living cells, DNA serves as an "informational" molecule and genetic material that is inherited. Its known functions include:

- Carrier of information for RNA structures.
- Carrier of information for protein structures.
- Replication, so that it can be passed down from one generation to the next.

❖ *Polysaccharides*: These large, polymeric carbohydrates occur in different types and serve various functions. Examples are as follows.

- *Cellulose*: It is a common material that provides structure for plant cell walls.
- *Starch*: It is a combination of two polysaccharides (amylose and amylopectin) and is made by plants to store excess glucose.
- *Glycogen* ("animal starch"): This polysaccharide is the main storage form of glucose in animal and human cells.

Examples of Thermoplastics

❖ Acrylonitrile butadiene styrene (ABS)
❖ Celluloid
❖ Cellulose acetate
❖ Ethylene vinyl acetate (EVA)
❖ Ethylene vinyl alcohol (EVAL)
❖ Fluoroplastics (including polytetrafluoroethylene (PTFE), or Teflon®)
❖ Ionomers

- Kydex™, an acrylic/PVC alloy
- Liquid crystal polymer (LCP)
- Polyacetal (POM or Acetal)
- Polyacrylates (Acrylic or Acrylates)
- Polyacrylonitrile (PAN or Acrylonitrile)
- Polyamide (PA) (including nylon and Kevlar®)
- Polyamide-imide (PAI)
- Polyaryletherketone (PAEK or Ketone)
- Polybutadiene (PBD)
- Polybutylene (PB)
- Polycyclohexylene dimethylene terephthalate (PCT)
- Polyhydroxyalkanoates (PHAs)
- Polyketone (PK)
- Polyester (including polycarbonate (PC), polyethylene terephthalate (PET), polybutylene terephthalate (PBT), polylactic acid (PLA))
- Polyethylene (PE)
- Polyetheretherketone (PEEK)
- Polyetherimide (PEI)
- Polyethersulfone (PES)—see Polysulfone
- Polyethylenechlorinates (PEC)
- Polyimide (PI)
- Polymethylpentene (PMP)
- Polyphenylene oxide (PPO)
- Polyphenylene sulfide (PPS)
- Polyphthalamide (PPA)
- Polypropylene (PP)
- Polystyrene (PS)
- Polysulfone (PSU)
- Polyvinyl chloride (PVC)
- Spectralon

Examples of Thermosets

- Vulcanized rubber
- Bakelite (a phenol formaldehyde resin, used in electrical insulators and plastic wear)
- Duroplast
- Urea-formaldehyde foam (used in plywood, particleboard, and medium-density fibreboard)
- Melamine resin (used on worktop surfaces)
- Polyester resin (used in glass-reinforced plastics/fiberglass)

- Epoxy resin (used as an adhesive and in fibre-reinforced plastics such as glass-reinforced plastic and graphite-reinforced plastic)

Examples of Elastomers

Unsaturated rubbers that can be cured by sulfur vulcanization:

- Natural rubber (NR)
- Polyisoprene (IR):
  - Butyl rubber (copolymer of isobutylene and isoprene, IIR)
  - Halogenated butyl rubbers: chloro butyl rubber (CIIR), bromo butyl rubber (BIIR)
- Polybutadiene (BR):
  - Styrene-butadiene rubber (SBR, copolymer of polystyrene and polybutadiene)
  - Nitrile rubber (NBR, copolymer of polybutadiene and acrylonitrile), also called buna N rubbers
  - Hydrated nitrile rubbers (HNBR): Therban® and Zetpol®
- Chloroprene rubber (CR): polychloroprene, Neoprene, Baypren Saturated rubbers that cannot be cured by sulfur vulcanization.
- Ethylene propylene rubber (EPM, a copolymer of polyethylene and polypropylene)
- Ethylene propylene diene rubber (EPDM, a combination of polyethylene, polypropylene, and a diene)
- Epichlorohydrin rubber (ECO)
- Polyacrylic rubber (ACM, ABR)
- Silicone rubber (SI, Q, VMQ)
- Fluorosilicone rubber (FVMQ)
- Fluoroelastomers (FKM, FPM): Viton®, Tecnoflon®, Fluorel®, Dai-El®
- Perfluoroelastomers (FFKM)
- Tetrafluoro ethylene/propylene rubbers (FEPM)
- Chlorosulfonated polyethylene (CSM): Hypalon®
- Ethylene-vinyl acetate (EVA)

Other Types of Elastomers

- Thermoplastic Elastomers (TPE): Hytrel®, Santoprene®
- Polyurethane rubber
- Resilin, Elastin
- Polysulfide rubber

1.6 Synthesis and Polymer Technology

Our core competencies include the development of innovative processes and technologies for the synthesis of polymers and copolymers with new, improved or tailor-made performance characteristics and the setting of foundations for the scale-up of these processes and technologies.

Microcapsules consist of a solid, liquid or gaseous nucleus of active substance and a polymer shell of varying permeability or density. Matrix particles are solid spheres of one or more polymers. The physical structure of matrix particles can range from dense to highly porous. We vary the molecular and macroscopic properties of the particles to tailor the geometric and morphological structure and functionality to specific requirements. Microcapsules and matrix particles can be produced both by non-reactive polymer dripping processes and by reactive processes from suitable monomers or low-molecular-weight prepolymers.

Proven applications include:

- Controlled-release systems for pharmaceutical, cosmetic or agrochemical active ingredients,
- Compatibilization of plastics additives,
- Microparticles with sensor functionalities,
- Microparticles for applications in electroplating.

Tailored materials with defined requirement profiles often consist of composites made from different material classes. To replace metals by lightweight, stable, easily processable polymer materials, the matrix reinforcement of plastics must be improved. Research work worldwide is concentrating on replacement of glass fibers by organic polymer fibers. Our optimized polymer fiber reinforcement attains the strength of glass-fiber-reinforced materials. In this way, component weight and wear and tear on compounding and processing machines can be reduced. Waste disposal is also simplified. Through reactive compounding, it is possible to integrate chemical modification into the extrusion process. In equipment with high mixing efficiency, the heat deflection resistance and modulus of thermoplastics can be increased by reactive compounding with network-forming polymers such as amino or phenolic resins. With heterochain polymers such as polyesters, polyamides or polycarbonates, we also use reactive compounding to produce non-random copolymers.

Industrial polymerization and modification processes are optimized with the help of reaction and process analysis, theoretical models and simulation methods. We have developed special know-how on polycondensation processes and polysaccharide derivatization. Examples include:

- Production of polyesters and polyamides
- Polylactic acid
- New amino resin products
- Hydrophobically modified starches
- Polymer-fiber-reinforced composite materials.

2

Basic Understanding of Polymers and Polymer Nomenclature

2.1 Introduction

A polymer is a large molecule (macromolecule) composed of repeating structural units connected by covalent chemical bonds. The word is derived from the Greek words ðïëõ (poly), meaning "many"; and ττολ (meros), meaning "part". Well known examples of polymers include plastics, DNA and proteins. A simple example is polypropylene whose repeating unit structure is shown at right.

While "polymer" in popular usage suggests "plastic", the term actually refers to a large class of natural and synthetic materials with a variety of properties and purposes. Natural polymer materials such as shellac and amber have been in use for centuries. Biopolymers such as proteins and nucleic acids play crucial roles in biological processes. A variety of other natural polymers exist, such as cellulose, which is the main constituent of wood and paper. Some common synthetic polymers are Bakelite, neoprene, nylon, PVC (polyvinyl chloride), polystyrene, polyacrylonitrile and PVB (polyvinyl butyral). Polymers are studied in the fields of polymer chemistry, polymer physics, and polymer science.

Historical Development

Starting in 1811 Henri Braconnot did pioneering work in derivative cellulose compounds, perhaps the earliest important work in polymer science. The term polymer was coined in 1833 by Jöns Jakob Berzelius. The development of vulcanization later in the nineteenth century improved the durability of the natural polymer rubber, signifying the first popularized semi-synthetic polymer. In 1907, Leo Baekeland created the first completely synthetic polymer by reacting phenol and formaldehyde at precisely controlled temperature and pressure. Bakelite was then publicly introduced in 1909.

Despite significant advances in synthesis and characterization of polymers, a correct understanding of polymer molecular structure did not emerge until the 1920s. Before that, scientists believed that polymers were clusters of small molecules (called colloids), without definite molecular weights, held together by an unknown force, a concept known as association theory. In 1922, Hermann Staudinger proposed that polymers consisted of long chains of atoms held together by covalent bonds, an idea which did not gain wide acceptance for over a decade, and for which Staudinger was ultimately awarded the Nobel Prize. Work by Wallace Carothers in the 1920s also demonstrated that polymers could be synthesized rationally from their constituent monomers. An important contribution to synthetic polymer science was made by the Italian chemist Giulio Natta and the German chemist Karl Ziegler who won the Nobel Prize in Chemistry in 1963 for the development of the Ziegler-Natta catalyst. In the intervening century, synthetic polymer materials such as Nylon, polyethylene, Teflon, and silicone have formed the basis for a burgeoning polymer industry. These years have also shown significant developments in rational polymer synthesis. Most commercially important polymers today are entirely synthetic and produced in high volume, on appropriately scaled organic synthetic techniques.

Synthetic polymers today find application in nearly every industry and area of life. Polymers are widely used as adhesives and lubricants, as well as structural components for products ranging from children's toys to aircraft. They have been employed in a variety of biomedical applications ranging from implantable devices to controlled drug delivery. Polymers such as poly (methyl methacrylate) find application as photoresist materials used in semiconductor manufacturing and low-k dielectrics for use in high-performance microprocessors. Recently polymers have also been employed in the development of flexible polymer-based substrates for electronic displays.

Polymer Structure

The structural properties of a polymer relate to the physical arrangement of monomer residues along the backbone of the chain. Structure has a strong influence on the other properties of a polymer. For example, a linear chain polymer may be soluble or insoluble in water depending on whether it is composed of polar monomers (such as ethylene oxide) or nonpolar monomers (such as styrene). On the other hand, two samples of natural rubber may exhibit different durability even though their molecules comprise the same monomers. Polymer scientists have developed terminology to precisely describe both the nature of the monomers as well as their relative arrangement:

Monomer Identity

The identity of the monomers comprising the polymer is generally the first and most important attribute of a polymer. The repeat unit is the constantly repeated unit of the chain, and is also characteristic of the polymer. Polymer nomenclature is generally based upon the type of monomers comprising the polymer. Polymers that contain only a single type of monomer are known as *homopolymers*, while polymers containing a mixture of monomers are known as *copolymers*. Poly(styrene), for example, is composed only of styrene monomers, and is therefore is classified as a homopolymer. Ethylene-vinyl acetate, on the other hand, contains more than one variety of monomer and is thus a copolymer. Some biological polymers are composed of a variety of different but structurally related monomers, such as polynucleotides composed of nucleotide subunits.

A very common error is to use the term "monomer" to refer to the repeating units of the polymer. In fact, these two things are different. The monomer is the stable molecule that will be used as the polymerization reaction starts. Then, a loss of a minimum of two chemical groups of the monomer forms the repeating unit. A simple example is polyethylene. The monomer is the ethylene (ethene) molecule, while the repeating unit is -C-C-.

A polymer molecule containing ionizable subunits is known as a *polyelectrolyte*. An *ionomer* is a subclass of polyelectrolyte with a low fraction of ionizable subunit.

Tacticity

This property describes the relative stereochemistry of chiral centers in neighboring structural units within a macromolecule. There are three types: isotactic, atactic, and syndiotactic.

Chain Linearity

The simplest form of polymer molecule is a straight chain or *linear* polymer, composed of a single main chain. The flexibility of an unbranched chain polymer is characterized by its persistence length. A *branched polymer* molecule is composed of a main chain with one or more substituent side chains or branches. Special types of branched polymers include star polymers, comb polymers, and brush polymers. If the polymer contains a side chain that has a different composition or configuration than the main chain, the polymer is called a graft or grafted polymer. A *cross-link* suggests a branch point from which four or more distinct chains emanate. A polymer molecule with a high degree of crosslinking is referred to as a *polymer network*. Sufficiently high crosslink concentrations may lead to the formation of an 'infinite network', also known as a 'gel', in which networks of chains are of unlimited extend—essentially all chains have linked into one molecule.

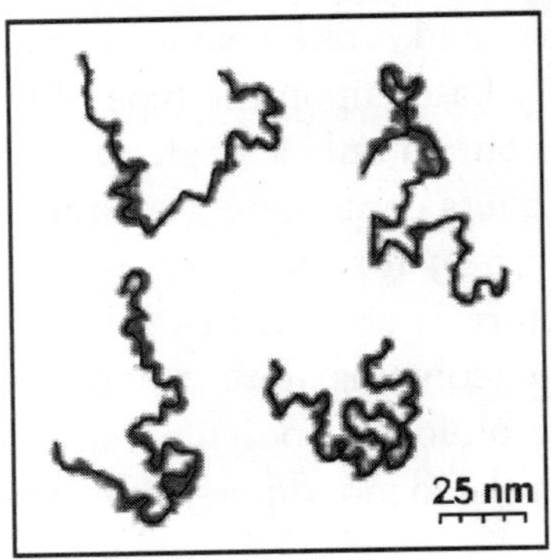

Chain Length

Polymer bulk properties may be strongly dependent on the size of the polymer chain. Like any molecule, a polymer molecule's size may be described in terms of molecular weight or mass. In polymers, however, the molecular mass may be expressed in terms of degree of polymerization, essentially the number of monomer units which comprise the polymer. For synthetic polymers, the molecular weight is expressed statistically to describe the distribution of molecular weights in the sample. This is because of the fact that almost all industrial processes produce a distribution of polymer chain sizes. Examples of such statistics include the number average molecular weight and weight average molecular weight. The ratio of these two values is the polydispersity index, commonly used to express the "width" of the molecular weight distribution.

The maximum length of a polymer chain is its contour length.

Monomer Arrangement in Copolymers

Monomers within a copolymer may be organized along the backbone in a variety of ways:

- *Alternating copolymers* possess regularly alternating monomer residues
- *Periodic copolymers* have monomer residue types arranged in a repeating sequence
- *Random copolymers* have a random sequence of monomer residue types
- *Statistical copolymers* have monomer residues arranged according to a known statistical rule
- *Block copolymers* have two or more homopolymer subunits linked by covalent

bonds. Block copolymers with two or three distinct blocks are called diblock copolymers and triblock copolymers, respectively.

Polymer Properties

Types of polymer 'properties' can be broadly divided into several categories based upon scale. At the nano-micro scale are properties that directly describe the chain itself. These can be thought of as polymer structure. At an intermediate mesoscopic level are properties that describe the morphology of the polymer matrix in space. At the macroscopic level are properties that describe the bulk behavior of the polymer.

The bulk properties of a polymer are those most often of end-use interest. These are the properties that dictate how the polymer actually behaves on a macroscopic scale.

Relationship between Chain Length and Polymer Properties

Polymer bulk properties are strongly dependent upon their structure and mesoscopic behavior. A number of qualitative relationships between structure and properties are known.

Increasing chain length tends to decrease chain mobility, increase strength and toughness, and increase the glass transition temperature (Tg). This is a result of the increase in chain interactions such as Van der Waals attractions and entanglements that come with increased chain length. These interactions tend to fix the individual chains more strongly in position and resist deformations and matrix breakup, both at higher stresses and higher temperatures. Chain length is related to melt viscosity roughly as $1:10^{3.2}$, so that a tenfold increase in polymer chain length results in a viscosity increase of over 1000 times.

Crystallinity

When applied to polymers, the term crystalline has a somewhat ambiguous usage. In some cases, the term crystalline finds identical usage to that used in conventional crystallography. For example, the structure of a crystalline protein or polynucleotide, such as a sample prepared for x-ray crystallography, may be defined in terms of a conventional unit cell composed of one or more polymer molecules with cell dimensions of hundreds of angstroms or more.

A synthetic polymer may be lightly described as crystalline if it contains regions of three-dimensional ordering on atomic (rather than macromolecular) length scales, usually arising from intramolecular folding and/or stacking of adjacent chains. Synthetic polymers may consist of both crystalline and amorphous regions; the *degree of crystallinity* may be expressed in terms of a weight fraction or volume fraction of crystalline material. Few synthetic polymers are entirely crystalline.

The crystallinity of polymers is characterized by their degree of crystallinity, ranging from zero for a completely noncrystalline polymer to one for a theoretical completely crystalline polymer. Increasing degree of crystallinity tends to make a polymer more rigid. It can also lead to greater brittleness. Polymers with a degree of crystallinity approaching zero or one will tend to be transparent, while polymers with intermediate degrees of crystallinity will tend to be opaque due to light scattering by crystalline/ glassy regions.

Tensile Strength

The tensile strength of a material quantifies how much stress the material will endure before failing. This is very important in applications that rely upon polymer's physical strength or durability. For example, a rubber band with a higher tensile strength will hold a greater weight before snapping. In general tensile strength increases with polymer chain length.

Young's Modulus of Elasticity

Young's Modulus quantifies the elasticity of the polymer. It is defined, for small strains, as the ratio of rate of change of stress to strain. Like tensile strength this is highly relevant in polymer applications involving the physical properties of polymers, such as rubber bands.

Transport Properties

Transport properties such as diffusivity relate to how rapidly molecules move through the polymer matrix. These are very important in many applications of polymers for films and membranes.

Melting Point

The term "melting point" when applied to polymers suggests not a solid-liquid phase transition but a transition from a crystalline or semi-crystalline phase to a solid amorphous phase. Though abbreviated as simply "T_m", the property in question is more properly called the "crystalline melting temperature". Among synthetic polymers, crystalline melting is only discussed with regards to thermoplastics, as thermosetting polymers will decompose at high temperatures rather than melt.

Boiling Point

The boiling point of a polymer substance is never defined because polymers will decompose before reaching theoretical boiling temperatures.

Glass Transition Temperature

A parameter of particular interest in synthetic polymer manufacturing is the glass transition temperature (T_g), which describes the temperature at which amorphous polymers undergo a second order phase transition from a rubbery, viscous amorphous solid to a brittle, glassy amorphous solid. The glass transition temperature may be engineered by altering the degree of branching or cross-linking in the polymer or by the addition of plasticizer.

Mixing Behavior

In general, polymeric mixtures are far less miscible than mixtures of small molecule materials. This effect is a result of the fact that the driving force for mixing is usually entropics, not energetics. In other words, miscible materials usually form a solution not because their interaction with each other is more favorable than their self-interaction but because of an increase in entropy and hence free energy associated with increasing the amount of volume available to each component. This increase in entropy scales with the number of particles (or moles) being mixed. Since polymeric molecules are much larger and hence generally have much higher specific volumes than small molecules, the number of molecules involved in a polymeric mixture are far less than the number in a small molecule mixture of equal volume. The energetics of mixing, on the other hand, are comparable on a per volume basis for polymeric and small molecule mixtures. This tends to increase the free energy of mixing for polymer solutions and thus make solvation less favorable. Thus, concentrated solutions of polymers are far rarer than those of small molecules.

In dilute solution, the properties of the polymer are characterized by the interaction between the solvent and the polymer. In a *good solvent*, the polymer appears swollen and occupies a large volume. In this scenario, intermolecular forces between the solvent and monomer subunits dominate over intramolecular interactions. In a *bad solvent* or *poor solvent*, intramolecular forces dominate and the chain contracts. In the *theta solvent*, or the state of the polymer solution where the value of the second virial coefficient becomes 0, the intermolecular polymer-solvent repulsion balances exactly the intramolecular monomer-monomer attraction. Under the theta condition (also called the Flory condition) the polymer behaves like an ideal random coil.

Chain Conformation

The space occupied by a polymer molecule is generally expressed in terms of radius of gyration, which is an average distance from the center of mass of the chain to the chain itself. Alternatively, it may be expressed in terms of pervaded volume, which is the volume of solution spanned by the polymer chain and scales with the cube of the radius of gyration.

Branching

Branching of polymer chains also affect the bulk properties of polymers. Long chain branches may increase polymer strength, toughness, and Tg due to an increase in the number of entanglements per chain. Random length and atactic short chains, on the other hand, may reduce polymer strength due to disruption of organization. Short side chains may likewise reduce crystallinity due to disruption of the crystal structure. Reduced crystallinity may also be associated with increased transparency due to light scattering by small crystalline regions. A good example of this effect is related to the range of physical attributes of polyethylene. High density polyethylene (HDPE) has a very low degree of branching, is quite stiff, and is used in applications such as milk jugs. Low density polyethylene (LDPE), on the other hand, has significant numbers of short branches, is quite flexible, and is used in applications such as plastic films. The branching index of the polymer is a parameter that characterizes the effect of long-chain branches on the size of a branched macromolecule in solution. Dendrimers are a special case of polymer where every monomer unit is branched. This tends to reduce intermolecular chain entanglement and crystallization. Alternatively, dendritic polymers are not perfectly branched, but share similar properties to dendrimers due to their high degree of branching.

Chemical Cross-Linking

Cross linking tends to increase T_g and increase strength and toughness. Cross linking consists of the formation of chemical bonds between chains. Among other applications, this process is used to strengthen rubbers in a process known as vulcanization, which is based on cross linking by sulphur. Car tires, for example, are highly cross linked in order to reduce the leaking of air out of the tire and to toughen their durability. Eraser rubber, on the other hand, is not cross linked to allow flaking of the rubber and prevent damage to the paper.

Inclusion of Plasticizers

Inclusion of plasticizers tends to lower Tg and increase polymer flexibility.

Plasticizers are generally small molecules that are chemically similar to the polymer and create gaps between polymer chains for greater mobility and reduced interchain interactions. A good example of the action of plasticizers is related to polyvinylchlorides or PVCs. A uPVC or unplasticized polyvinylchloride is used for things such as pipes. A pipe has no plasticizers in it because it needs to remain strong and heat resistant. Plasticized PVC is used for clothing for a flexible quality. Plasticizers are also put in some types of cling film to make the polymer more flexible.

Chemical Properties of Polymers

The attractive forces between polymer chains play a large part in determining a polymer's properties. Because polymer chains are so long, these interchain forces are amplified far beyond the attractions between conventional molecules. Different side groups on the polymer can lend the polymer to ionic bonding or hydrogen bonding between its own chains. These stronger forces typically result in higher tensile strength and melting points.

The intermolecular forces in polymers can be affected by dipoles in the monomer units. Polymers containing amide or groups can form hydrogen bonds between adjacent chains; the partially positively charged hydrogen atoms in N-H groups of one chain are strongly attracted to the partially negatively charged oxygen atoms in C = O groups on another. These strong hydrogen bonds, for example, result in the high tensile strength and melting point of polymers containing urethane or urea linkages. Polyesters have dipole-dipole bonding between the oxygen atoms in C = O groups and the hydrogen atoms in H-C groups. Dipole bonding is not as strong as hydrogen bonding, so a polyester's melting point and strength are lower than Kevlar's (Twaron), but polyesters have greater flexibility.

Ethene, however, has no permanent dipole. The attractive forces between polyethylene chains arise from weak van der Waals forces. Molecules can be thought of as being surrounded by a cloud of negative electrons. As two polymer chains approach, their electron clouds repel one another. This has the effect of lowering the electron density on one side of a polymer chain, creating a slight positive dipole on this side. This charge is enough to actually attract the second polymer chain. Van der Waals forces are quite weak, however, so polyethene can have a lower melting temperature compared to other polymers.

Polymer Characterization

The characterization of a polymer requires several parameters which need to be specified. This is because a polymer actually consists of a statistical distribution of chains of varying lengths, and each chain consists of monomer residues which affect its properties.

A variety of lab techniques are used to determine the properties of polymers. Techniques such as wide angle X-ray scattering, small angle X-ray scattering, and small angle neutron scattering are used to determine the crystalline structure of polymers. Gel permeation chromatography is used to determine the number average molecular weight, weight average molecular weight, and polydispersity. FTIR, Raman and NMR can be used to determine composition. Thermal properties such as the glass transition temperature and melting point can be determined by differential scanning calorimetry and dynamic mechanical analysis. Pyrolysis followed by analysis of the fragments is one more technique for determining the possible structure of the polymer. The Thermogravimetry is an useful technique to evaluate the thermal stability of the polymer. Detailed analyses of TG curves also allow to know a bit of the phase segregation in polymers.

Polymer Degradation

A plastic item with thirty years of exposure to heat and cold, brake fluid, and sunlight. Notice the discoloration, swollen dimensions, and tiny splits running through the material. Polymer degradation is a change in the properties—tensile strength, colour, shape, etc. of a polymer or polymer based product under the influence of one or more environmental factors such as heat, light or chemicals. It is often due to the hydrolysis of the bonds connecting the polymer chain, which in turn leads to a decrease in the molecular mass of the polymer. These changes may be undesirable, such as changes during use, or desirable, as in biodegradation or deliberately lowering the molecular mass of a polymer. Such changes occur primarily because of the effect of these factors on the chemical composition of the polymer. Ozone cracking and UV degradation are specific failure modes for certain polymers.

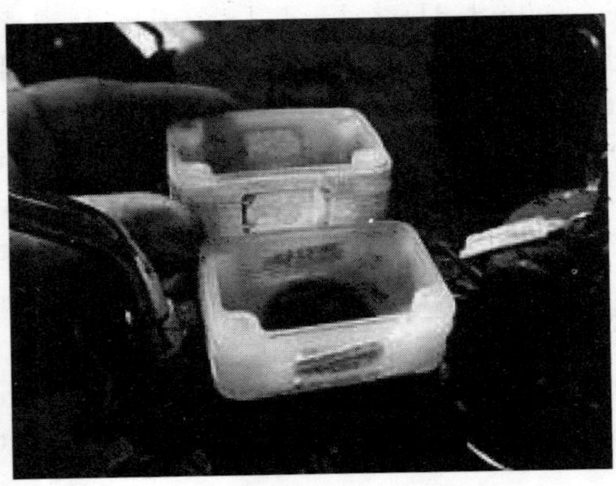

The degradation of polymers to form smaller molecules may proceed by *random scission* or *specific scission*. The degradation of polyethylene occurs by *random scission*—that is by a random breakage of the linkages (bonds) that hold the atoms of the polymer together. When heated above 450 Celsius it degrades to form a mixture of hydrocarbons. Other polymers—like polyalphamethylstyrene—undergo 'specific' chain scission with breakage occurring only at the ends. They literally unzip or depolymerize to become the constituent monomer.

However the degradation process can be useful from the view points of understanding the structure of a polymer or recycling/reusing the polymer waste to prevent or reduce environmental pollution. Polylactic acid and Polyglycolic acid, for example, are two polymers that are useful for their ability to degrade under aqueous conditions. A copolymer of these polymers is used for biomedical applications such as hydrolysable stitches that degrade over time after they are applied to a wound. These materials can also be used for plastics that will degrade over time after they are used and will therefore not remain as litter.

Product Failure

In a finished product such a change is to be prevented or delayed. Failure of safety-critical polymer components can cause serious accidents, such as fire in the case of cracked and degraded polymer fuel lines. Chlorine-induced cracking of acetal resin plumbing joints and polybutylene pipes has caused many serious floods in domestic properties, especially in the USA in the 1990s—2000. Traces of chlorine in the water supply attacked vulnerable polymers in the plastic plumbing, a problem which occurs faster if any of the parts have been poorly extruded or injection moulded. Attack of the acetal joint occurred because of faulty moulding leading to cracking along the threads of the fitting, which are serious stress concentrations.

Polymer oxidation leads to cracking and failure of the parts affected, and has caused accidents involving medical devices. One of the oldest known failure modes is ozone cracking caused by chain scission when ozone gas attacks susceptible elastomers such as natural rubber and nitrile rubber. They possess double bonds in their repeat units which are cleaved during ozonolysis. Cracks in fuel lines can penetrate the bore of the tube, and so cause fuel leakage. If cracking occurs in the engine compartment, electric sparks can ignite the gasoline and can cause a serious fire.

Fuel lines can also be attacked by another form of degradation: hydrolysis. Nylon 6,6 is susceptible to acid hydrolysis and in one accident, a fractured fuel line led to a spillage of diesel into the road. If diesel fuel leaks onto the road, accidents to following cars can be caused by the slippery nature of the deposit, which is like black ice.

2.2 Polymer Fundamentals: Fiber Applications

Polymer science addresses the chemistry and physics of large, chain-like molecules. As with the molecules themselves, this technical pursuit is diverse and complicated. The following discussion provides an introduction to the manufacture and use of synthetic organic polymers for those with some knowledge of basic science. More advanced tutorial information on polymers is contained in the links displayed at the end. Click here to go directly there.

What is a Polymer?

The term "polymer" is derived from the Greek "poly", meaning "many", and "mer", meaning "parts"—thus polymers are substances made of "many parts". In most cases the parts are small molecules which react together hundreds, or thousands, or millions of times. A molecule used in producing a polymer is a "monomer"—mono is Greek for single, thus a monomer is a "single part". A polymer made entirely from molecules of one monomer is referred to as a "homopolymer". Chains that contain two or more different repeating monomers are "copolymers".

The resulting molecules may be long, straight chains, or they may be branched, with small chains extending out from the molecular "backbone". The branches also may grow until they join with other branches to form a huge, three-dimensional matrix. Variants of these molecular shapes are among the most important factors in determining the properties of the polymers created.

The size of polymer molecules is important. This is usually expressed in terms of molecular weight. Since a polymeric material contains many chains with the same repeating units, but with different chain lengths, average molecular weight must be used. In general, higher molecular weights lead to higher strength. But as polymer chains get bigger, their solutions, or melts, become more viscous and difficult to process.

Proteins and Carbohydrates

Life as we know it could not exist without polymers. Proteins, with large numbers of amino acids joined by amide linkages, perform a wide variety of vital roles in plants and animals. Carbohydrates, with chains made up of repeating units derived from simple sugars, are among the most plentiful compounds in plants and animals. Both of these natural polymers are important for fibers. Proteins are the basis for wool, silk and other animal-derived filaments. Cellulose as a carbohydrate occurs as cotton, linen and other vegetable fibers. The properties of these fibers are limited by the form provided in their natural state. Some, like linen and silk, are difficult to isolate from their sources, which makes them scarce and expensive. There are, of course, many other sources of proteins and cellulose, such as wood pulp, but natural polymers tend to be very difficult to work with and form into fibers or other useful structures. The inter-chain forces tend to be strong because of the large number of polar groups in the molecular chains. Thus, natural polymers usually have melting points that are so high that they degrade before they liquefy.

The most useful molecules for fibers are long chains with few branches and a very regular, extended structure. Thus, cellulose is a good fiber-former. It has few side chains or linkages between the sugar units forcing its chains into extended configurations. However, starches, which contain the same basic sugar units, do not form useful fibers because their chains are branched and coiled into almost spherical configurations.

Synthetic Polymers

Synthetic polymers offer more possibilities, since they can be designed with molecular structures that impart properties for desired end uses. Many of these polymers are capable of dissolving or melting, allowing them to be extruded into the long, thin filaments needed to make most textile products. Synthetic polymer fibers can be made with regular structures that allow the chains to pack together tightly, a characteristic that gives filaments good strength. Thus, filaments can be made from some synthetic polymers that are much lighter and stronger than steel. Bullet-proof vests are made from synthetic fibers.

There are two basic chemical processes for the creation of synthetic polymers from small molecules (1) condensation, or step-growth polymerization, and (2) addition, or chain-growth polymerization.

Step-Growth Polymerization

In step-growth polymerization, monomers with two reactive ends join to form dimers (two "parts" joined together), then "trimers" (three "parts"), and so on. However, since each of the newly formed oligomers (short chains containing only a few parts) also has two reactive ends, they can join together; so a dimer and a trimer would form a pentamer (five repeating "parts"). In this way the chains may quickly great length achieve large size. This form of step-growth polymerization is used for the manufacture of two of the most important classes of polymers used for textile fibers, polyamide (commonly known as nylon), and polyester.

There are many different commercial versions of polyester in a wide variety of applications, including plastics, coatings, films, paints, and countless other products. The polymer usually used for textile fibers is poly(ethylene terephthalate), or PET, which is formed by reacting ethylene glycol with either terephthalic acid or dimethyl terephthalate. Antimony oxide is usually added as a catalyst, and high vacuum is used to remove the water or methanol byproducts. High temperature (>250°C) is necessary to provide the energy for the reaction, and to keep the resultant polymer in a molten state.

PET molecules are regular and straight, so their inter-chain forces are strong—but not strong enough to prevent melting. Thus, PET is a "thermoplastic" material; that is, it can be melted and then solidified to form specific products. Since its melting point is high, it does not soften or melt at temperatures normally encountered in laundering or drying. Another important property of PET is its T_g, or "glass transition temperature". When a polymer is above its glass transition temperature, it is easy to change its shape. Below its T_g, the material is dimensionally stable and it resists changes in shape. This property is very important for textile applications because it allows some fibers, and the fabrics made from them, to be texturized or heat-set into a given shape. This can provide bulk to the yarn, or wrinkle resistance to the fabric. These set-in shapes remain permanent as long as the polymer is not heated above its T_g. Because its chains are closely packed and its ester groups do not form good hydrogen bonds, polyesters are also hydrophobic (i.e., they do not absorb water). This property also requires special dyeing techniques.

There are also many important classes of synthetic polyamides (nylons) and they have a wide variety of commercial uses. These are usually distinguished from each other by names based on the number of carbon atoms contained in their monomer units. As with polyesters, polyamides are formed by step-growth polymerization of monomers possessing two reactive groups. Here, the reactive functions are acids and amines. The monomers used may have their two reactive functions of the same chemical type (both acids, or both amines), or of different types. Thus, nylon 6,6—a very common fiber polymer—is made by reacting molecules of adipic acid (containing six carbons in a chain, with an acid function at each end) with hexamethylene diamine (also six carbon atoms, with amine functions at each end). In another variant the diamine contains ten carbons atoms, the product designated nylon 6,10.

The other common polyamide fiber polymer is nylon 6. Its monomer has six carbons in the chain, with an amine at one end and an acid at the other. Thus only one form of monomer is needed to conduct the reaction.

Commercial production of nylon 6 makes use of caprolactam, a derivative which provides the same result.

As with the polyesters, nylons have regular structures to allow good inter-chain forces that impart high strength. Both nylon 6 and nylon 6,6 have melting points similar to PET but they have a lower T_g. Also, since the amide functions in nylon chains are good at hydrogen bonding, nylons can be penetrated by water molecules. This allows them to be dyed from aqueous media, unlike their polyester counterparts.

In addition to nylon, there is another commercially important group of synthetic polyamides. These are the aramids, which contain aromatic rings as part of their polymer chain backbone. Due to the stability of their aromatic structures and their conjugated amide linkages, the aramids are characterized by exceptionally high strength and thermal stability. Their usefulness for common textile applications is limited by their high melting points and by their insolubility in common solvents. They are expensive to fabricate, and they carry an intrinsic color that ranges from light yellow to deep gold.

Other step-growth polymers—the polyurethanes—are produced by the reaction of polyols and polyisocyanates. For fiber purposes, this class of linear polymers is formed from glycols and diisocyanates. Usually, the reactions are carried out to form block copolymers containing at least two different chemical structures—one rigid, and the other flexible. The flexible segments stretch, while the rigid sections act as molecular anchors to allow the material to recover its original shape when the stretching force is removed. Varying the properties of the segments, and the ratio of flexible to rigid segments controls the amount of stretch. Fibers made in this way are classified as spandex and they are used widely in apparel where stretch is desirable.

Chain-Growth Polymerization

Chain-growth polymerization occurs when an activated site on a chemical, such as a free radical or ion, adds to a double bond, producing a new bond and a new by activated location. That location then attacks another double bond, adding another unit to the chain, and a new reactive end. The process may be repeated thousands, or millions, of times, to produce very large molecules. This is usually a high energy process and the intermediate species are so reactive that, in addition to attacking available monomer, they also may attack other chains, producing highly branched structures. Since these branches prevent the molecules from forming regular structures with other molecules, their inter-chain forces are weak. The resulting polymers tend to be low-melting and waxy.

The breakthrough in making chain-growth polymers useful for fibers and for most commercial plastics came with the development of special selective catalysts that drive the production of long, straight polymer chains from monomers containing basic carbon-to-carbon double bonds.

Ethylene and propylene form the simplest chain-growth polymers. Since their polymer chains contain no polar groups, these polyolefins must rely on close contact between the molecular chains for strength. Thus, the physical characteristics of polyethylene are very sensitive to even a small number of chain branches. Very straight chains of polyethylene can form strong crystalline structures which exhibit exceptional strength. Protective fabrics made from this type of highly structured polyethylene are virtually impossible to penetrate or cut.

Polypropylene is more complicated. Even without chain branching, each monomer unit adds one methyl group pendant to the chain. The arrangement of these side groups is described as the "tacticity" of the polymer. A random arrangement is considered "atactic", or without tacticity. Regular arrangement with all side groups on one side of the chain is "isotactic", and a regular alternating structure is "syndiotactic". Polypropylene molecules can only pack closely in an isotactic arrangement. Synthesis of these polymers was a major challenge, but several stereoselective catalysts are now available, and high-density polypropylene has become a commodity product. Fibers made from it are lightweight, hydrophobic and highly crystalline. Their resistance to wetting gives them good moisture wicking and anti-staining properties. This also makes them virtually undyeable, except when the dye is applied to the polymer in its molten state—a process know as "solution dyeing".

By contrast, the pendant nitrile functions in polyacrylonitrile are sufficiently polar to produce very strong inter-chain forces. Pure homopolymers from acrylonitrile are non-thermoplastic and difficult to dissolve or dye. Thus, for most commercial acrylonitrile polymers, small amounts of other monomers with bulky side chains are introduced to force the chains apart, to reduce the inter-chain forces. Common co-monomers for these fiber applications include vinyl chloride, vinyl acetate, acrylic acid, and methyl acrylate.

Specialty Fiber Polymers

There are also a number of complex, specialty fiber polymers with methods of synthesis that are not easily classified. These materials are occasionally used in high performance materials where the complex structures impart exceptional strength, thermal stability, electrical conductivity, and others desirable properties. They include PBI (polybenzimidazole) and sulfar. Their chemistry is beyond the scope of this introductory discussion.

2.3 Polymer Science: Overview

In the late 1960s, IUPAC formally established the Commission on Macromolecular Nomenclature (Commission IV. 1), which has since become the leading nomenclature body in the polymer field; it is now superseded by the Subcommittee on Macromolecular Terminolgy. This group promulgates rules and definitions on polymer terminology and nomenclature.

The Subcommittee (*sic* Commission) has developed rules for naming regular, single-strand organic and inorganic macromolecules, copolymer molecules, irregular macromolecules, ladder and spiro macromolecules, and nonlinear and network macromolecules. It has developed rules for representing the structures of macromolecules. It also works to standardize the terminology used in polymer science and has developed definitions for the basic terms dealing with polymer molecules, assemblies of polymer molecules and non-linear macromolecules, polymer solutions, polymer crystals, types of polymerization, and the degradation and aging, and mechanical behavior.

Since its beginning, this group has attracted the participation of a number of scientists from academia, the publishing industry, and the polymer industry.

2.4 Polymer Nomenclature

The classification system used in the rubber industry for naming elastomers is based on that described in ISO 1629-1976. The last letter of the identification code defines the basic group to which the polymer belongs whilst the earlier ones provide more specific information and in many cases define the polymer absolutely. The list below is not exhaustive but covers many of the more common elastomers or rubbers and some "rubber-like" materials.

'M' Group: Rubbers with a Saturated -C-C- Main Chain

IM : Polyisobutylene (e.g. VISTANEX), a soft inert polymer.

EPM : Copolymer of ethylene and propylene, rubber-like material.

EPDM: A terpolymer of ethylene, propylene and a di- or polyene giving pendent olefin groups as crosslinking sites (e.g. NORDEL).

CSM : Chlorosulphonated polyethylene (e.g. HYPALON), containing both C-Cl and C-SO2CI groups. Cl content 20-45 per cent; S content 0.5-2.5 per cent. Optimum properties 30 per cent Cl, 1.5 per cent S; ozone-resistant rubber also used in varnishes.

FPM : Fluoro/fluoroalkyl groups on C-C backbone (e.g. VITON, FLUOREL—copolymers of hexafluoropropylene and vinylidene fluoride).

(e.g. TECHNOFLON copolymer of vinylidene fluoride and 1-hydropentafluoropropylene.

CFM : As FPM, but containing Cl as well as F; vinylidene fluoride (VF): chlorotrifluoroethylene (CTFE) copolymer (e.g. VOLTALEF, KEL F).

'O' Group: Rubbers with Carbon and Oxygen in the Main Chain

These materials have good heat resistance and good low temperature properties.

COPoly(epichlorohydrin) (HERCLOR H)—the parent material from which comes:

ECO : Copolymer of epichlorohydrin and ethylene oxide (HERCLOR C).
GPO : Copolymer of propylene oxide and allyl glycidyl ether (PAREL).

'Q' Group: Silicone Rubbers

These are all relatively stable thermally and because of their cold cure characteristics may be used as electrical insulants, seals, moulds, etc.

MQ : Polydimethylsiloxane; depending on the molar mass this can be an oil, wax or rubber.
MPQ : As MQ with the addition of phenylmethylsiloxane.
MPVQ: As above but with *vinyl* groups.
MFQ : As MQ but fluorinated.

'R' Group: Rubbers with an Unsaturated Carbon Backbone

ABR : Refers to copolymers of butadiene and methyl methacrylate (e.g. BUTAKON ML) used to impregnate paper but also includes the terpolymer with acrylonitrile (primer, before adhesive layer applied) and tetrapolymer with styrene (used as a synthetic rubber).

BR : Poly(butadiene)—available as high cis (98%+), high trans (98%+) and anywhere in between. Can also have vinyl groups present at any level. General-purpose rubbers usually 90 per cent + cis or about 45 per cent cis 45 per cent trans 10 per cent vinyl. High vinyls have some specialist uses.

CR : Poly(ß-chlorobutadiene) (e.g. CHLOROPRENE, NEOPRENE). Two main types, 'G', amber in colour with large molar mass range centred at about 100 000; 'W', white, molar mass of narrower range and centred about 200 000. Used as an adhesive or where oil or ozone resistance required; gaskets, subaqua suits, etc.

IIR : Copolymer of isobutylene and diene such as butadiene or isoprene

(BUTYL). Only a small amount of diene added (ca 2-5%) to give crosslinkable sites. Has low gas permeability, hence uses in inflatable products, and as general-purpose rubber.

CIIR : Chlorinated IIR} with 2-3 per cent w/w halogen to decrease gas.

BIIR : Brominated IIR} permeability and improve self-adhesion on building.

IR : Synthetic *cis*-poly(isoprene) (e.g. CARIFLEX, NATSYN, SKI3) *cis* level 90-99 per cent, remainder *trans* and *vinyl*. General-purpose rubber.

NBR : Copolymer; acrylonitrile and butadiene (e.g. KRYNAC, NITRILE) available with a wide range of ACN loadings to alter hardness; oil-resistant applications. Also available is terpolymer (see ABR) and tetrapolymer with styrene.

NR : *Cis*-poly(isoprene) natural rubber, essentially 100 per cent *cis*, *trans/vinyl* <0.1 per cent. Contains about 95 per cent polyisoprene. Various grades available RSS, SMR, SIR, SLR, NIG with number identifying grade—5, 10, 20. Also modified NR—PA, SP, OENR, ENR, DPNR. NR was the original general purpose (GP) rubber.

SBR : Random copolymer of styrene and butadiene. Styrene level varies from 10 to 80 per cent but the general purpose level is 23.5 per cent. Many types available and the exact type identified by a numeric code. General purpose rubber. Vast amounts used in tyres. Also available as ter/tetra polymer systems (see ABR & NBR).

'T' Group: Rubbers with Carbon, Oxygen and Sulphur in the Main Chain

OT : Polymer of bischloroalkylether (or formal), with sulphur. Most common one uses bis-2-chloroethylformal; $CH_2(OCH_2CH_2Cl)_2$ (with a little 1,2,3-trichloropropane for crosslinking) THIOKOL ST.

EOT : As above, but copolymerized with ethylene dichloride. All of these smell strongly of sulphur and are used for oil and solvent seals. The liquid polymers cold cure and are used as sealants in the building trade. Popular ones include Poly(ethylene disulphide) and Poly(butyl ether disulphide).

'U' Group: Polymer Chain Contains Carbon, Oxygen and Nitrogen

A wide range of materials used as oil-resistant materials, in oxidation-resisting applications and as lightweight shoe soling.

AU : Polyesterurethanes.

EU : Polyether urethanes.

Although not elastomers, certain other polymeric materials can exhibit "rubber-like" behaviour:

PVC : Poly(vinylchloride); hard brittle material (d = 1.4) often copolymerised with vinylidine chloride, vinyl acetate, styrene, ABR, ethylene vinyl acetate etc. for a wide range of applications. When plasticized, usually with esters such as phthalates, it becomes quite 'rubbery', used in conveyer belts, paints, varnishes, floor coverings, erasers (rubbers), flexible tubing, wellington boots and many cheap "rubber" goods. Thermoplastic.

PE : Polyethylene; a wide range of types available—HDPE (high density PE) and LDPE (low density PE). Numerous applications—medical implants to polythene bags, blended with elastomers such as EPDM to produce thermoplastic elastomers. Type distinguished by their melting points.—LDPE <110°C, HDPE up to 136°C. A reclaimed mixture shows a spread of melting points.

PP : Polypropylene; similar applications to PE but higher melting (165°C). Also used to make thermoplastic elastomers.

PS : Polystyrene; occasionally met as a reinforcing plastic within a continuous elastomeric phase (e.g.shoe soling) but can be considered to be present in some thermoplastic elastomers such as the block copolymers:

1. SIS styrene-isoprene-styrene
2. SBS styrene-butadiene-styrene.

- *Chlorinated rubber*: Refers specifically to chlorinated natural rubber. Used for paints and adhesives. The theoretical level for (C5H8Cl2)n is 51 per cent but commercial chlorinated rubber contains 65 per cent Cl.
- *Rubber hydrochloride*: Again refers specifically to hydrochlorinated natural rubber—usually with about 90 per cent of the double bonds hydrochlorinated (30% Cl). Plasticized material produced as film (e.g. PLIOFILM) was used for packaging.
- *M. G. Rubber*: Natural rubber to which methyl methacrylate has been grafted, commercial materials generally contain 30 per cent or 49 per cent w/w methacrylate.
- *GuttaPercha and Balata*: Polyisoprene, with 100 per cent of the units trans; pure material not unlike PVC in feel and, when plasticized, can have similar uses.
- *Chicle*: A naturally occurring mixture of cis and trans polyisoprene (25:75), with resins, used in chewing-gum.

- *Guayule*: Natural *cis*-polyisoprene isolated from the shrub Parthenium argentatum by solvent extraction. Uses and properties as for NR, but smell reminiscent of gin. Efforts to develop commercial exploitation have not been particularly successful.

3

Polymer Characterization

3.1 Introduction

Polymer characterization is the analytical branch of polymer science. The discipline is concerned with the characterization of polymeric materials on a variety of levels. The characterization typically has as a goal to improve the performance of the material which implies that the characterization should ideally also be linked to the parameters that are related to the desirable properties of the material such as strength, impermeability, toughness, optical poperties and the like

Chemical Structure

The chemical structure of many polymers is rather complex because the polymerization reaction does not necessarily produce identical molecules as in the case in the DNA-coded synthesis of biopolymers. A polymeric material typically consists of a distribution of molecular sizes and sometimes also of shapes. Chromatographic methods like Size Exclusion Chromatography often in combination with LALLS and or Viscometry can be used to determine the molecular weight distribution as well as the degree of long chain branching of a polymer, provided a suitable solvent can be found. Copolymers with short chain branching such as Linear Low-density Polyethylene (a copolymer of ethylene and and higher alkene such as hexene of octene) require a different approach. There are fractional elution technique (ATREF) that can reveal how the short chain branches are distributed over the various molecular weight.

Thermal Properties

A true workhorse for polymer characterization is thermal analysis, particularly Differential scanning calorimetry. Changes in the compositional and structural parameters of the material usually affect its melting transitions or glass transitions and these in trun can be linked to many performance parameters. For semicrystalline polymers it is an important method to measure crystallinity.

Mechanical and Dielectric Spectroscopy

Dynamic mechanical spectroscopy and Dielectric spectroscopy are essentially extensions of thermal analysis that can reveal more subtle transitions with temperature as they affect the complex modulus or the dielectric function of the material.

Morphology

Morphological parameters, particularly on a mesoscale (nanometers to microns) are very important for the mechanical properties of many materials. Transmission Electron Microscopy in combination with staining techniques, but also Scanning Electron Microscopy, Scanning Probe Microscopy and other forms of microscopy are important tools to optimize the morphology of materials like polybutadiene-polystyrene polymers and many polymer blends.

X-ray diffraction is generally not as powerful for this class of materials as they are either amorphous or poorly crystallized. The small angle range (Small Angle X-ray Scattering: SAXS can be used to measure the long periods of semicrystalline polymers.

Other Techniques

- Solid state NMR
- Spectroscopic techniques: FTIR etc.

3.2 Size Exclusion Chromatography

Size exclusion chromatography (SEC) is a chromatographic method in which particles are separated based on their size, or in more technical terms, their hydrodynamic volume. It is usually applied to large molecules or macromolecular complexes such as proteins and industrial polymers. When an aqueous solution is used to transport the sample through the column, the technique is known as gel filtration chromatography. The name gel permeation chromatography is used when an organic solvent is used as a mobile phase. The main application of gel filtration chromatography is the fractionation of proteins and other water-soluble polymers, while gel permeation chromatography is used to analyze the molecular weight distribution of organicsoluble polymers. Either technique should not be confused with gel electrophoresis, where an electric field is used to “pull” or “push” molecules through the gel depending on their electrical charges.

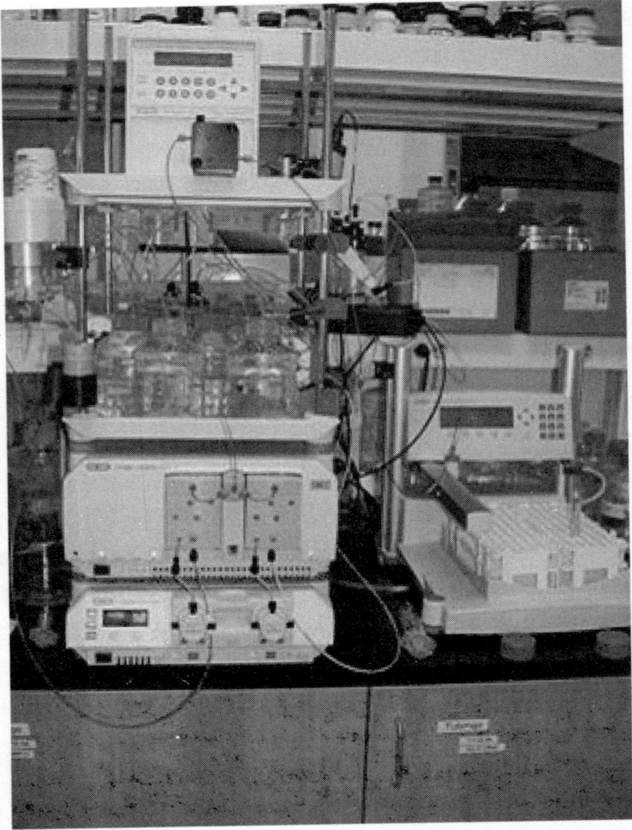

Fig. 3.1: Equipment for running size exclusion chromatography. The buffer is pumped through the column (right) by a computer controlled device.

SEC is a widely used technique for the purification and analysis of synthetic and biological polymers, such as proteins, polysaccharides and nucleic acids. Biologists and biochemists typically use a gel medium—usually polyacrylamide, dextran or agarose—and filter under low pressure. Polymer chemists typically use either a silica or crosslinked polystyrene medium under a higher pressure. These media are known as the stationary phase.

The advantage of this method is that the various solutions can be applied without interfering with the filtration process, while preserving the biological activity of the particles to be separated. The technique is generally combined with others that further separate molecules by other characteristics, such as acidity, basicity, charge, and affinity for certain compounds.

Discovery

The technique was invented by Grant Henry Lathe and Colin R Ruthven, working at Queen Charlotte's Hospital, London. They later received the John Scott Award for this invention. While Lathe and Ruthven used starch gels as the matrix, Porath and Flodin later introduced dextran gels; other gels with size fractionation properties include agarose and polyacrylamide. A short review of these developments has appeared

Theory and Method

The underlying principle of SEC is that particles of different sizes will elute (filter) through a stationary phase at different rates. This results in the separation of a solution of particles based on size. Provided that all the particles are loaded simultaneously or near simultaneously, particles of the same size should elute together.

This is usually achieved with an apparatus called a column, which consists of a hollow tube tightly packed with extremely small porous polymer beads designed to have pores of different sizes. These pores may be depressions on the surface or channels through the bead. As the solution travels down the column some particles enter into the pores. Larger particles cannot enter into as many pores. The larger the particles, the less overall volume to traverse over the length of the column, and the faster the elution.

The filtered solution that is collected at the end is known as the eluate. The *void volume* includes any particles too large to enter the medium, and the solvent volume is known as the *column volume*.

Factors Affecting Filtration

In real life situations particles in solution do not have a constant, fixed size, resulting in the probability that a particle which would otherwise be hampered by a pore may pass right by it. Also, the stationary phase particles are not ideally defined; both particles and pores may vary in size. Elution curves therefore resemble Gaussian distributions. The stationary phase may also interact in undesirable ways with a particle and influence retention times, though great care is taken by column manufacturers to use stationary phases which are inert and minimize this issue.

Like other forms of chromatography, increasing the column length will enhance the resolution, and increasing the column diameter increases the capacity of the column. Proper column packing is important to maximize resolution: an overpacked column can collapse the pores in the beads, resulting in a loss of resolution. An underpacked column can reduce the relative surface area of the stationary phase accessible to smaller species, resulting in those species spending less time trapped in pores.

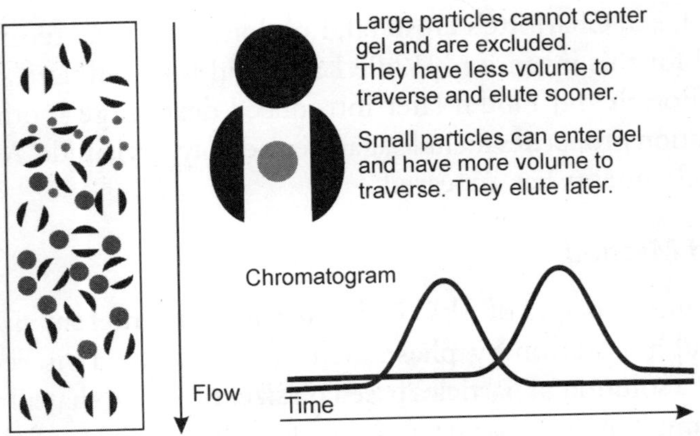

Fig. 3.2: A cartoon illustrating the theory behind size exclusion chromatography.

Analysis

In simple manual columns the eluent is collected in constant volumes, known as fractions. The more similar the particles are in size, the more likely they will be in the same fraction and not detected separately. More advanced columns overcome this problem by constantly monitoring the eluent.

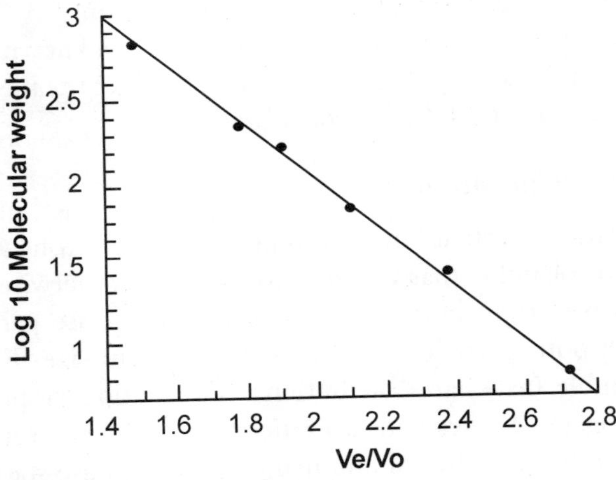

Fig. 3.3: Standardization of a size exclusion column.

The collected fractions are often examined by spectroscopic techniques to determine the concentration of the particles eluted. Three common spectroscopy detection techniques are refractive index (RI), evaporative light scattering (ELS), and ultraviolet (UV). When eluting spectroscopically similar species (such as during biological purification) other techniques may be necessary to identify the contents of each fraction. It is also possible to analyse the eluent flow continuously with RI, LALLS UV and or viscosity measurements.

The elution volume (Ve) decreases roughly linearly with the logarithm of the molecular hydrodynamic volume (for globular proteins this is proportional to molecular weight). Columns are often calibrated using 4-5 standard samples (e.g., folded proteins of known molecular weight), and a sample of the very large molecule blue dextran to determine the void volume. The elution volumes of the standards are divided by the elution volume of the blue dextran (Ve/Vo) and plotted against the log of the standards' molecular weights.

Applications

Proteomics

SEC is generally considered a low resolution chromatography as it does not discern similar species very well, and is therefore often reserved for the final "polishing" step of a purification. The technique can determine the quaternary structure of purified proteins which have slow exchange times, since it can be carried out under native solution conditions, preserving macromolecular interactions. SEC can also assay protein tertiary structure as it measures the hydrodynamic volume (not molecular weight), allowing folded and unfolded versions of the same protein to be distinguished. For example, the apparent hydrodynamic radius of a typical protein domain might be 14 Å and 36 Å for the folded and unfolded forms respectively. SEC allows the separation of these two forms as the folded form will elute much later due to its smaller size. Alternatively, folded and unfolded versions of the same metalloproteins can be separated according to their different isoelectric points by using quantitative preparative native continuous polyacrylamide gel electrophoresis (QPNC-PAGE).

Polymer Synthesis

SEC can be used as a measure of both the size and the polydispersity of a synthesised polymer—that is, the ability to be able to find the distribution of the sizes of polymer molecules. If standards of a known size are run previously, then a calibration curve can be created to determine the sizes of polymer molecules of interest. Alternatively, techniques such as light scattering and/or viscometry can be used online with SEC to yield absolute molecular weights that do not rely on calibration with standards of known molecular weight. Due to the difference in size of two polymers with identical molecular weights, the absolute determination methods are generally more desirable. A typical SEC system can quickly (in about half an hour) give polymer chemists information on the size and polydispersity of the sample.

3.3 Viscometry, Viscometer and Viscosity

The basis for determination of molecular weight according to the Staudinger method (since replaced by the more general Mark-Houwink equation) is the fact that relative viscosity of suspensions depends on volumetric proportion of solid particles. A viscometer (also called *viscosimeter*) is an instrument used to measure the viscosity of a fluid. For liquids with viscosities which vary with flow conditions, an instrument called a rheometer is used. Viscometers only measure under one flow condition.

In general, either the fluid remains stationary and an object moves through it, or the object is stationary and the fluid moves past it. The drag caused by relative motion of the fluid and a surface is a measure of the viscosity. The flow conditions must have a sufficiently small value of Reynolds number for there to be laminar flow.

At 20.00 degrees Celsius the viscosity of water is 1.002 mPa·s and its kinematic viscosity (ratio of viscosity to density) is 1.0038 mm^2/s. These values are used for calibrating certain types of viscometer.

Standard Laboratory Viscometers for Liquids

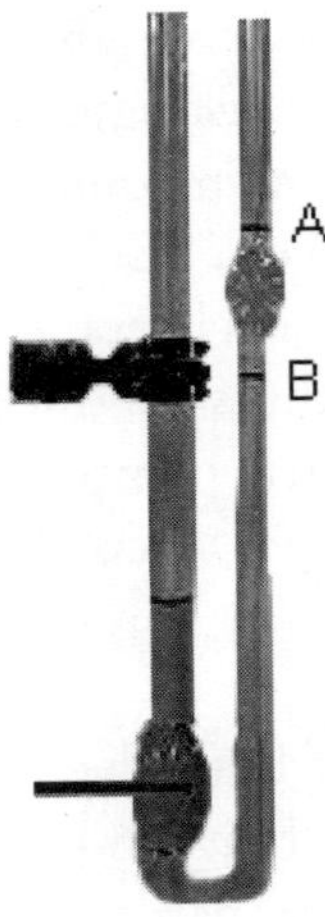

Fig. 3.4: These viscometers measure the viscosity of a fluid with a known density.

U-tube Viscometers

These are also known as Ostwald viscometers named after Wilhelm Ostwald or glass capillary viscometers. Another type is the Ubbelohde viscometer. They basically consist of a glass tube in the shape of a U held vertically in a controlled temperature bath. In one arm of the U is a vertical section of precise narrow bore (the capillary). Above this is a bulb, there is another bulb lower down in the other arm. In use, liquid is drawn into the upper bulb by suction, then allowed to flow down through the capillary into the lower bulb. Two marks (one above and one below the upper bulb) indicate a known volume. The time taken for the level of the liquid to pass between these marks is proportional to the kinematic viscosity. Most commercial units are provided with a conversion factor, or can be calibrated by a fluid of known properties.

The time it takes for the test liquid to flow through a capillary of a known diameter of a certain factor between 2 marked points is measured. By multiplying the time taken by the factor of the viscometer, the kinematic viscosity is obtained. The viscometers are usually placed in a constant temperature water bath as temperature affects viscosity.

Such viscometers are also classified as direct flow or reverse flow. Reverse flow viscometers have the reservoir above the markings and direct flow are those with the reservoir below the markings. Such classifications exists so that the level can be determined even when opaque or staining liquids are measured, otherwise the liquid will cover the markings and make it impossible to gauge the time the level passes the mark. This also allows the viscometer to have more than 1 set of marks to allow for an immediate timing of the time it takes to reach the 3rd mark, therefore yielding 2 timings and allowing for subsequent calculation of Determinability to ensure accurate results.

Falling Sphere Viscometers

Stokes' law is the basis of the falling sphere viscometer, in which the fluid is stationary in a vertical glass tube. A sphere of known size and density is allowed to descend through the liquid. If correctly selected, it reaches terminal velocity, which can be measured by the time it takes to pass two marks on the tube. Electronic sensing can be used for opaque fluids. Knowing the terminal velocity, the size and density of the sphere, and the density of the liquid, Stokes' law can be used to calculate the viscosity of the fluid. A series of steel ball bearings of different diameter is normally used in the classic experiment to improve the accuracy of the calculation. The school experiment uses glycerine as the fluid, and the technique is used industrially to check the viscosity of fluids used in processes. It includes many different oils, and polymer liquids such as solutions.

In 1851, George Gabriel Stokes derived an expression for the frictional force (also called drag force) exerted on spherical objects with very small Reynolds numbers (e.g., very small particles) in a continuous viscous fluid by solving the small fluid-mass limit of the generally unsolvable Navier-Stokes equations:

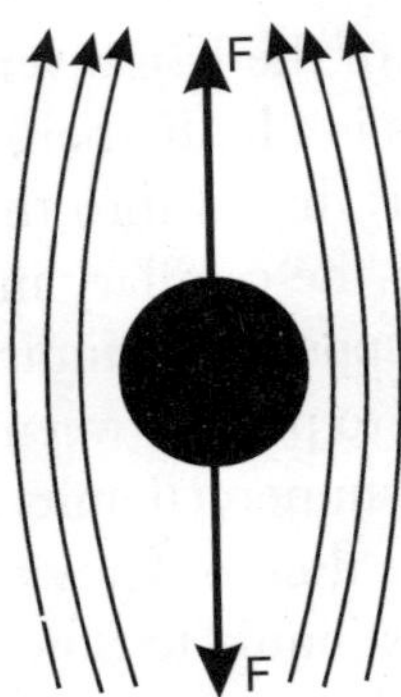

Fig. 3.5: Creeping Flow Past a Sphere.

$$F = 6\pi r \eta \upsilon$$

where:

F is the frictional force,
r is the radius of the spherical object,
η is the fluid viscosity, and
υ is the particle's velocity.

If the particles are falling in the viscous fluid by their own weight, then a terminal velocity, also known as the settling velocity, is reached when this frictional force combined with the buoyant force exactly balance the gravitational force. The resulting settling velocity (or terminal velocity) is given by:

$$V_s = \frac{2}{9}\frac{r^2 g(\rho_p - \rho_f)}{\mu}$$

where:

V_s is the particles' settling velocity (m/s) (vertically downwards if $\rho_p > \rho_f$, upwards if $\rho_p < \rho_f$),
r is the Stokes radius of the particle (m),
g is the gravitational acceleration (m/s^2),
ρ_p is the density of the particles (kg/m^3),
ρ_f is the density of the fluid (kg/m^3), and
μ is the (dynamic) fluid viscosity (Pa s).

Note that Stokes flow is assumed, so the Reynolds number must be small.

Vibrational Viscometers

Vibrational viscometers date back to the 1950s Bendix instrument, which is of a class that operates by measuring the damping of an oscillating electromechanical resonator immersed in a fluid whose viscosity is to be determined. The resonator generally oscillates in torsion or transversely (as a cantilever beam or tuning fork). The higher the viscosity, the larger the damping imposed on the resonator. The resonator's damping may be measured by one of several methods:

1. Measuring the power input necessary to keep the oscillator vibrating at a constant amplitude. The higher the viscosity, the more power is needed to maintain the amplitude of oscillation.
2. Measuring the decay time of the oscillation once the excitation is switched off. The higher the viscosity, the faster the signal decays.
3. Measuring the frequency of the resonator as a function of phase angle between excitation and response waveforms. The higher the viscosity, the larger the frequency change for a given phase change.

The vibrational instrument also suffers from a lack of a defined shear field, which makes it unsuited to measuring the viscosity of a fluid whose flow behaviour is not known before hand. Vibrating viscometers are rugged industrial systems used to measure viscosity in the process condition. The active part of the sensor is a vibrating rod. The vibration amplitude varies according to the viscosity of the fluid in which the rod is immersed. These viscosity meters are suitable for measuring clogging fluid and high-viscosity fluids even with fibers (up to 1,000 Pa·s). Currently, many industries around the world consider these viscometers as the most efficient system to measure viscosity of any fluid, contrasted to rotational viscometers, which require more maintenance, inability to measure clogging fluid, and frequent calibration after intensive use. Vibrating viscometers has no moving parts, no weak parts and the sensitive part is very small. Actually even the very basic or acid fluid can be measured by adding a special coating or by changing the material of the sensor to a material such as 316L, SUS316, Hastelloy, or enamel.

Rotational Viscometers

Rotational viscometers use the idea that the torque required to turn an object in a fluid, can indicate the viscosity of that fluid. The common Brookfield-type viscometer determines the required torque for rotating a disk or bob in a fluid at known speed. 'Cup and bob' viscometers work by defining the exact volume of sample which is to be sheared within a test cell, the torque required to achieve a certain rotational speed is measured and plotted. There are two classical geometries in "cup and bob" viscometers, known as either the "Couette" or "Searle" systems—distinguished by whether the cup or bob rotates. The rotating cup is preferred in some cases, because it reduces the onset of Taylor vortices, but is more difficult to thermostat accurately.

'Cone and Plate' viscometers use a cone of very shallow angle in bare contact with a flat plate. With this system the shear rate beneath the plate is constant to a modest degree of precision and deconvolution of a flow curve; a graph of shear stress (torque) against shear rate (angular velocity) yields the viscosity in a straightforward manner.

Stabinger Viscometer

Fig. 3.6: Stabinger viscometer (SVM 3000).

By modifying the classic Couette rotational viscometer, an accuracy comparable to that of kinematic viscosity determination is achieved. The internal cylinder in the Stabinger Viscometer is hollow and specifically lighter than the sample, thus floats freely in the sample, centered by centrifugal forces. The formerly inevitable bearing friction is thus fully avoided. The speed and torque measurement is implemented without direct contact, by a rotating magnetic field and an eddy current brake. This allows for a previously unprecedented torque resolution of 50 pN·m and an exceedingly large measuring range from 0.2 to 20,000 mPa·s with a single measuring system. A built-in density measurement based on the oscillating U-tube principle allows the determination of kinematic viscosity from the measured dynamic viscosity employing the relation:

$$\upsilon = \frac{\eta}{\rho}$$

The Stabinger Viscometer was presented for the first time by Anton Paar GmbH at the ACHEMA in the year 2000. The measuring principle is named after its inventor Dr. Hans Stabinger.

Stormer Viscometer

The *Stormer viscometer* is a rotation instrument used to determine the viscosity of paints, commonly used in paint industries. It consists of a paddle-type rotor that is spun by an internal motor, submerged into a cylinder of viscous substance. The rotor speed can be adjusted by changing the amount of load supplied onto the rotor. For example, in one brand of viscometers, pushing the level upwards decreases the load and speed, downwards increases the load and speed. The viscosity can be found by adjusting the load until the rotation velocity is 200 rotations per minute. By examining the load applied and comparing tables found on ASTM D 562, one can find the viscosity in Krebs units (KU), unique only to the Stormer type viscometer. This method is intended for paints applied by brush or roller.

Bubble Viscometer

Bubble viscometers are used to quickly determine kinematic viscosity of known liquids such as resins and varnishes. The time required for an air bubble to rise is directly proportional to the visosity of the liquid, so the faster the bubble rises, the lower the viscosity. The Alphabetical Comparison Method uses 4 sets of lettered reference tubes, A5 through Z10, of known viscosity to cover a viscosity range from 0.005 to 1,000 stokes. The Direct Time Method uses a single 3-line times tube for determining the "bubble seconds", which may then be converted to stokes.

Miscellaneous Viscometer Types

Other viscometer types use balls or other objects. Viscometers that can characterize non-Newtonian fluids are usually called *rheometers* or *plastometers*. In the I.C.I "Oscar" viscometer, a sealed can of fluid was oscillated torsionally, and by clever measurement techniques it was possible to measure both viscosity and elasticity in the sample. Viscosity is a measure of the resistance of a fluid which is being deformed by either shear stress or extensional stress. In general terms it is the resistance of a liquid to flow, or its "thickness". Viscosity describes a fluid's internal resistance to flow and may be thought of as a measure of fluid friction. Thus, water is "thin", having a lower viscosity, while vegetable oil is "thick" having a higher viscosity. All real fluids (except superfluids) have some resistance to stress, but a fluid which has no resistance to shear stress is known as an ideal fluid or inviscid fluid. For example, a high viscosity magma will create a tall volcano, because it cannot spread fast enough; low viscosity lava will create a shield volcano, which is large and wide. The study of viscosity is known as rheology.

Etymology

The word "viscosity" derives from the Latin word "viscum" for mistletoe. A viscous glue was made from mistletoe berries and used for lime-twigs to catch birds.

Viscosity Coefficients

When looking at a value for viscosity, the number that one most often sees is the coefficient of viscosity. There are several different viscosity coefficients depending on the nature of applied stress and nature of the fluid. They are introduced in the main books on hydrodynamics and rheology.

- Dynamic viscosity determines the dynamics of an incompressible Newtonian fluid;
- Kinematic viscosity is the *dynamic viscosity* divided by the density for a Newtonian fluid;
- Volume viscosity (or bulk viscosity) determines the dynamics of a compressible Newtonian fluid;
- Shear viscosity is the viscosity coefficient when the applied stress is a shear stress (valid for non-Newtonian fluids); and
- Extensional viscosity is the viscosity coefficient when the applied stress is an extensional stress (valid for non-Newtonian fluids).

Shear viscosity and *dynamic viscosity* are much better known than the others. That is why they are often referred to as simply *viscosity*. Simply put, this quantity is the ratio between the pressure exerted on the surface of a fluid, in the lateral or horizontal direction, to the change in velocity of the fluid as you move down in the fluid (this is what is referred to as a velocity gradient). For example, at room temperature, water has a nominal viscosity of 1.0×10^{-3} Pa·s and motor oil has a nominal apparent viscosity of 250×10^{-3} Pa·s.

Extensional viscosity is widely used for characterizing polymers.

Volume viscosity is essential for Acoustics in fluids, see Stokes' law (sound attenuation).

Newton's Theory

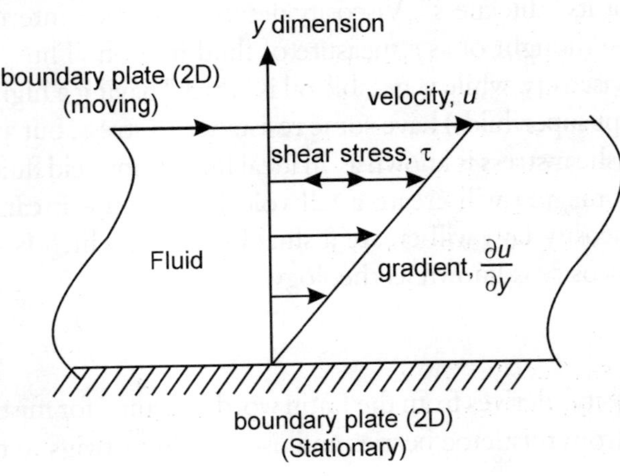

Laminar shear of fluid between two plates. Friction between the fluid and the moving boundaries causes the fluid to shear. The force required for this action is a measure of the fluid's viscosity. This type of flow is known as a Couette flow.

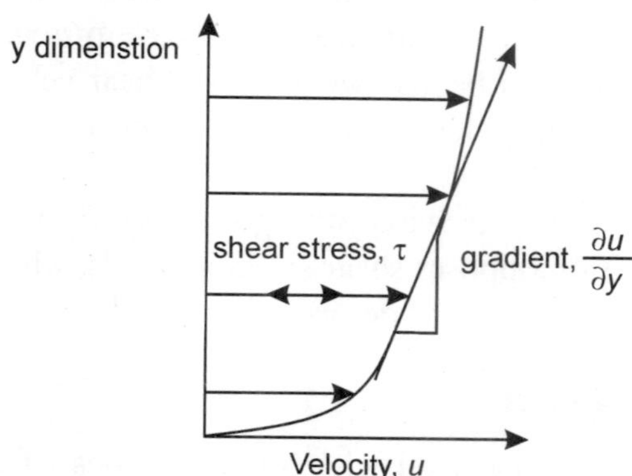

Laminar shear, the non-constant gradient, is a result of the geometry the fluid is flowing through (e.g. a pipe). In general, in any flow, layers move at different velocities and the fluid's viscosity arises from the shear stress between the layers that ultimately opposes any applied force. Isaac Newton postulated that, for straight, parallel and uniform flow, the shear stress, ∂, between layers is proportional to the velocity gradient, $\partial u/\partial y$, in the direction perpendicular to the layers.

$$\tau = \eta \frac{\partial u}{\partial y}$$

Here, the constant ç is known as the *coefficient of viscosity*, the *viscosity*, the *dynamic viscosity*, or the *Newtonian viscosity*. Many fluids, such as water and most gases, satisfy Newton's criterion and are known as Newtonian fluids. Non-Newtonian fluids exhibit a more complicated relationship between shear stress and velocity gradient than simple linearity.

The relationship between the shear stress and the velocity gradient can also be obtained by considering two plates closely spaced apart at a distance y, and separated by a homogeneous substance. Assuming that the plates are very large, with a large area A, such that edge effects may be ignored, and that the lower plate is fixed, let a force F be applied to the upper plate. If this force causes the substance between the plates to undergo shear flow (as opposed to just shearing elastically until the shear stress in the substance balances the applied force), the substance is called a fluid. The applied force is proportional to the area and velocity of the plate and inversely proportional to the distance between the plates. Combining these three relations results in the equation $F = \eta(Au/y)$, where η is the proportionality factor called the *absolute viscosity* [with units Pa·s = kg/(m·s) or slugs/(ft·s)]. The absolute viscosity is also known as the *dynamic viscosity*, and is often shortened to simply *viscosity*. The equation can be expressed in terms of shear stress; $\tau = F/A = \eta(u/y)$. The rate of shear deformation is u/y and can be also written as a shear velocity, du/dy. Hence, through this method, the relation between the shear stress and the velocity gradient can be obtained.

James Clerk Maxwell called viscosity *fugitive elasticity* because of the analogy that elastic deformation opposes shear stress in solids, while in viscous fluids, shear stress is opposed by *rate* of deformation.

Viscosity Measurement

Dynamic viscosity is measured with various types of rheometer. Close temperature control of the fluid is essential to accurate measurements, particularly in materials like lubricants, whose viscosity can double with a change of only 5°C. For some fluids, it is a constant over a wide range of shear rates. These are Newtonian fluids. The fluids without a constant viscosity are called Non-Newtonian fluids. Their viscosity cannot be described by a single number. Non-Newtonian fluids exhibit a variety of different correlations between shear stress and shear rate. One of the most common instruments for measuring kinematic viscosity is the glass capillary viscometer.

In paint industries, viscosity is commonly measured with a Zahn cup, in which the efflux time is determined and given to customers. The efflux time can also be converted to kinematic viscosities (cSt) through the conversion equations. Also used in paint, a Stormer viscometer uses load-based rotation in order to determine viscosity. The viscosity is reported in Krebs units (KU), which are unique to Stormer viscometers. Vibrating viscometers can also be used to measure viscosity. These models such as the *Dynatrol* use vibration rather than rotation to measure viscosity. *Extensional viscosity* can be measured with various rheometers that apply extensional stress Volume viscosity can be measured with acoustic rheometer.

Units of Measure

Viscosity (Dynamic/Absolute Viscosity)

Dynamic viscosity and absolute viscosity are synonymous. The IUPAC symbol for dynamic viscosity is the Greek letter eta (η), but it is also commonly referred to using the Greek symbol mu (μ). The SI physical unit of dynamic viscosity is the -second (Pa·s), which is identical to $kg \cdot m^{-1} \cdot s^{-1}$. If a fluid with a viscosity of one Pa·s is placed between two plates, and one plate is pushed sideways with a shear stress of one pascal, it moves a distance equal to the thickness of the layer between the plates in one second.

The name poiseuille (Pl) was proposed for this unit (after Jean Louis Marie Poiseuille who formulated Poiseuille's law of viscous flow), but not accepted internationally. Care must be taken in not confusing the poiseuille with the poise named after the same person.

The cgs physical unit for dynamic viscosity is the *poise* (P), named after Jean Louis Marie Poiseuille. It is more commonly expressed, particularly in ASTM standards, as *centipoise* (cP). Water at 20 °C has a viscosity of 1.0020 cP.

$$1\ \text{P} = 1\ \text{g}\cdot\text{cm}^{-1}\cdot\text{s}^{-1}$$

The relation between poise and pascal-seconds is:

$$10\ \text{P} = 1\ \text{kg}\cdot\text{m}^{-1}\cdot\text{s}^{-1} = 1\ \text{Pa}\cdot\text{s}$$

$$1\ \text{cP} = 0.001\ \text{Pa}\cdot\text{s} = 1\ \text{mPa}\cdot\text{s}$$

Kinematic Viscosity

In many situations, we are concerned with the ratio of the viscous force to the inertial force, the latter characterised by the fluid density ρ. This ratio is characterised by the *kinematic viscosity* (ν), defined as follows:

$$\nu = \frac{\mu}{\rho},$$

or

$$\nu = \frac{\eta}{\rho}.$$

where μ or η is the (dynamic or absolute) viscosity (in centipoise cP), and ñ is the density (in grams/cm^3), and í is the kinematic viscosity (in centistokes cSt). Kinematic viscosity (Greek symbol: ν) has SI units Pa.s/(kg/m^3) = m^2·s^{-1}. The cgs physical unit for kinematic viscosity is the *stokes* (St), named after George Gabriel Stokes. It is sometimes expressed in terms of *centistokes* (cSt or ctsk). In U.S. usage, *stoke* is sometimes used as the singular form.

1 stokes = 100 centistokes = 1 cm^2·s^{-1} = 0.0001 m^2·s^{-1}.
1 centistokes = 1 mm^2·s^{-1} = 10^{-6}m^2·s^{-1}.

Saybolt Universal Viscosity

At one time the petroleum industry relied on measuring kinematic viscosity by means of the Saybolt viscometer, and expressing kinematic viscosity in units of Saybolt Universal Seconds (SUS). Kinematic viscosity in centistoke can be converted from SUS according to the arithmetic and the reference tabel provided in ASTM D 2161. It can also be converted in computerized method, or vice versa.

Relation to Mean Free Path of Diffusing Particles

In relation to diffusion, the kinematic viscosity provides a better understanding of the behavior of mass transport of a dilute species. Viscosity is related to shear stress and the rate of shear in a fluid, which illustrates its dependence on the mean free path, λ, of the diffusing particles.

From fluid mechanics, shear stress, τ, is the rate of change of velocity with distance perpendicular to the direction of movement.

$$\tau = \mu \frac{du}{dx}$$

Interpreting shear stress as the time rate of change of momentum, p, per unit area (rate of momentum flux) of an arbitrary control surface gives,

$$\tau = \frac{p}{A} = \frac{mu}{A}$$

Further manipulation will show:

$$\frac{p}{u} = m = \rho \bar{u} A \Rightarrow \tau = 2\rho \bar{u} \lambda . \frac{du}{dx} \Rightarrow \nu = \frac{\mu}{\rho} = 2\bar{u}\lambda$$

where

m is the rate of change of mass
ρ is the density of the fluid
$\bar{u}$ is the average molecular speed
μ is the dynamic viscosity.

Dynamic Versus Kinematic Viscosity

Conversion between kinematic and dynamic viscosity is given by íñ = ì.

For example,

if $\nu = 0.0001\ m^2 \cdot s^{-1}$ and $\rho = 1000$ kg m^{-3} then $\mu = \nu\rho = 0.1\ kg \cdot m^{-1} \cdot s^{-1} = 0.1$ Pa·s
if $\nu = 1$ St ($= 1\ cm^2 \cdot s^{-1}$) and $\rho = 1$ g cm^{-3} then $\mu = \nu\rho = 1\ g \cdot cm^{-1} \cdot s^{-1} = 1$ P

A plot of the kinematic viscosity of air as a function of absolute temperature is available on the Internet.

Example: Viscosity of Water

Because of its density of $\rho = 1$ g/cm^3 (varies slightly with temperature), and its dynamic viscosity is near 1 mPa·s, the viscosity values of water are, to rough precision, all powers of ten:

Dynamic Viscosity

$$\mu = 1 \text{ mPa·s} = 10^{-3} \text{ Pa·s} = 1 \text{ cP} = 10^{-2} \text{ poise}$$

Kinematic viscosity:

$$\nu = 1 \text{ cSt} = 10^{-2} \text{ stokes} = 1 \text{ mm}^2\text{/s}$$

Molecular Origins

The viscosity of a system is determined by how molecules constituting the system interact. There are no simple but correct expressions for the viscosity of a fluid. The simplest exact expressions are the Green-Kubo relations for the linear shear viscosity or the Transient Time Correlation Function expressions derived by Evans and Morriss in 1985. Although these expressions are each exact in order to calculate the viscosity of a dense fluid, using these relations requires the use of molecular dynamics computer simulations.

Fig. 3.7: Pitch has a viscosity approximately 100 billion times that of water.

Gases

Viscosity in gases arises principally from the molecular diffusion that transports momentum between layers of flow. The kinetic theory of gases allows accurate prediction of the behavior of gaseous viscosity.

Within the regime where the theory is applicable:

- Viscosity is independent of pressure and
- Viscosity increases as temperature increases.

James Clerk Maxwell published a famous paper in 1866 using the kinetic theory of gases to study gaseous viscosity.

Effect of Temperature on the Viscosity of a Gas

Sutherland's formula can be used to derive the dynamic viscosity of an ideal gas as a function of the temperature:

$$\eta = \eta_0 \frac{T_0 + C}{T + C}\left(\frac{T}{T_0}\right)^{3/2}$$

where:

η = viscosity in (Pa·s) at input temperature T
η_0 = reference viscosity in (Pa·s) at reference temperature T_0
T = input temperature in kelvin
T_0 = reference temperature in kelvin
C = Sutherland's constant for the gaseous material in question

Valid for temperatures between $0 < T < 555$ K with an error due to pressure less than 10 per cent below 3.45 MPa

Sutherland's constant and reference temperature for some gases:

Gas	*C* [K]	T_0 [K]	η_0 [10^{-6} Pa s]
air	120	291.15	18.27
nitrogen	111	300.55	17.81
oxygen	127	292.25	20.18
carbon dioxide	240	293.15	14.8
carbon monoxide	118	288.15	17.2
hydrogen	72	293.85	8.76
ammonia	370	293.15	9.82
sulfur dioxide	416	293.65	12.54
helium	79.4	273	19

Viscosity of a Dilute Gas

The Chapman-Enskog equation may be used to estimate viscosity for a dilute gas. This equation is based on semi-theorethical assumption by Chapman and Enskoq. The equation requires three empirically determined parameters: the collision diameter (σ), the maximum energy of attraction divided by the Boltzmann constant (ε/κ) and the collision integral [ω(T*)].

$$\eta_0 \times 10^7 = 266.93\frac{(MT)^{1/2}}{\sigma^2\omega(T^*)}$$

T^* = êT/å Reduced temperature (dimensionless)
η_0 = viscosity for dilute gas (uP)
M = molecular mass (g/mol)
T = temperature (K)
σ = the collision diameter (Å)
ε/κ = the maximum energy of attraction divided by the Boltzmann constant (K)
ω_η = the collision integral.

Liquids

In liquids, the additional forces between molecules become important. This leads to an additional contribution to the shear stress though the exact mechanics of this are still controversial. Thus, in liquids:

- Viscosity is independent of pressure (except at very high pressure); and
- Viscosity tends to fall as temperature increases (for example, water viscosity goes from 1.79 cP to 0.28 cP in the temperature range from 0 °C to 100 °C); see temperature dependence of liquid viscosity for more details.

The dynamic viscosities of liquids are typically several orders of magnitude higher than dynamic viscosities of gases.

Viscosity of Blends of Liquids

The viscosity of the blend of two or more liquids can be estimated using the Refutas equation. The calculation is carried out in three steps. The first step is to calculate the Viscosity Blending Number (VBN) (also called the Viscosity Blending Index) of each component of the blend:

$$\text{VBN} = 14.534 \times ln\,[ln\,(v + 0.8)] + 10.975 \qquad (1)$$

where v is the kinematic viscosity in centistokes (cSt). It is important that the kinematic viscosity of each component of the blend be obtained at the same temperature.

The next step is to calculate the VBN of the blend, using this equation:

$$VBN_{Blend} = [x_A \times VBN_A] + [x_B \times VBN_B] + \ldots + [x_N \times VBN_N] \quad (2)$$

where x_X is the mass fraction of each component of the blend.

Once the viscosity blending number of a blend has been calculated using equation (2), the final step is to determine the kinematic viscosity of the blend by solving equation (1) for v:

$$\upsilon = e^{e^{\frac{VBN_{Blend}-10.975}{14.534}}} - 0.8 \quad (3)$$

where VBN_{Blend} is the viscosity blending number of the blend.

Viscosity of Selected Substances

The viscosity of air and water are by far the two most important materials for aviation aerodynamics and shipping fluid dynamics. Temperature plays the main role in determining viscosity.

Viscosity of Air

The viscosity of air depends mostly on the temperature. At 15.0 °C, the viscosity of air is 1.78×10^{-5} kg/(m·s) or 1.78×10^{-4} P. One can get the viscosity of air as a function of temperature from the Gas Viscosity Calculator

Viscosity of Water

The viscosity of water is 8.90×10^{-4} Pa·s or 8.90×10^{-3} dyn·s/cm² or 0.890 cP at about 25 °C.

As a function of temperature T (K): $\mu(\text{Pa·s}) = A \times 10^{B/(T-C)}$

where

$$A = 2.414 \times 10^{-5} \text{ Pa·s} ; B = 247.8 \text{ K} ; \text{and } C = 140 \text{ K}.$$

Viscosity of water at different temperatures is listed below:

Temperature [°C]	*Viscosity [Pa·s]*	*Temperature [°C]*	*Viscosity [Pa·s]*
10	$1.308 \times 10^{?3}$	60	$4.668 \times 10^{?4}$
20	$1.003 \times 10^{?3}$	70	$4.044 \times 10^{?4}$
30	$7.978 \times 10^{?4}$	80	$3.550 \times 10^{?4}$
40	$6.531 \times 10^{?4}$	90	$3.150 \times 10^{?4}$
50	$5.471 \times 10^{?4}$	100	$2.822 \times 10^{?4}$

Viscosity of Various Materials

Some dynamic viscosities of Newtonian fluids are listed below:

	Viscosity [Pa·s]
hydrogen	$8.4 \times 10^{?6}$
air	$17.4 \times 10^{?6}$
xenon	$2.12 \times 10^{?5}$

Liquids (at 25 °C)

	Viscosity [Pa·s]	*Viscosity [cP]*
liquid nitrogen @ 77K	$1.58 \times 10^{?4}$	0.158
acetone*	$3.06 \times 10^{?4}$	0.306
methanol*	$5.44 \times 10^{?4}$	0.544
benzene*	$6.04 \times 10^{?4}$	0.604
water	$8.94 \times 10^{?4}$	0.894
ethanol*	$1.074 \times 10^{?3}$	1.074
mercury*	$1.526 \times 10^{?3}$	1.526
nitrobenzene*	$1.863 \times 10^{?3}$	1.863
propanol*	$1.945 \times 10^{?3}$	1.945
Ethylene glycol	$1.61 \times 10^{?2}$	16.1
sulfuric acid*	$2.42 \times 10^{?2}$	24.2
olive oil	.081	81
glycerol	1.5	1500
castor oil*	.985	985
corn syrup*	1.3806	1380.6
HFO-380	2.022	2022
pitch	2.3×10^{8}	2.3×10^{11}

*Note:** Data from CRC Handbook of Chemistry and Physics, 73rd edition, 1992-1993.

Fluids with variable compositions, such as honey, can have a wide range of viscosities. A more complete table can be found at Transwiki, including the following:

	Viscosity [cP]
honey	2,000–10,000
molasses	5,000–10,000
molten glass	10,000–1,000,000
chocolate syrup	10,000–25,000
molten chocolate*	45,000–130,000
ketchup*	50,000–100,000
peanut butter	~250,000
shortening*	~250,000

Note: * These materials are highly non-Newtonian.

Viscosity of Solids

On the basis that all solids, such as granite flow to a small extent in response to shear stress some researchers have contended that substances known as amorphous solids, such as glass and many polymers, may be considered to have viscosity. This has led some to the view that solids are simply liquids with a very high viscosity, typically greater than 10^{12} Pa·s. This position is often adopted by supporters of the widely held misconception that glass flow can be observed in old buildings. This distortion is more likely the result of the glass making process rather than the viscosity of glass.

However, others argue that solids are, in general, elastic for small stresses while fluids are not. Even if solids flow at higher stresses, they are characterized by their low-stress behavior. Viscosity may be an appropriate characteristic for solids in a plastic regime. The situation becomes somewhat confused as the term *viscosity* is sometimes used for solid materials, for example Maxwell materials, to describe the relationship between stress and the rate of change of strain, rather than rate of shear. These distinctions may be largely resolved by considering the constitutive equations of the material in question, which take into account both its viscous and elastic behaviors. Materials for which both their viscosity and their elasticity are important in a particular range of deformation and deformation rate are called *viscoelastic*. In geology, earth materials that exhibit viscous deformation at least three times greater than their elastic deformation are sometimes called rheids.

Viscosity of Amorphous Materials

Viscous flow in amorphous materials (e.g. in glasses and melts) is a thermally activated process:

$$\eta = A \,.\, e^{QR/T}$$

where Q is activation energy, T is temperature, R is the molar gas constant and A is approximately a constant.

The viscous flow in amorphous materials is characterized by a deviation from the Arrhenius-type behavior: Q changes from a high value Q_H at low temperatures (in the glassy state) to a low value Q_L at high temperatures (in the liquid state). Depending on this change, amorphous materials are classified as either:

- *strong when*: $Q_H - Q_L < Q_L$ or
- *fragile when*: $Q_H - Q_L \geq Q_L$

The fragility of amorphous materials is numerically characterized by the Doremus' fragility ratio:

$$R_D = Q_H/Q_L$$

and strong material have $R_D < 2$ whereas fragile materials have RD ≥ 2.

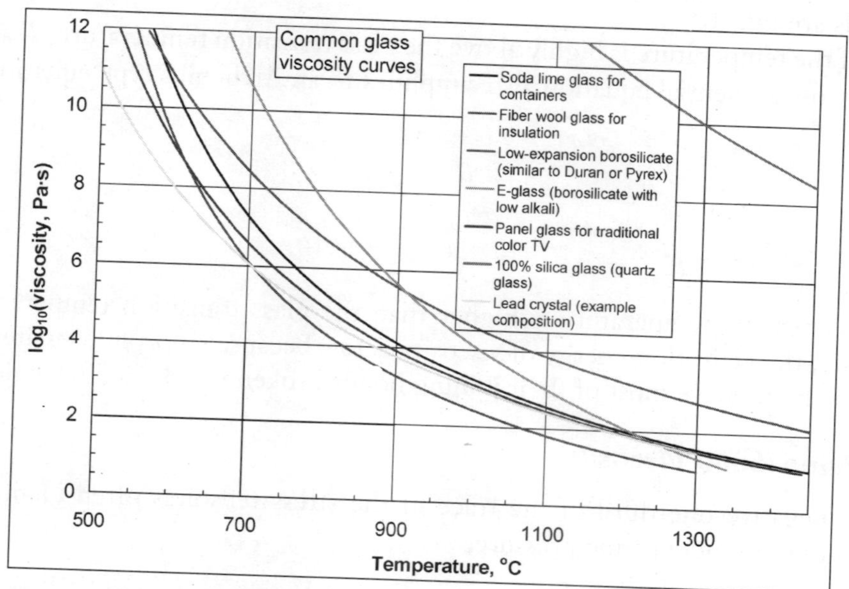

Fig. 3.8: Common glass viscosity curves.

The viscosity of amorphous materials is quite exactly described by a two-exponential equation:

$$\eta = A_1 . T [1 + A_2 . e^{B/RT}] . [1 + C . e^{D/RT}]$$

with constants A_1, A_2, B, C and D related to thermodynamic parameters of joining bonds of an amorphous material.

Not very far from the glass transition temperature, T_g, this equation can be approximated by a Vogel-Tammann-Fulcher (VTF) equation or a Kohlrausch-type stretched-exponential law.

If the temperature is significantly lower than the glass transition temperature, $T < T_g$, then the two-exponential equation simplifies to an Arrhenius type equation:

$$\eta = A_L T . e^{QH/RT}$$

with:

$$Q_H = H_d + H_m$$

where H_d is the enthalpy of formation of broken bonds (termed configurons) and H_m is the enthalpy of their motion. When the temperature is less than the glass transition temperature, $T < T_g$, the activation energy of viscosity is high because the amorphous materials are in the glassy state and most of their joining bonds are intact.

If the temperature is highly above the glass transition temperature, $T > T_g$, the two-exponential equation also simplifies to an Arrhenius type equation:

$$\eta = A_H T \,.\, e^{QL/RT}$$

with:

$$Q_L = H_m$$

When the temperature is higher than the glass transition temperature, $T > T_g$, the activation energy of viscosity is low because amorphous materials are melt and have most of their joining bonds broken which facilitates flow.

Volume (bulk) Viscosity

The negative-one-third of the trace of the stress tensor is often identified with the thermodynamic pressure,

$$-\frac{1}{3}T_a^a = p\,,$$

which only depends upon the equilibrium state potentials like temperature and density (equation of state). In general, the trace of the stress tensor is the sum of thermodynamic pressure contribution plus another contribution which is proportional to the divergence of the velocity field. This constant of proportionality is called the volume viscosity.

Eddy Viscosity

In the study of turbulence in fluids, a common practical strategy for calculation is to ignore the small-scale *vortices* (or *eddies*) in the motion and to calculate a large-scale motion with an *eddy viscosity* that characterizes the transport and dissipation of energy in the smaller-scale flow (see *large eddy simulation*). Values of eddy viscosity used in modeling ocean circulation may be from 5×10^4 to 10^6 Pa·s depending upon the resolution of the numerical grid.

Fluidity

The reciprocal of viscosity is *fluidity*, usually symbolized by $\varphi = 1/\eta$ or $F = 1/\eta$, depending on the convention used, measured in *reciprocal poise* ($cm \cdot s \cdot g^{-1}$), sometimes called the *rhe*. *Fluidity* is seldom used in engineering practice.

The concept of fluidity can be used to determine the viscosity of an ideal solution. For two components *a* and *b*, the fluidity when *a* and *b* are mixed is

$$F \approx \chi_a F_a + \chi_b F_b$$

which is only slightly simpler than the equivalent equation in terms of viscosity:

$$\eta = \frac{1}{\chi_a \eta_a + \chi_b / \eta_b}$$

where χ_a and χ_b is the mole fraction of component *a* and *b* respectively, and η_a and η_b are the components pure viscosities.

The Linear Viscous Stress Tensor

Viscous forces in a fluid are a function of the rate at which the fluid velocity is changing over distance. The velocity at any point r is specified by the velocity field v(r). The velocity at a small distance dr from point r may be written as a Taylor series:

$$v(\mathrm{r} + d\mathrm{r}) = v(\mathrm{r}) + \frac{d\mathrm{v}}{d\mathrm{r}} d\mathrm{r} + \ldots$$

where $\frac{d\mathrm{v}}{d\mathrm{r}}$ is shorthand for the dyadic product of the del operator and the velocity:

$$\frac{d\mathrm{v}}{d\mathrm{r}} = \begin{bmatrix} \frac{\partial v_x}{\partial x} & \frac{\partial v_x}{\partial y} & \frac{\partial v_x}{\partial z} \\ \frac{\partial v_y}{\partial x} & \frac{\partial v_y}{\partial y} & \frac{\partial v_y}{\partial z} \\ \frac{\partial v_z}{\partial x} & \frac{\partial v_z}{\partial y} & \frac{\partial v_z}{\partial z} \end{bmatrix}$$

This is just the Jacobian of the velocity field. Viscous forces are the result of relative motion between elements of the fluid, and so are expressible as a function of the velocity field. In other words, the forces at r are a function of v(r) and all derivatives of v(r) at that point. In the case of linear viscosity, the viscous force will be a function of the Jacobian tensor alone. For almost all practical situations, the linear approximation is sufficient.

If we represent x, y, and z by indices 1, 2, and 3 respectively, the i, j component of the Jacobian may be written as $\partial_i v_j$ where ∂_i is shorthand for $\partial / \partial x_i$. Note that when the first and higher derivative terms are zero, the velocity of all fluid elements is parallel, and there are no viscous forces.

Any matrix may be written as the sum of an antisymmetric matrix and a symmetric matrix, and this decomposition is independent of coordinate system, and so has physical significance. The velocity field may be approximated as:

$$v_i\,(\mathrm{r} + d\mathrm{r}) = v_i(\mathrm{r}) + \frac{1}{2}\left(\partial_i v_j - \partial_i v_j - \partial_j v_i\right) dr_i + \frac{1}{2}\left(\partial_i v_j + \partial_j v_i\right) dr_i$$

where Einstein notation is now being used in which repeated indices in a product are implicitly summed. The second term from the right is the asymmetric part of the first derivative term, and it represents a rigid rotation of the fluid about r with angular velocity ∂ where:

$$\omega = \frac{1}{2}\nabla \times v = \frac{1}{2}\begin{bmatrix} \partial_2 v_3 - \partial_3 v_2 \\ \partial_3 v_1 - \partial_1 v_3 \\ \partial_1 v_2 - \partial_2 v_1 \end{bmatrix}$$

For such a rigid rotation, there is no change in the relative positions of the fluid elements, and so there is no viscous force associated with this term. The remaining symmetric term is responsible for the viscous forces in the fluid. Assuming the fluid is isotropic (i.e. its properties are the same in all directions), then the most general way that the symmetric term (the rate-of-strain tensor) can be broken down in a coordinate-independent (and therefore physically real) way is as the sum of a constant tensor (the rate-of-expansion tensor) and a traceless symmetric tensor (the rate-of-shear tensor):

$$\frac{1}{2}\left(\partial_i v_j + \partial_j v_i\right) = \underbrace{\frac{1}{3}\partial_k v_k \delta_{ij}}_{\text{rate-of-expansion tensor}} + \underbrace{\left(\frac{1}{2}\left(\partial_i v_j + \partial_j v_i\right) - \frac{1}{3}\partial_k v_k \delta_{ij}\right)}_{\text{rate-or-shear tensor}}$$

where δ_{ij} is the unit tensor. The most general linear relationship between the stress tensor σ and the rate-of-strain tensor is then a linear combination of these two tensors:

$$\sigma\ visc;\ i, j = \varsigma \partial_k v_k \delta_{ij} + \eta\left(\partial_i v_j + \partial_j v_i - \frac{2}{3}\partial_k v_k \delta_{ij}\right)$$

where æ is the coefficient of bulk viscosity (or "second viscosity") and ç is the coefficient of (shear) viscosity.

The forces in the fluid are due to the velocities of the individual molecules. The velocity of a molecule may be thought of as the sum of the fluid velocity and the thermal velocity. The viscous stress tensor described above gives the force due to the fluid velocity only. The force on an area element in the fluid due to the thermal velocities of the molecules is just the hydrostatic pressure. This pressure term ($-p\delta_{ij}$) must be added to the viscous stress tensor to obtain the total stress tensor for the fluid.

$$\sigma_{ij} = -pd_{ij} \ \sigma visc;ij$$

The infinitesimal force dF_i on an infinitesimal area dA_i is then given by the usual relationship.

3.4 Polymer Characterization

In addition to PDI's excellent capabilities in the determination of the composition of material, it has several labs dedicated to characterizing various behaviors such as rheology, failure mechanisms, particle size, molecular weight and thermal properties. These labs play an important role in understanding why materials behave in a certain manner, and they compliment the labs looking at composition to provide complete and thorough understanding of polymer materials.

4

Natural Polymers and their Sources

4.1 Back to Nature

When Wallace Carothers invented nylon, he made a polymer with a structure similar to that of silk. Natural materials inspired his work, and now we're going to take a look at some of those natural polymers.

```
 ┌ CH2      CH2 ┐
 │   \      /   │
 │    C == C    │
 │   /      \   │
 │  H        CH3│
 └              ┘n
```

polyisoprene

Rubber

Natural rubber is one of the more unusual materials found in nature. It can be stretched and it snaps back to its original shape. It's also waterproof. What more could anyone want? A lot of different plants produce rubber, but the most important is the hevea tree (*hevea brasiliensis*). Natural rubber's scientific name is polyisoprene, and it has a repeat unit like you see on your right.

Starch and Cellulose

Starch and cellulose are both made from the same monomer, glucose. Glucose is a simple kind of sugar that your body uses for fuel. Starch is found in foods like bread, potatoes, corn, and rice. Your body breaks it down to get the glucose it needs.

On the other hand, cellulose is a polymer we can't digest. It is the main component of cotton and wood (and therefore paper), and it makes very strong fibers. Also, even though we can't digest cellulose, certain animals can. Termites can digest the cellulose in wood, while cattle and sheep happily get their nourishment from the cellulose in grass.

Proteins

Proteins are the family of polymers that inspired nylon. Wallace Carothers suspected that silk, one kind of protein, was a polyamide. So he decided to make a synthetic polyamide when he set out to make a synthetic fiber. But silk is just one kind of protein. Your skin, hair, and fingernails are made of a protein called keratin. In your blood, a protein called hemoglobin carries oxygen to your cells, while other proteins called antibodies fight disease. One amazing family of proteins are the enzymes. Enzymes are catalysts; that is, they speed up chemical reactions in your body. Some of them can make chemical reactions take place a million times faster than they would without an enzyme.

$$H_2N-\underset{R}{\overset{H}{C}}-\overset{O}{\overset{\|}{C}}-OH \longrightarrow \left[-\underset{R}{\overset{H}{C}}-\overset{O}{C}-\underset{H}{N}-\right]_n$$

an amino acid **a protein**

The body polymerizes amino acids to make proteins.

Like nylon, proteins are polyamides. They have a repeat structure like you see above. However, unlike nylon, each repeat unit in a protein may be different. Your body builds proteins from twenty different amino acids. The group *R* in the diagram above is different in each amino acid. A given protein may contain all twenty, in a precise sequence consisting of thousands of amino acid repeat units. The number of possible combinations and sequences is endless. With this much possible variety, its easy to see why there are so many different proteins and why they can do so many different things.

Since proteins do so many different jobs in the body, it's no wonder that Linneaus Dorman decided to see if it was possible to make synthetic proteins to make the body work even better.

DNA and RNA

Deoxyribonucleic acid (DNA) and ribonucleic acid (RNA) are polymers we can't live without. The DNA is the blueprint that determines everything about your body. DNA is passed from parents to their children, so children inherit the physical characteristics of their parents. RNA helps the blueprints of DNA to be carried out.

4.2 SNP a Synthetic Natural Polymers

SNP is ... a privately owned chemical company that specializes in synthetic and natural water-soluble polymers. For over 20 years SNP, Inc. (formerly Syn-Chem, Inc.) has been a manufacturer and distributor of natural and synthetic water-soluble polymers to the paper, textile, adhesives, and other specialty industries. SNP is committed to providing its customers with high quality products, superior technical support, and economical pricing.

SNP's Mission is ... to work closely with customers to improve their product performance and reduce their manufacturing costs.

SNP designs and manufactures products to match the customer's specifications. SNP listens to the customer's needs, wants, and opportunities. Through flexible business practices and technical innovation, SNP meets the customers needs, satisfies the customer's wants, and explores opportunities, which are mutually beneficial.

SNP is committed to product development by accepting the challenge of an initial concept, conducting laboratory screening and optimization designed experiments, facilitating pilot trial evaluations, expanding laboratory studies if necessary, and finally, implementing new products into production.

The primary SNP facilities, including the technical laboratory, primary manufacturing site and warehousing are conveniently located in the Durham, NC near the Research Triangle Park. SNP's location offers the advantage of proximity to North Carolina State University and close academic relationships in the areas paper and textile science and engineering.

4.3 History of Polymers and Plastics

Plastics are polymers. What is a polymer? The most simple definition of a polymer is something made of many units. Think of a polymer as a chain. Each link of the chain is the "mer" or basic unit that is made of carbon, hydrogen, oxygen, and/or silicon. To make the chain, many links or "mers" are hooked or polymerized together. Polymerization can be demonstrated by linking strips of construction paper together to make paper garlands or hooking together hundreds of paper clips to form chains.

Polymers have been with us since the beginning of time. Natural polymers include such things as tar and shellac, tortoise shell and horns, as well as tree saps that produce amber and latex. These polymers were processed with heat and pressure into useful articles like hair ornaments and jewelry. Natural polymers began to be chemically modified during the 1800s to produce many materials. The most famous of these were vulcanized rubber, gun cotton, and celluloid. The first semi-synthetic polymer produced was Bakelite in 1909 and was soon followed by the first synthetic fiber, rayon, which was developed in 1911.

Even with these developments, it was not until World War II that significant changes took place in the polymer industry. Prior to World War II, natural substances were generally available; therefore, synthetics that were being developed were not a necessity. Once the world went to war, our natural sources of latex, wool, silk, and other materials were cut off, making the use of synthetics critical. During this time period, we saw the use of nylon, acrylic, neoprene, SBR, polyethylene, and many more polymers take the place of natural materials that were no longer available. Since then, the polymer industry has continued to grow and has evolved into one of the fastest growing industries in the U.S. and in the world.

The Structure of Polymers

Many common classes of polymers are composed of hydrocarbons. These polymers are specifically made of small units bonded into long chains. Carbon makes up the backbone of the molecule and hydrogen atoms are bonded along the backbone. Below is a diagram of polyethylene, the simplest polymer structure.

```
    H     H        H        H        H     H
    |  H  |  H     |  H     |  H     |  H  |
    |  |  |  |     |  |     |  |     |  |  |
 ---C  |  C  |     C  |     C  |     C  |  C- -
     \ C /   \ C  /   \ C  /   \ C  /  \ C /
    |  |  |  |     |  |     |  |     |  |  |
    H  |  H  |     H  |     H  |     H  |  H
       H     H        H        H        H
```

There are polymers that contain only carbon and hydrogen. Polypropylene, polybutylene, polystyrene, and polymethylpentene are examples of these. Even though the basic makeup of many polymers is carbon and hydrogen, other elements can also be involved. Oxygen, chlorine, fluorine, nitrogen, silicon, phosphorous, and sulfur are other elements that are found in the molecular makeup of polymers. Polyvinyl chloride (PVC) contains chlorine. Nylon contains nitrogen and oxygen. Teflon contains fluorine. Polyester and polycarbonates contain oxygen. Vulcanized rubber and thiokol contain sulfur. There are also some polymers that, instead of having a carbon backbone, have a silicon or silicon-oxygen backbone. These are considered inorganic polymers. One of the most famous silicon-based polymers is Silly Putty.

Molecular Arrangement of Polymers

Think of how spaghetti noodles look on a plate. This is similar to how polymers can be arranged if they are amorphous. An amorphous arrangement of molecules has no long-range order or form in which the polymer chains arrange themselves. Amorphous polymers are generally transparent. This is an important characteristic for many applications such as food wrap, Plexiglas™, headlights, and contact lenses. Controlling and quenching the polymerization process can result in amorphous organization.

Fig. 4.1: How spaghetti noodles look on a plate is similar to how polymers can be arranged if they are amorphous.

Obviously, not all polymers are transparent. The polymer chains in objects that are translucent and opaque are in a more crystalline arrangement. By definition a crystalline arrangement has atoms, ions, or in this case, molecules in a distinct pattern. You generally think of crystalline structures in salt and gemstones, but not in plastics. Just as quenching can produce amorphous arrangements, processing can control the degree of crystallinity. The higher the degree of crystallinity, the less light can pass through the polymer. Therefore, the degree of translucence or opaqueness of the polymer is directly affected by its crystallinity.

Engineers are always producing better materials by manipulating the molecular structure that affects the final polymer produced. Manufacturers and processors introduce various fillers, reinforcements, and additives into the base polymers to expand product possibilities.

Characteristics of Polymers

Polymers are divided into two distinct groups: thermoplastics and thermosets. The majority of polymers are thermoplastic, meaning that once the polymer is formed it can be heated and reformed over and over again. This property allows for easy processing and recycling. The other group, the thermosets, can not be remelted. Once these polymers are formed, reheating will cause the material to scorch. Every polymer has very distinct characteristics, but most polymers have the following general attributes.

1. *Polymers can be very resistant to chemicals:* Consider all the cleaning fluids in your home that are packaged in plastic. Reading the warning labels that describe what happens when the chemical comes in contact with skin or eyes or is ingested will emphasize the chemical resistance of these materials.
2. *Polymers can be both thermal and electrical insulators:* A walk through your house will reinforce this concept, as you consider all the appliances, cords, electrical outlets, and wiring, that are made or covered with polymeric materials. Thermal resistance is evident in the kitchen with pot and pan handles made of polymers, the coffee pot handle, the foam core of refrigerators and freezers, insulated cups, coolers, and microwave cookware. The thermal underwear that many skiers wear is made of polypropylene and the fiberfill in winter jackets is acrylic.
3. *Generally, polymers are very light in mass with varying degrees of strength:* Consider the range of applications, from dime store toys to the frame structure of space stations, or from delicate nylon fiber in pantyhose to Kevlar™, which is used in bulletproof vests.
4. *Polymers can be processed in various ways to produce thin fibers or very intricate parts:* Plastics can be molded into bottles or the body of a car or be mixed with solvents to become an adhesive or a paint. Elastomers and some plastics stretch and are very flexible. Other polymers can be foamed like polystyrene (Styrofoam™) and urethane, to name just two examples. Polymers are materials with a seemingly limitless range of characteristics and colors. Polymers have many inherent properties that can be further enhanced by a wide range of additives to broaden their uses and applications.

In addressing all the superior attributes of polymers, it is equally important to discuss some of the difficulties associated with the material. Plastics deteriorate but never decompose completely, but neither does glass, paper, or aluminum. Plastics make up 9.9 per cent of our trash by weight compared to paper, which constitutes 39 per cent. Glass and metals make up 13 per cent by weight. In 1997, Americans produced 217 million tons of trash.

Applications for recycled plastics are growing every day. Plastics can be blended with virgin plastic (plastic that has not been processed before) to reduce cost without sacrificing properties. Recycled plastics are used to make polymeric timbers for use in picnic tables, fences, and outdoor toys, thus saving natural lumber. Plastic from 2-liter bottles is even being spun into fiber for the production of carpet.

A solution for plastics that are not recycled, especially those that are soiled, such as used microwave food wrap or diapers, can be a waste-to-energy system (WTE). Incineration of polymers produces heat energy. The heat energy produced by the burning plastics not only can be converted to electrical energy but helps burn the wet trash that is present. Paper also produces heat when burned, but not as much as plastics. On the other hand, glass, aluminum and other metals do not release any energy when burned.

Polymers affect every day of our life. These materials have so many varied characteristics and applications that their usefulness can only be measured by our imagination. Polymers are the materials of past, present, and future generations.

4.4 Introduction to Plastics

Plastics have become a universal material, used for everything from throwaway bags to wings for combat aircraft. Plastics are cheap, lightweight, strong, often attractive, and can be synthesized with a wide range of properties. This document provides a short introduction to plastics technology.

Natural Polymers

People have been using natural organic polymers for centuries in the form of waxes and shellacs, as well as fabrics and ropes, which are based on a plant polymer named "cellulose". By the early 19th century natural rubber, based on a polymer now known as "isoprene" and tapped from rubber trees, was in widespread use.

Eventually, inventors learned to improve the properties of natural polymers. Natural rubber was sensitive to temperature, becoming sticky and smelly in hot weather, and brittle in cold weather. In 1834, two inventors, Friedrich Ludersdorf of Germany and Nathaniel Hayward of the US, independently discovered that adding sulfur to raw rubber helped prevent the material from becoming sticky.

In 1839, the American inventor Charles Goodyear was experimenting with the sulfur treatment of natural rubber when, according to legend, he dropped a piece of sulfur-treated rubber on a stove. The rubber seemed to have improved properties, and Goodyear followed up with further experiments, developing a process known as "vulcanization" that involved cooking the rubber with sulfur. Compared to untreated natural rubber, Goodyear's "vulcanized rubber" was stronger and more resistant to abrasion; more elastic, much less sensitive to temperature; impermeable to gases; and highly resistant to chemicals and electric current.

Vulcanization still remains an important industrial process for the manufacture of rubber in both natural and artificial forms. Vulcanization creates sulfur bonds that link separate rubber polymers together, improving the material's structural integrity and its other properties.

Cellulose-Based Plastics: Celluloid & Rayon

Inventors were particularly interested in developing synthetic substitutes for natural materials that were expensive and in short supply, since that meant a profitable market to exploit. Ivory was a particularly attractive target for a synthetic replacement.

An Englishman named Alexander Parkes developed a "synthetic ivory" named "pyroxlin", which he marketed under the trade name "Parkesine", and which won a bronze medal at the 1862 World's Fair in London. Parkesine was made from cellulose treated with nitric acid and a solvent. The output of the process hardened into a hard, ivory-like material that could be molded when heated.

However, Parkes was not able to scale up the process to an industrial level, and products made from Parkesine quickly warped and cracked after a short time. An American printer and amateur inventor named John Wesley Hyatt took up where Parkes left off. Parkes had failed for lack of a proper solvent, but Hyatt discovered that camphor would do the job nicely.

Hyatt was something of an industrial genius who understood what could be done with such a shapeable, or "plastic", material, and proceeded to design much of the basic industrial machinery needed to produce quality plastic materials in quantity. Since cellulose was the main constituent used in the synthesis of his new material, Hyatt named it "celluloid". It was introduced in 1863.

One of the first products were dental pieces. Sets of false teeth built around celluloid proved cheaper than existing rubber dentures. However, celluloid dentures tended to soften when hot, making tea drinking tricky, and the camphor taste proved to be hard to eliminate.

Celluloid's real breakthrough products were waterproof shirt collars, cuffs, and the false shirt fronts known as "dickies", whose willingness to pop up unpredictably became a stock joke in silent-movie comedies. Such celluloid items didn't wilt and didn't stain easily, and Hyatt sold them by trainloads. Corsets made with celluloid stays also proved popular, since perspiration didn't rust the stays as it did metal stays.

Celluloid proved extremely versatile in its field of application, providing a cheap and attractive replacement for ivory, tortoise-shell, and bone. Not only was celluloid cheaper in itself, but products that had been made with such traditional materials could now be molded in large batches, instead of produced by expensive hand craftsmanship. Some of the 19th-century items made with cellulose were beautifully designed and made. For example, celluloid combs made to tie up the long tresses of hair fashionable at the time are now jewel-like museum pieces. Such pretty trinkets were no longer only for the rich.

Celluloid could also be used in entirely new applications. Hyatt figured out how to fabricate the material in a strip format for movie film. By the year 1900, movie film was a major market for celluloid.

However, celluloid still tended to yellow and crack over time, and it had another, more dangerous defect: it burned very easily and spectacularly, unsurprising given that mixtures of nitric acid and cellulose are also used to synthesize smokeless powder.

Ping-pong balls, onse of the few products still made with celluloid, sizzle and burn if set on fire, and Hyatt liked to tell stories about celluloid billiard balls exploding when struck very hard. These stories might have had a basis in fact, since the billiard balls were often made of celluloid and covered with paints based on another, even more flammable, nitrocellulose product known as "collodion". The paints might have acted as primer to set the rest of the ball off with a bang.

Cellulose was also used to produce cloth. While the men who developed celluloid were interested in replacing ivory, those who developed the new fibers were interested in replacing another expensive material, silk.

In 1884, a French chemist, the Comte de Chardonnay, introduced a cellulose-based fabric that became known as "Chardonnay silk". It was an attractive cloth, but like celluloid it was very flammable, a property completely unacceptable in clothing. After some ghastly accidents, Chardonnay silk was taken off the market.

In 1894, three British inventors, Charles Cross, Edward Bevan, and Clayton Beadle, patented a new "artificial silk" or "art silk" that was much safer. The three men sold the rights for the new fabric to the French Courtald company, a major silk manufacturer, who put it into production in 1905, using cellulose from wood pulp as the "feedstock" material.

Art silk became well known under the trade name "rayon", and was produced in great quantities through the 1930s, when it was supplanted by better artificial fabrics. It still remains in production today, often in blends with other natural and artificial fibers. It is cheap and feels smooth on the skin, though it is weak when wet and creases easily. It could also be produced in a transparent sheet form known as "cellophane".

Bakelite (Phenolic)

The limitations of celluloid led to the next major advance, known as "phenolic" or "phenol-formaldehyde" plastics. A chemist named Leo Hendrik Baekelund, a Belgian-born American living in New York state, was searching for an insulating shellac to coat wires in electric motors and generators. Baekelund found that mixtures of phenol (C6H5OH) and formaldehyde (HCOH) formed a sticky mass when mixed together and heated, and the mass became extremely hard if allowed to cool and dry.

He continued his investigations and found that the material could be mixed with wood flour, asbestos, or slate dust to create "composite" materials with various improved properties. Most of these compositions were strong and fire-resistant. The only problem was that the material tended to foam during synthesis, and the resulting product was of unacceptable quality.

Baekelund built pressure vessels force out the bubbles and provide a smooth, uniform product. He publicly announced his discovery in 1909, naming it "bakelite". It was originally used for electrical and mechanical parts, finally coming into widespread use in consumer goods in the 1920s.

Bakelite was the first true plastic. It was a purely synthetic material, not based on any material or molecule found in nature. It was also the first "thermoset" plastic. Conventional "thermoplastics" can be molded and then melted again, but thermoset plastics form bonds between polymers when "cured", creating a tangled matrix that cannot be undone without destroying the plastic. Thermoplastics are tough and temperature resistant.

Bakelite was cheap, strong, and durable. It was molded into thousands of forms, such as radios, telephones, clocks, and of course billiard balls. Phenolic plastics are still in widespread use. For example, electronic circuit boards are made of sheets of paper or cloth impregnated with phenolic resin.

Polystyrene and PVC

After the First World War, improvements in chemical technology led to an explosion in new forms of plastics. Among the earliest examples in the wave of new plastics were "polystyrene (PS)" and "polyvinyl chloride (PVC)", developed by the I.G. Farben company of Germany.

Polystyrene is a rigid, brittle plastic that is now used to make plastic model kits, disposable eating utensils, and similar knicknacks. It would also be the basis for one of the most popular "foamed" plastics, under the name "styrene foam" or "styrofoam". Foam plastics can be synthesized in an "open cell" form, in which the foam bubbles are interconnected, as in an absorbent sponge, or "closed cell" form, in which all the bubbles are distinct, like tiny balloons, as in gas-filled foam insulation and floatation devices.

PVC has side chains incorporating chlorine molecules, which form strong bonds. PVC in its normal form is stiff, strong, heat and weather resistant, and is now used for making pipe, gutters, house siding, enclosures for computers and other electronics gear, and compact-disk media. PVC can also be softened with chemical processing, and in this form it is now used for shrink-wrap, food packaging, and raingear.

Nylon

The real star of the plastics industry in the 1930s was "polyamide (PA)", far better known by its trade name, "nylon". Nylon was the first purely synthetic fiber, introduced by Du Pont Corporation at the 1939 World's Fair in New York City. In 1927, Du Pont had begun a secret development project designated "Fiber66", under the direction of a Harvard chemist named Wallace Carothers. Carothers had been hired to perform pure research, and not only investigated new materials, but worked to understand their molecular structure and how it related to material properties. He took some of the first steps on the road to "molecular design" of materials.

His work led to the discovery of synthetic nylon fiber, which was very strong but still very flexible. The first application was for bristles for toothbrushes. However, Du Pont's real target was silk, particularly silk stockings. It took Du Pont twelve years and $27 million USD to refine nylon and develop the industrial processes for its manufacture in bulk. With such a major investment, it was no surprise that Du Pont spared little expense to promote nylon after its introduction, creating a public sensation that was called "nylon mania".

Nylon mania came to an abrupt stop at the end of 1941, when America entered World War II. The production capacity that had been built up to produce nylon stockings, or just "nylons", for American women, was taken over to manufacture numbers of parachutes for fliers and paratroopers. After the war ended, Du Pont went back to selling nylon to the public, engaging in another promotional campaign in 1946 that resulted in an even bigger craze, triggering off "nylon riots".

Nylon still remains an important plastic, and not just for use in fabrics. In its bulk form it is very wear-resistant, and so is used to build gears, bearings, bushings, and other mechanical parts.

Synthetic Rubber

Another plastic that was critical to the war effort was "synthetic rubber", which was produced in a variety of forms. Practical synthetic rubber grew out of studies published in 1930 written independently by Carothers and the German scientist Hermann Staudinger. These studies led in 1931 to one of the first successful synthetic rubbers, known as "neoprene". Neoprene is highly resistant to heat and chemicals such as oil and gasoline, and is used in fuel hoses and as an insulating material in machinery.

In 1935, German chemists synthesized the first of a series of synthetic rubbers known as "Buna rubbers". These were "copolymers", meaning that their polymers were made up from not one but two monomers, in alternating sequence. One such Buna rubber, known as "GR-S (Government Rubber Styrene)", is a copolymer of butadiene and styrene, became the basis for US synthetic rubber production during World War II.

Worldwide natural rubber supplies were limited, and by mid-1942 most of the rubber-producing regions were under Japanese control. Military trucks needed rubber for tires, and rubber was used in almost every other war machine. The US government launched a major effort to ramp up synthetic rubber production, and by 1944 a total of 50 factories were manufacturing it, pouring out a volume of the material twice that of the world's natural rubber production before the beginning of the war.

After the war, natural rubber plantations no longer had a stranglehold on rubber supplies, particularly after chemists learned to synthesize isoprene. GR-S remains the primary synthetic rubber for the manufacture of tires.

* Synthetic rubber would also play an important part in the space race and nuclear arms race. Solid-fuel rockets used during World War II used nitrocellulose explosives for propellants, but it was impractical and dangerous to make such rockets very big.

During the war, California Institute of Technology (CalTech) researchers came up with a new solid fuel, based on asphault fuel mixed with an oxidizer, such as potassium or ammonium percholorate, plus aluminum powder, which burns very hot. This new solid fuel burned more slowly and evenly than nitrocellulose explosives, and was much less dangerous to store and use, though it tended to flow slowly out of the rocket in storage and the rockets using it had to be stockpiled nose-down.

After the war, the CalTech researchers began to investigate the use of synthetic rubbers instead of asphault as the fuel in the mixture. By the mid-1950s, large missiles were being built using solid fuels based on synthetic rubber, mixed with ammonium perchlorate and high proportions of aluminum powder. Such solid fuels could be cast into large, uniform blocks that had no cracks or other defects that would cause nonuniform burning. Ultimately, most large military rockets and missiles would use solid fuels based on synthetic rubbers, and they would also play a significant part in the civilian space effort.

Plastics Explosion: Acrylic, Polyethylene, and etc.

Other plastics emerged in the prewar period, though some wouldn't come into widespread use until after the war. By 1936, American, British, and German companies were producing "polymethyl methacrylate (PMMA)", better known as "acrylic". Although acrylics are now well-known for the use in paints and synthetic fibers, such as "fake furs", in their bulk form they are actually very hard and more transparent than glass, and are sold as glass replacements under trade names such as "plexiglas" and "lucite". Plexiglas was used to built aircraft canopies during the war, and it is also now used as a marble replacement for countertops.

Another important plastic, "polyethylene (PE)", sometimes known as "polythene", was discovered in 1933 by the Reginald Gibson and Eric Fawcett at the British industrial giant Imperial Chemical Industries (ICI). This material evolved into two forms, "low density polyethylene (LDPE)", and "high density polyethylene (HDPE)".

PEs are cheap, flexible, durable, and chemically resistant. LDPE is used to make films and packaging materials, while HDPE is used to containers, pipes, and automotive fittings. While PE has low resistance to chemical attack, it was found later that a PE container could be made much more robust by exposing it to fluorine gas, which modified the surface layer of the container into the much tougher "polyfluoroethylene".

Polyethylene would lead after the war to an improved material, "polypropylene (PP)", which was discovered in the early 1950s. It is common in modern science and technology that the growth of the general body of knowledge can lead to the same inventions in different places at about the same time, but polypropylene was an extreme case of this phenomenon, being separately invented about nine times. It was a patent attorney's dream scenario, and litigation wasn't resolved until 1989.

Polypropylene managed to survive the legal process, and two American chemists working for Phillips Petroleum of the Netherlands, Paul Hogan and

Robert Banks, are now generally credited as the "official" inventors of the material. Polypropylene is similar to its ancestor, polyethylene, and shares polyethylene's low cost, but it is much more robust. It is used in everything from plastic bottles to carpets to plastic furniture, and is very heavily used in automobiles.

Polyurethane was invented by Friedrich Bayer & Company of Germany in 1937, and would come into use after the war in blown form for mattresses, furniture padding, and thermal insulation. It is also used in non-blown form for sports wear as "lycra".

In 1939, I.G. Farben Industrie of Germany filed a patent for "polyepoxide" or "epoxy". Epoxies are a class of thermoset plastic that form cross-links and "cure" when a catalyzing agent, or "hardener", is added. After the war they would come into wide use for coatings, "super glues", and composite materials.

Composites using epoxy as a matrix include "fiberglass", where the structural element is glass fiber, and "carbon-epoxy composites", in which the structural element is carbon fiber. Fiberglass is now often used to build sport boats, and carbon-epoxy composites are an increasingly important structural element in aircraft, as they are lightweight, strong, and heat-resistant.

Two chemists named Rex Whinfield and James Dickson, working at a small English company with the quaint name of the "Calico Printer's Association" in Manchester, developed "polyethylene terephthalate (PET or PETE)" in 1941, and it would be used for synthetic fibers in the postwar era, with names such as "polyester", "dacron", and "terylene".

PET is more impermeable than other low-cost plastics and so is a popular material for making bottles for Coke and other "fizzy drinks", since carbonation tends to attack other plastics; and for acidic drinks such as fruit or vegetable juices. PET is also strong and abrasion resistant, and is used for making mechanical parts, food trays, and other items that have to endure abuse. PET films, tradenamed "mylar", are used to make recording tape.

One of the most impressive plastics used in the war, and a top secret, was "polytetrafluoroethylene (PTFE)", better known as "teflon", which could be deposited on metal surfaces as a scratchproof and corrosion-resistant, low-friction protective coating. The polyfluoroethylene surface layer created by exposing a polyethylene container to fluorine gas is very similar to teflon.

A Du Pont chemist name Roy Plunkett discovered teflon by accident in 1938. During the war, it was used in gaseous-diffusion processes to refine uranium for the atomic bomb, as the process was highly corrosive. By the early 1960s, teflon "non-stick" frying pans were a hot consumer item.

Teflon was later used to synthesize the miracle fabric "GoreTex", which can be used to build raingear that in principle "breathes" to keep the wearer's moisture from building up. GoreTex is also used for surgical implants; teflon strand is used to make dental floss; and teflon mixed with fluorine compounds is used to make "decoy" flares dropped by aircraft to distract heat-seeking missiles.

After the war, the new plastics that had been developed entered the consumer mainstream in a flood. New manufacturing were developed, using various forming, molding, casting, and extrusion processes, to churn out plastic products in vast quantities. American consumers enthusiastically adopted the endless range of colorful, cheap, and durable plastic gimmicks being produced for new suburban home life.

One of the most visible parts of this plastics invasion was Earl Tupper's "tupperware", a complete line of sealable polyethylene food containers that Tupper cleverly promoted through a network of housewives who sold Tupperware as a means of bringing some money. The tupperware line of products was well thought out and highly effective, greatly reducing spoilage of foods in storage. Thin-film "plastic wrap" that could be purchased in rolls also helped keep food fresh.

Another prominent element in 1950s homes was "formica", a plastic laminate that was used to surface furniture and cabinetry. Formica was durable and attractive. It was particularly useful in kitchens, as it did not absorb, and could be easily cleaned of, stains from food preparation, such as blood or grease. With formica, a very attractive and well-built table could be built using low-cost and lightweight plywood with formica covering, rather than expensive and heavy hardwoods like oak or mahogany.

Composite materials like fiberglass came into use for building boats and, in some cases, cars. Polyurethane foam was used to fill mattresses, and styrofoam was used to line ice coolers and make float toys.

* Plastics continue to be improved. General Electric introduced "lexan", a high-impact "polycarbonate" plastic, in the 1970s. Du Pont developed "kevlar", an extremely strong synthetic fiber that was best-known for its use in bullet-proof vests and combat helmets. Kevlar was so remarkable that Du Pont officials actually had to release statements to deny rumors that the company had received the recipe for it from space aliens.

One of the most potentially important new developments in plastic are circuits out of plastics is conductive polymers. Electronic circuitry fabricated using plastics or other materials that could be simply printed on a substrate could be incredibly cheap, opening the door to throwaway electronic devices that would cost pennies, or to applications hardly dreamed of now.

So far, electronic devices made with such materials have not been acceptable for production, but in 2001, prototypes of flat-panel displays based on such technologies were being publicly demonstrations, with predictions of commercial introduction in two or three years.

Plastics and the Environment

Although plastics have had a remarkable impact on our culture, it has become increasingly obvious that there is a price to be paid for their use. The first controversy arose in the late 1950s and early 1960s. There were a number of incidents where small children suffocated after crawling into plastic bags used by launderers to cover clothing. The plastics industry managed to fend off trouble by launching a massive public-education campaign.

By the late 1960s, plastics were increasingly seen as a symbol of an outdated 1950s consumer culture. The term "plastic" became an insult, used to describe someone thought of as soulless. At the end of the 1960s, the Beatles would even sing of "Polyethylene Pam", a "go-getter" who would do anything to get ahead.

This was partly just a fashion statement, since plastics remained in widespread use anyway, and in many cases were much more effective and environmentally benign than the alternatives. However, this led to a problem as well, since the consumption of massive amounts of plastic goods created an equally massive problem with litter and waste disposal.

Plastic was almost too good, as it was durable and degraded very slowly. In some cases burning it could release toxic fumes. There were also the problems that manufacturing plastics often created large quantities of nasty chemical pollutants, and depleted the Earth's bounded supply of fossil fuels.

By the 1990s, plastic recycling programs were common in the United States and elsewhere. Thermoplastics can be remelted and reused, and thermoset plastics can be ground up and used as filler, though the purity of the material tends to degrade with each reuse cycle. There are methods by which plastics can be broken back down to a feedstock state.

Products such as automobiles are now being designed to make recycling of their large plastic parts easier. To assist recycling of disposable items, the Plastic Bottle Institute of the Society of the Plastics Industry devised a now-familiar scheme to mark plastic bottles by plastic type. A recyclable plastic container using this scheme is marked with a triangle with three "chasing arrows" inside of it, which enclose a number giving the plastic type:

1. PETE
2. HDPE
3. PVC
4. LDPE

5. PP
6. PS
7. OTHER

Unfortunately, recycling plastics proved difficult. The biggest problem with plastics recycling is that it is difficult to automate the sorting of plastic waste, and so it is labor-intensive. While containers are usually made from a single type and color of plastic, making them relatively easy to sort, a consumer toy like a cellphone may have many small parts consisting of over a dozen different types and colors of plastics. As the value of the material is low, recycling plastics is unprofitable. For this reason, the percentage of plastics recycled in the US is very small, somewhere around 5 per cent.

Research has been done on "biodegradable" plastics that break down with exposure to sunlight. Starch can be mixed with plastic to allow it to degrade more easily, but it still doesn't lead to complete breakdown of the plastic. Some researchers have actually genetically engineered bacteria that synthesize a completely biodegradable plastic.

So far, these plastics have proven too costly and limited for general use, and critics have pointed out that they only real problem they address is roadside litter, which is regarded as a secondary issue. When such plastic materials are dumped into landfills, they can become "mummified" and persist for decades even if they are supposed to be biodegradable.

There have been some success stories. The Courtald concern, the original producer of rayon, came up with a revised process for the material in the mid-1980s to produce "tencel". Tencel has much superior properties to rayon, but is still produced from "biomass" feedstocks, and its manufacture is extraordinarily clean by the standards of plastic production. Whether the use of plastics can be made completely consistent with environmental quality still remains to be seen.

Comments and Sources

I think the dullness has a lot to do with the way chemistry is typically taught, based on the memorization of lists of dry facts that are then promptly forgotten, and with astonishingly little attempt to relate the science to its pervasive everyday application.

Anyway, after writing this article I've picked up the habit of checking for recycling symbols on plastic objects around the house to see what they're made of. It turns out that the little plastic chair and table set I bought is made out of PP, for example.

The chair and table are an example of the utility of plastics. While they're not really meant for heavy use by any means, they are perfectly functional, durable and even, given a little open-mindedness, reasonably attractive. This at a cost less than that of a meal at a mid-priced restaurant. They would be a marvel to a citizen from several centuries ago, who would not only be astounded at the material itself and its cost, but at the vision of limitless numbers of such items "stamped out" by machines.

Although this document follows the outline of the MODERN MARVELS installment, it is much more detailed. I fleshed it out with a few print sources:

- MOLECULES AT AN EXHIBITION by John Emsley, Oxford University Press, 1998. Emsley is a professor at the University of London and a remarkably good science writer. I have to recommend this book, as it is very entertaining and witty.
- "Plastics Get Wired" by Philip Yam, SCIENTIFIC AMERICAN, July 1995, 82:87.
- "Disappearing Act", by Tim Beardsley, SCIENTIFIC AMERICAN, November 1988.

I also found some useful information in the from the MicroSoft ENCARTA online encyclopedia's article on the subject. As usual, sources contradicted each other, for example with the show claiming that most synthetic rubber production during the war was of neoprene, while Encarta said, somewhat more convincingly, that it was GR-S.

5

Synthetic Polymers and their Sources

5.1 Macromolecules: Natural and Synthetic Polymers

Objectives

In this labortaory you will become familiar with the classifications of polymers by synthesizing and examining several of the following:

1. A linear condensation copolymer (GlyptalTM resin),
2. A branched addition polymer (polymethylmethacrylate),
3. A cross-link a natural linear polymer (cellulose),
4. A loosely cross-linked silicon-based condensation polymer (a polymethylsiloxane), and
5. A cross-linked polyvinyl alcohol.

Introduction

Approximately 50 per cent of the industrial chemists in the United States work in some area of polymer chemistry, a fact that illustrates just how important polymers are to our economy and standard of living. These polymers are essential to the production of goods ranging from toys to roofing materials. So what exactly are polymers? Polymers are substances composed of extremely large molecules termed macromolecules, with molecular masses ranging from 104 to 108 amu. The macromolecules consist of many smaller molecular units, monomers joined together through covalent bonds. The molar mass of the polymer is quoted as an average molar mass.

Both natural and synthetic polymers are ubiquitous in our lives: elastomers (polymers with elastic, rubber-like properties), plastics (the first plastic was used in 1843 to make buttons), textile fibers, resins, and adhesives. The more common polymers include acrylics, alkyds, cellulosics, epoxy resins, phenolics, polycarbonates, polyamides, polyesters, polyfluorocarbons, polyolefins, polystyrenes, silicones, and vinyl plastics, to name but a few.

Naturally occurring macromolecules are obviously derived from living things ñ wood, wool, paper, cotton, starch, silk, rubber ñ and have provided us for centuries with materials for clothing, food, and housing. Starch, glycogen, and cellulose are all polymeric versions of the monomer glucose. Proteins are macromolecules composed of monomeric units of alpha amino acids; nucleic acids are composed of subunits (nucleotides) containing a nitrogeneous base, sugar and phosphate groups. Natural rubber is a latex exudate of certain trees and composed of monomers called isoprene units. The usefulness of latex was first discovered by Lord Mackintosh in Malayasia in the last century and provided the foundation of his waterproof rainwear empire.

The temptation to improve upon nature has always been great and has rarely been resisted. When scientists linked the special properties of these substances (physical properties such as tensile strength and flexibility) to the sizes of their molecules the next logical step involved chemical modifications of naturally occurring polymers.

Synthetic celluloid derives from natural cellulose and stems from an accident that Christian Schoenbein, a chemistry professor had in 1846—the age of plastic had begun although, initially the interest in cellulose nitrate was more for their explosive properties. When cellulose (from wood chips or fiber) is treated with a mixture of nitric acid, camphor, and alcohol, the resultant product is called CelluloidTM and bears very little resemblance to the starting material. CelluloidTM possess the ability to be molded into hard, smooth billiard balls (replacing the original, very expensive ivory balls) and into thin sheets for making movie pictures. CelluloidTM is highly flammable and today has been replaced by greatly improved synthetic polymers such as bakelite discovered in 1907 by the Belgian-American Chemist, Leo H. Baekeland.

When cellulose is treated with sodium hydroxide and carbon disulfide (CS2), cellulose xanthate is formed. A viscous (thick) solution of cellulose.xanthate, forced

through fine holes into dilute sulfuric acid, regenerates the cellulose as fine, continuous cylindrical threads called rayon. If the solution is forced instead through a narrow slit, a thin transparent film or sheet is obtained called cellophane.

Thermoplasts Versus Thermosetting

Polymers generally are classified into two broad groups in accordance with their behavior upon heating. Polymers that can be repeatedly melted and solidified (without damage) are said to be thermoplasts; those that solidify once but will not melt again with damage are said to be thermosets. Technically, though only thermoplasts are true plastics, even though the term "plastic" is commonly applied to all synthetic polymers.

The thermoplastic substance contains long, thin molecules which form tangled chains and is rigid at lower temperatures but gradually softens upon the application of heat and after it passes a characteristic temperature known as its glass transition temperature (Tg). Below Tg, the substance is brittle, having the characteristic properties of a glass; above Tg, the substance becomes flexible and soft. Chewing gum is a thermoplast that becomes extremely brittle when the outside temperatures drop below its glass transition temperature—this is an useful property to use in order to remove chewing gum from your clothes. Once warmed above Tg, however, the gum quickly softens and regains its flexibility. Some thermoplasts, such as polystyrene, melt before reaching their glass transition temperatures and remain rigid materials up to their melting points. Thermoplastic polymers are used frequently for injection molding of such items as food storage containers and toys that are not exposed to high temperatures. Additionally, thermoplastic polymers can be molded, pressed and extruded.

Thermosetting substances contain large, cross-linked molecules and are also rigid at lower temperatures, undergo irreversible chemical and physical changes (including decomposition) upon heating. Such substances remain solids at higher temperatures than do thermoplastic materials, and they do not melt. Thermosets are often employed in high-temperature environments, such as for electrical insulation in electric motors and gasoline engines.

Thermoplastic polymers are composed of small monomers covalently bonded end-to-end in a long chain but without covalent bonds joining adjacent chains. Such macromolecules constitute the linear polymers. Shorter side groups attached to the long chains at periodic intervals, cause the polymers to be termed branched polymers. The chains, having average molecular masses up to one million amu, may be independent of each other (as in polyethylene) or loosely lined through hydrogen bonding (as in nylon). If the long chains are linked by covalent bonds, the polymeric network becomes two- or three-dimensional, resulting in an infusible (nonmelting) and insoluble material. Such macromolecules make up the cross-linked polymers, and are found in thermosetting materials. Polymers of all types in which the long chains are produced by joining two or more different kinds of monomers are termed copolymers.

The process by which the polymerization reaction occurs permits classification of polymers into two categories: addition and condensation polymers. Addition polymers are those in which the monomers join at unsaturated carbon atoms; several are summarized in theTable. During polymerisation, the double bonds between the pairs of carbon atoms 'open up' and the carbon atoms of separate ethylene molecules join together to form a molecule of polyethylene. The first polymersiation of ethylene was accomplished in 1933 by the use of very high pressure (1000 atm) and oxygen as a catalyst. Nowadays, with the development of the use of powerful catalysts, addition can occur at atmospheric pressure. Polymethyl methacrylate, also called Lucite™ or Plexiglas™ (originally developed as an unbreakable substitute for glass in airplane canopies), belongs to this group of addition polymers. The polymerization is initiated by a variety of substances (such as benzoyl peroxide) that can form a free radical with the unsaturated carbon atom. The resulting addition polymer is described as a branched polymer.

The second way to make a polymer is by condensation polymerisation. In this process, two compounds with reactive atoms at the end of their molecules react together, usually with the release of a small molecular unit such as water or hydrogen chloride. The presence of two or more functional groups in the monomer usually leads to the production of a cross-linked polymer. Glyptal resin, formed by the reaction between phthalic acid and glycerol, is a condensation copolymer, a copolymer since 2 different types of monomers combine to form the chain.

Monomers are linked through ester ($-C(=O)-O-CH_2-$) bonds; the resin is termed a polyester. The more reactive phthalic anhydride is often used in place of phthalic acid in this reaction.

Paper is composed of naturally occurring cellulose, the polymeric structural material of plants. The cellulose chains are composed of linear glucose units. Parchment paper is made by cross-linking the linear cellulose polymer to form a sheet-like structure. Adjacent chains are cross-linked by ether ($-CH_2-O-CH_2-$) bonds resulting from dehydration of the alcohol groups (through the use of sulfuric acid as the dehydrating agent). Resulting reactions that form the macromolecules required for polymerization by either addition or condensation processes must be capable of proceeding indefinitely.

By far the most interesting uses of polymers involves replacement of diseased, worn out, or missing parts of the human body including flexible replacements for major blood vessels, replacement valves for hearts, temporary skin, and artificial joints. Artificial ball-and-socket pelvic (hip) joints made of steel (ball) and plastic (socket) are installed at the rate of 25,000 per year. People with crippling arthritis, debilitating coronary and circulatory problems, and burns all benefit from the development of biomedical polymers. Preventive and cosmetic dentistry, as well, benefit from polymers used to seal porous teeth and reconstruct missing enamel.

Linear silicones, or polysiloxanes, are comparatively new polymers based upon silicon-oxygen-silicon linkages. These polymers may be cross-linked to various degrees by additional -Si-O-Si- bonding between adjacent chains:

The R group is generally a hydrocarbon group such as $-CH_3$ (methyl), $-CH_2CH_3$ (ethyl), or $-C_6H_5$ (phenyl). Silicones are stable at much higher temperatures than carbon-based polymers, yet they remain flexible even at exceedingly low temperatures. Among such silicones is Silly PuttyTM, the cross-linked polymerization product of dimethyldichlorosilane, $(CH_3)_2Si(OH)_2$ with the release of hydrogen chloride (HCl):

$$Cl-\underset{CH_3}{\overset{CH_3}{|}}{Si}-Cl + H_2O \longrightarrow HO-\underset{CH_3}{\overset{CH_3}{|}}{Si}-OH + 2HCl$$

The unstable dimethyldihydroxysilane condenses rapidly to form a low molecular mass linear polymethylsiloxane polymer with the elimination of water:

$$Cl-\underset{CH_3}{\overset{CH_3}{|}}{Si}-Cl \longrightarrow HO-\underset{CH_3}{\overset{CH_3}{|}}{Si}-OH + (n-1)H_2O$$

This material is an oily liquid. Additional heating continues the polymerization to form longer chains. The addition of boron trioxide allows three such chains to be joined to form Silly PuttyTM:

O—Si—O—Si—
—Si—O—Si—O—B—
O—Si—O—Si—

This medium molecular mass polymer has physical properties between those of a fluid and an elastomer. Although it is resistant to rapid deformation, it flows easily with slowly applied stress.

5.2 Synthetic Polymer

Man-made polymers are used in a wide array of applications: food packaging, films, fibers, tubing, pipes, etc. The personal care industry also uses polymers to aid in texture of products, binding, and moisture retention (e.g. in hair gel and conditioners).

Table 5.1: Addition Polymers.

Example	*Monomer(s)*	*Polymer*	*Use*
Polyethylene	$CH_2{=}CH_2$	$-CH_2-CH_2-$	Common polymer: bags, wire insulat-ions, squeeze bottles
Polypropylene	$CH_2{=}CH_2$ with CH_3	$-CH_2-CH_2-$ with CH_3	Fibers, indoor-outdoor carpet, bottles, rope
Polystyrene	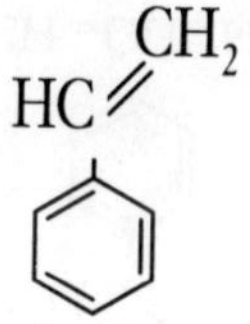 $HC{=}CH_2$ on benzene ring	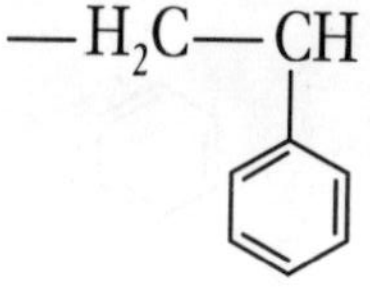 $-H_2C-CH$ on benzene ring	Styrofoam™; drinking cups, building insulation, packing materials
Polyvinylchloride (PVC)	$-H_2C{=}CH$ with α	$-CH_2-CH_2-$ with α	synthetic leathers, clear bottles, floor coverings, phonograph records, water pipes
Polytetrafluoroethylene (Teflon™)	$CF_2{=}CF_2$	—	nonstick surfaces, chemically resistant films, cookware coatings
Polymethylmethacrylate (Lucite™, plexiglass™)	CH_2-C with CO_2CH_3 and CH_3	CH_2-C- with CO_2CH_3 and CH_3	unbreakable glass, latex paints
Polyacrylonitrile (Orlon™, Acrilan™, Creslan™)	$H_2C{=}CH$ with CN	$-H_2C-CH-$ with CN	fibers for sweaters, blankets, carpets

(*Contd.*)

(Contd.)

Example	Monomer(s)	Polymer	Use
Polyvinylacetate (PVA)	$CH_2{=}CH_2$ with pendant OCCH_3 ($C{=}O$)	$—CH_2—CH_2—$ with pendant OCCH_3 ($C{=}O$)	adhesives, latex paints, chewing gum, textile coatings
natural rubber	$CH_2{=}C(CH_3)CH{=}CH_2$	$—CH_2—C(CH_3){=}CH—CH_2—$	The polymer is cross-linked with sulfur (vulcanization).
Polychloroprene (neoprene rubber)	$CH_2{=}C(Cl)CH{=}CH_2$	$—CH_2—C(Cl){=}CH—CH_2—$	cross-linked with zinc oxide; resistant to oil, gasoline
Styrene-Butadiene Rubber (SBR)	$HC(C_6H_5){=}CH_2$; $CH_2{=}CHCH{=}CH_2$	$—CH_2—CH(C_6H_5)—H_2C—$	cross-linked with peroxides; most commonly used for tires, 25% styrene, 75% butadiene
Polyvinylacetate (PVA)	$CH_2{=}CH_2$ with pendant OCCH_3 ($C{=}O$)	$—CH_2—C(CH_3){=}CH—CH_2—$	adhesives, latex paints, chewing gum, textile coatings

Table 5.2: Condensation Polymers.

Example	*Monomer(s)*	*Polymer*	*Use*
Polyamides (nylon)	$HO{-}C(=O){-}(CH_2)n{-}C(=O){-}OH$ $H_2N{-}(CH_2)n{-}NH_2$	$-C(=O){-}(CH_2)n{-}C(=O){-}N(H){-}(CH_2)n{-}N(H)-$	fibers, molded objects
Polyesters (Dacron™, Mylar™, Fortrel™)	$HOC(=O){-}C_6H_4{-}C(=O)OH$ $HO{-}(CH_2)n{-}OH$	$-C(=O){-}C_6H_4{-}C(=O){-}O{-}(CH_2)n{-}O-$	linear polyesters, fibers, recording tape
Polyesters (Glyptal™ resin)	phthalic anhydride (C_6H_4, $C(=O){-}O{-}C(=O)$ ring) $HO{-}CH_2CHCH_2{-}OH$ (with OH on the middle carbon)	C_6H_4 bearing $-C(=O)-$ and $-C(=O){-}O{-}CH_2CHCH_2O-$, with $-O-$ on the middle carbon	crosslinked polyester; paints
Polyesters (casting resin)	$HO{-}C(=O){-}CH{=}CH{-}C(=O){-}HO$ $HO{-}(CH_2)n{-}OH$	$-C(=O){-}CH{=}CH{-}C(=O){-}(CH_2)n{-}O-$	crosslinked with styrene and peroxide: fiberglass, boat resin
Cellulose Acetate	glucose ring unit with CH_2OH, O, $-O-$, OH, OH; CH_3COOH	ring unit with CH_2OAc, O, $-O-$, OAc, OAc	photographic film
Silicones	$Cl{-}Si(CH_3)_2{-}Cl$	$-O{-}Si(CH_3)_2{-}O-$	waterrepellant coatings, temperature resistant fluids, rubbers (CH_3SiCl_3 crosslinks in water)
Polyurethanes	$CH_3C_6H_3(N{=}C{=}O)_2$ $HO{-}(CH_2)n{-}OH$	$CH_3C_6H_3$ bearing $NHC(=O){-}O{-}(CH_2)n{-}O-$ and $NHC(=O){-}O{-}(CH_2)n{-}O-$	rigid and flexible foams, fibers

Examples

A non-exhaustive list of these ubiquitous materials includes:

- acrylonitrile butadiene styrene (ABS)
- polyamide (PA)
- polybutadiene
- poly(butylene terephthalate) (PBT)
- polycarbonate (PC)
- poly(ether sulphone) (PES, PES/PEES)
- poly(ether ether ketone)s (PEEK, PES/PEEK)
- polyethylene (PE)
- poly(ethylene glycol) (PEG)
- poly(ethylene terephthalate) (PET)
- polyimide
- polypropylene (PP)
- polytetrafluoroethylene (PTFE)
- polystyrene (PS)
- styrene acrylonitrile (SAN)
- poly(trimethylene terephthalate) (PTT)
- polyurethane (PU)
- polyvinylchloride (PVC)
- polyvinylidenedifluoride (PVDF)
- poly(vinyl pyrrolidone) (PVP)

Brand Names

These polymers are often better known through their brand names, for instance:

- Kevlar, Twaron, e.g. para-aramid
- Technora, e.g. copolyamid
- Kynar, e.g. PVDF
- Mylar, e.g. polyethylene terephthalate film
- Nylon, e.g. polyamide 6,6
- Rilsan, e.g. polyamide 11 & 12
- Teflon, e.g. PTFE
- Ultem, e.g. polyimide
- Vectran, aromatic polyamide
- Viton, e.g. poly-tetrafluoroethylene
- Zylon, e.g. poly-p-phenylene-2,6-benzobisoxazole (PBO).

5.3 Acrylonitrile Butadiene Styrene

Acrylonitrile butadiene styrene, or ABS, [chemical formula $(C_8H_8 \cdot C_4H_6 \cdot C_3H_3N)_n$] is a common thermoplastic used to make light, rigid, molded products such as piping, musical instruments (most notably recorders and plastic clarinets), golf club heads (used for its good shock absorbance), automotive body parts, wheel covers, enclosures, protective head gear, vballs [reusable paintballs], and toys including Lego bricks. In plumbing, ABS pipes are the black pipes (PVC pipes are white) and also in Plastic Pressure Pipe Systems. ABS plastic ground down to an average diameter of less than 1 micrometer is used as the colorant in some tattoo inks. Tattoo inks that use ABS are extremely vivid. This vividness is the most obvious indicator that the ink contains ABS, as tattoo inks rarely list their ingredients.

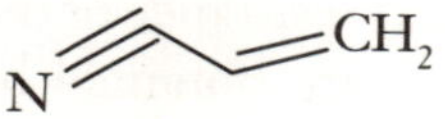

N CH_2 H_2C CH_2

acrylonitrile 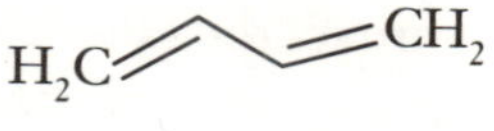   1,3-butadiene

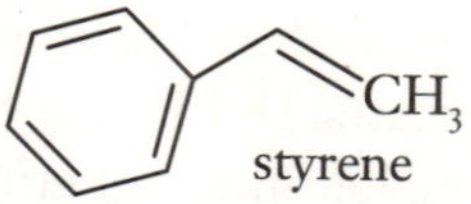

CH_3

styrene

It is a copolymer made by polymerizing styrene and acrylonitrile in the presence of polybutadiene. The proportions can vary from 15 to 35 per cent acrylonitrile, 5 to 30 per cent butadiene and 40 to 60 per cent styrene. The result is a long chain of polybutadiene criss-crossed with shorter chains of poly(styrene-co-acrylonitrile). The nitrile groups from neighboring chains, being polar, attract each other and bind the chains together, making ABS stronger than pure polystyrene. The styrene gives the plastic a shiny, impervious surface. The butadiene, a rubbery substance, provides resilience even at low temperatures. ABS can be used between -25 and 60 °C. The properties are created by rubber toughening, where fine particles of elastomer are distributed throughout the rigid matrix. Production of 1 kg of ABS requires the equivalent of about 2 kg of oil for raw materials and energy. It can also be recycled.

Properties

ABS is derived from acrylonitrile, butadiene, and styrene. Acrylonitrile is a synthetic monomer produced from propylene and ammonia; butadiene is a petroleum hydrocarbon obtained from butane; and styrene monomers, derived from coal, are commercially obtained from benzene and ethylene from coal. The advantage of ABS is that this material combines the strength and rigidity of the acrylonitrile and styrene polymers with the toughness of the polybutadiene rubber. The most important mechanical properties of ABS are resistance and toughness. A variety of modifications can be made to improve impact resistance, toughness, and heat resistance. The impact resistance can be amplified by increasing the proportions of polybutadiene in relation to styrene and acrylonitrile although this causes changes in other properties. Impact resistance does not fall off rapidly at lower temperatures. Stability under load is excellent with limited loads.

Even though ABS plastics are used largely for mechanical purposes, they also have good electrical properties that are fairly constant over a wide range of frequencies. These properties are little affected by temperature and atmospheric humidity in the acceptable operating range of temperatures. The final properties will be influenced to some extent by the conditions under which the material is processed to the final product; for example, molding at a high temperature improves the gloss and heat resistance of the product whereas the highest impact resistance and strength are obtained by molding at low temperature. ABS polymers are resistant to aqueous acids, alkalis, concentrated hydrochloric and phosphoric acids, alcohols and animal, vegetable and mineral oils, but they are swollen by glacial acetic acid, carbon tetrachloride and aromatic hydrocarbons and are attacked by concentrated sulfuric and nitric acids. They are soluble in esters, ketones and ethylene dichloride.

The aging characteristics of the polymers are largely influenced by the polybutadiene content, and it is normal to include antioxidants in the composition. On the other hand, while the cost of producing ABS is roughly twice the cost of producing polystyrene, ABS is considered superior for its hardness, gloss, toughness, and electrical insulation properties. However, it will be degraded (dissolve) when exposed to acetone. ABS is flammable when it is exposed to high temperatures, such as a wood fire. It will "boil", then burst spectacularly into intense, hot flames.

5.4 Nylon

Nylon is a generic designation for a family of synthetic polymers known generically as polyamides and first produced on February 28, 1935 by Wallace Carothers at DuPont. Nylon is one of the most commonly used polymers.

Density	*1.15 g/cm³*
Electrical conductivity (?)	10^{-12} S/m
Thermal conductivity	0.25 W/(m·K)
Melting point	463 K-624 K; 190°C-350°C; 374°F-663°F

Overview

Nylon is a thermoplastic silky material, first used commercially in a nylon-bristled toothbrush (1938), followed more famously by women's "nylons" stockings (1940). It is made of repeating units linked by peptide bonds (another name for amide bonds) and is frequently referred to as *polyamide* (PA). Nylon was the first commercially successful polymer and the first synthetic fiber to be made entirely from coal, water and air. These are formed into monomers of intermediate molecular weight, which are then reacted to form long polymer chains. Nylon was intended to be a synthetic replacement for silk and substituted for it in many different products after silk became scarce during World War II. It replaced silk in military applications such as parachutes, flak vests, and was used in many types of vehicle tires.

Nylon fibers are used in many applications, including fabrics, bridal veils, carpets, musical strings and rope. Solid nylon is used for mechanical parts such as gears and other low- to medium-stress components previously cast in metal. Engineering grade nylon is processed by extrusion, casting, and injection molding. Solid nylon is used in hair combs. Type 6/6 Nylon 101 is the most common commercial grade of nylon, and Nylon 6 is the most common commercial grade of cast nylon. Nylon is available in glass-filled and molybdenum sulfide-filled variants which increase structural and impact strength and rigidity or lubricity. Aramids are another type of polyamide with quite different chain structures which include aromatic groups in the main chain. Such polymers make excellent ballistic fibres.

Chemistry

Nylons are condensation copolymers formed by reacting equal parts of a diamine and a dicarboxylic acid, so that peptide bonds form at both ends of each monomer in a process analogous to polypeptide biopolymers. The numerical suffix specifies the numbers of carbons donated by the monomers; the diamine first and the diacid second. The most common variant is nylon 6-6 which refers to the fact that the diamine (hexamethylene diamine) and the diacid (adipic acid) each donate 6 carbons to the polymer chain. As with other regular copolymers like polyesters and polyurethanes, the "repeating unit" consists of one of each monomer, so that they alternate in the chain. Since each monomer in this copolymer has the same reactive group on both ends, the direction of the amide bond reverses between each monomer, unlike natural polyamide proteins which have overall directionality: C terminal? N terminal. In the laboratory, nylon 6,6 can also be made using adipoyl chloride instead of adipic It is difficult to get the proportions exactly correct, and deviations can lead to chain termination at molecular weights less than a desirable

10,000 daltons (u). To overcome this problem, a crystalline, solid "nylon salt" can be formed at room temperature, using an exact 1:1 ratio of the acid and the base to neutralize each other. Heated to 285 °C, the salt reacts to form nylon polymer. Above 20,000 daltons, it is impossible to spin the chains into yarn, so to combat this, some acetic acid is added to react with a free amine end group during polymer elongation to limit the molecular weight. In practice, and especially for 6,6, the monomers are often combined in a water solution. The water used to make the solution is evaporated under controlled conditions, and the increasing concentration of "salt" is polymerized to the final molecular weight.

DuPont patented nylon 6,6, so in order to compete, other companies (particularly the German BASF) developed the homopolymer nylon 6, or polycaprolactam—not a condensation polymer, but formed by a ring-opening polymerization (alternatively made by polymerizing aminocaproic acid). The peptide bond within the caprolactam is broken with the exposed active groups on each side being incorporated into two new bonds as the monomer becomes part of the polymer backbone. In this case, all amide bonds lie in the same direction, but the properties of nylon 6 are sometimes indistinguishable from those of nylon 6,6—except for melt temperature (N6 is lower) and some fiber properties in products like carpets and textiles. There is also nylon 9.

Nylon 5,10, made from pentamethylene diamine and sebacic acid, was studied by Carothers even before nylon 6,6 and has superior properties, but is more expensive to make. In keeping with this naming convention, "nylon 6,12" (N-6,12) or "PA-6,12" is a copolymer of a 6C diamine and a 12C diacid. Similarly for N-5,10 N-6,11; N-10,12, etc. Other nylons include copolymerized dicarboxylic acid/diamine products that are *not* based upon the monomers listed above. For example, some aromatic nylons are polymerized with the addition of diacids like terephthalic acid (→ Kevlar) or isophthalic acid (→ Nomex), more commonly associated with polyesters. There are copolymers of N-6,6/N6; copolymers of N-6,6/N-6/N-12; and others. Because of the way polyamides are formed, nylon would seem to be limited to unbranched, straight chains. But "star" branched nylon can be produced by the condensation of dicarboxylic acids with polyamines having three or more amino groups.

The general reaction is:

$$HO{-}\overset{O}{\overset{\|}{C}}{-}R{-}\overset{O}{\overset{\|}{C}}{-}OH + n\ H_2N{-}R{-}NH_2 \longrightarrow \left[\overset{O}{\overset{\|}{C}}{-}R{-}\overset{O}{\overset{\|}{C}}{-}\underset{H}{N}{-}R'{-}\underset{H}{N} \right]_n$$

A molecule of water is given off and the nylon is formed. Its properties are determined by the R and R' groups in the monomers. In nylon 6,6, R' = 6C and R = 4C alkanes, but one also has to include the two carboxyl carbons in the diacid to get the number it donates to the chain. In Kevlar, both R and R' are benzene rings.

Nylon Fiber

The Federal Trade Commission's definition for Nylon Fiber: A manufactured fiber in which the fiber forming substance is a long-chain synthetic polyamide in which less than 85 per cent of the amide-linkages are attached directly (-CO-NH-) to two aliphatic groups.

- A synthetic thermoplastic fiber (Nylon melts/glazes easily at relatively low temperatures)
- Round, smooth, and shiny filament fibers
- cross sections can be either:
  - trilobal to imitate silk
  - multilobal to increase staple like appearance and hand
- Its most widely used structures are multifilament, monofilament, staple or tow and is available as partially drawn or as finished filaments.
- Regular nylon has a round cross section and is perfectly uniform. The filaments are generally completely transparent unless they have been delustered or solution dyed. Thus, they are microscopically recognized as glass rods.
- Molecular chains of nylon are long and straight variations but have no side chains or linkages.
  - Cold drawing (step 18 on the model) can align the chains so they are oriented with the lengthwise direction and are highly crystalline.
- Nylon is related chemically to the protein fibers silk and wool.
  - They both have similar dye sites but nylon has many fewer dye sites than wool.

Basic Concepts of Nylon Production

- The first approach: combining molecules with an acid (COOH) group on each end are reacted with two chemicals that contain amine (NH_2) groups on each end.

  This process creates nylon 6,6, made of hexamethylene diamine with six carbon atoms and acidipic acid, as well as six carbon atoms.
- The second approach: a compound has an acid at one end and an amine at the other and is polymerized to form a chain with repeating units of (-NH-[CH2]n-CO-)x.

- In other words, nylon 6 is made from a single six-carbon substance called caprolactam.
- In this equation, if n = 5, then nylon 6 is the assigned name. (may also be referred to as polymer)

Nylon 6,6

- Pleats and creases can be heat-set at higher temperatures,
- Nylon is very easy to dye, but Nylon 6,6 is not.

Nylon 6

- Better dye Affinity;
- Softer Hand;
- Greater elasticity and elastic recovery;
- Better weathering properties; better sunlight resistance.

Producers The producers of nylon include: Honeywell Nylon Inc., Invista, Wellman Inc. among many others. The Dupont Company, is the most famous pioneer of the nylon we know today.

Characteristics

- Variation of luster: nylon has the ability to be very lustrous, semilustrous or dull.
- Durability: its high tenacity fibers are used for seatbelts, tire cords, ballistic cloth and other uses.
- High elongation
- Excellent abrasion resistance
- Highly resilient (nylon fabrics are heat-set)
- Paved the way for easy-care garments
- High resistance to:
    - insects, fungi and animals
    - molds, mildew, rot
    - many chemicals
- Used in carpets and nylon stockings
- Melts instead of burning
- Used in many military applications

Bulk Properties

Above their melting temperatures, T_m, thermoplastics like nylon are amorphous solids or viscous fluids in which the chains approximate random coils. Below

T_m, amorphous regions alternate with regions which are lamellar crystals. The amorphous regions contribute elasticity and the crystalline regions contribute strength and rigidity. The planar amide (-CO-NH-) groups are very polar, so nylon forms multiple hydrogen bonds among adjacent strands. Because the nylon backbone is so regular and symmetrical, especially if all the amide bonds are in the *trans* configuration, nylons often have high crystallinity and make excellent fibers. The amount of crystallinity depends on the details of formation, as well as on the kind of nylon. Apparently it can never be quenched from a melt as a completely amorphous solid.

Nylon 6,6 can have multiple parallel strands aligned with their neighboring peptide bonds at coordinated separations of exactly 6 and 4 carbons for considerable lengths, so the carbonyl oxygens and amide hydrogens can line up to form interchain hydrogen bonds repeatedly, without interruption. Nylon 5,10 can have coordinated runs of 5 and 8 carbons. Thus parallel (but not antiparallel) strands can participate in extended, unbroken, multi-chain â-pleated sheets, a strong and tough supermolecular structure similar to that found in natural silk fibroin and the â-keratins in feathers. (Proteins have only an amino acid á-carbon separating sequential -CO-NH- groups.) Nylon 6 will form uninterrupted H-bonded sheets with mixed directionalities, but the â-sheet wrinkling is somewhat different. The three-dimensional disposition of each alkane hydrocarbon chain depends on rotations about the 109.47° tetrahedral bonds of singly-bonded carbon atoms.

When extruded into fibers through pores in an industrial spinneret, the individual polymer chains tend to align because of viscous flow. If subjected to cold drawing afterwards, the fibers align further, increasing their crystallinity, and the material acquires additional tensile strength. In practice, nylon fibers are most often drawn using heated rolls at high speeds.

Block nylon tends to be less crystalline, except near the surfaces due to shearing stresses during formation. Nylon is clear and colorless, or milky, but is easily dyed. Multistranded nylon cord and rope is slippery and tends to unravel. The ends can be melted and fused with a heat source such as a flame or electrode to prevent this.

When dry, polyamide is a good electrical insulator. However, polyamide is hygroscopic. The absorption of water will change some of the material‘s properties such as its electrical resistance. Nylon is less absorbent than wool or cotton.

Historical Uses

Bill Pittendreigh, DuPont, and other individuals and corporations worked diligently during the first few months of World War II to find a way to replace

Asian silk with nylon in parachutes. It was also used to make tires, tents, ropes, ponchos, and other military supplies. It was even used in the production of a high-grade paper for U.S. currency. At the outset of the war, cotton accounted for more than 80 per cent of all fibers used and manufactured, and wool fibers accounted for the remaining 20 per cent. By August 1945, manufactured fibers had taken a market share of 25 per cent and cotton had dropped.

Some of the terpolymers based upon nylon are used every day in packaging. Nylon has been used for meat wrappings and sausage sheaths.

Use in Composites

Nylon can be used as the matrix material in composite materials, such as glass or carbon fiber, and yields a higher density than pure nylon.

Etymology

In 1940 John W. Eckelberry of DuPont stated that the letters “nyl” were arbitrary and the “on” was copied from the suffixes of other fibers such as cotton and rayon. A later publication by DuPont (*Context*, vol. 7, no. 2, 1978) explained that the name was originally intended to be “No-Run” (“run” meaning “unravel”), but was modified to avoid making such an unjustified claim and to make the word sound better. The story goes that Carothers changed one letter at a time until DuPont’s management was satisfied. But he was not involved in the nylon project during the last year of his life, and committed suicide before the name was coined.

Two theories about the origin of the name claim that it is an acronym of “Now you’ve lost, Old Nippon“ (N.Y.L.O.N.), or that it stands for “New York-London“. In the latter case, it is claimed that these were the two cities where the product was researched and developed, or that the inspiration came from a New York to London airplane ticket. There is no evidence for the ‘airline ticket’ theory, though some compelling evidence of the former from contemporary researchers at Oxford University who assisted in development.

Uses

- carpet fiber
- clothing
- fishing lines
- footwear
- nylon fiber
- pantyhose
- windpants

- toothbrush bristles
- velcro
- airbag fiber
- auto parts: intake manifolds, gas (petrol) tanks
- slings and rope used in climbing gear and slacklining
- machine parts, such as gears and bearings
- parachutes
- metallized nylon balloons
- classical and flamenco guitar strings
- paintball marker bolts
- racquetball, badminton, squash, and tennis racquet strings
- Strings for String instruments
- Drumstick heads
- As filter media in sterlizing grade filters
- Flexible tubing
- Basketball netting
- Sutures
- flags
- Rings for baby slings, etc.

5.5 Polybutadiene

Polybutadiene is a synthetic rubber that is a polymer formed from the polymerization of the monomer 1,3-butadiene. It has a high resistance to wear and is used especially in the manufacture of tires. It has also been used to coat or encapsulate electronic assemblies, offering extremely high electrical resistivity. It exhibits a recovery of 80 per cent after stress is applied, a value only exceeded by elastin and resilin.

Polymerization of Butadiene

1,3-Butadiene is an organic compound which is a rather simple conjugated diene hydrocarbon; the chemical structure is shown as a reactant in the diagram below. A hydrocarbon diene molecule has two carbon-carbon double bonds (i.e. between two sets of carbon atoms). Polybutadiene can be formed from many 1,3-butadiene monomers undergoing free radical polymerization to make a much longer polymer chain molecule.

A chain propagating step in this chemical reaction involves a free radical near the end of a growing polymer chain forming a covalent bond with the #1 carbon in a 1,3-butadiene monomer molecule being added, resulting in a polymer chain intermediate with a substituted allyl free radical at the end of the chain. This allyl free radical, formed from the butadiene just added, can further bond to another monomer molecule at either the #2 or #4 carbons of the previous butadiene monomer. Most of the time, the new monomer bonds to the #4 or terminal carbon of the previous butadiene, resulting in a 1,4-addition of the previous butadiene unit. In a 1,4-addition, the two double bonds of the previous butadiene unit are turned into single bonds and a new double bond is formed between the #2 and #3 carbons. This new double bond may have either a *cis* or a *trans* configuration. A smaller fraction of the time (perhaps 20%), the new monomer bonds to the #2 carbon of the previous butadiene, resulting in a 1,2-addition of the previous butadiene unit. The double bond between the #1 and #2 carbons turns into a single bond in the previous butadiene unit, and the double bond between the #3 and #4 carbons remains intact in a short vinyl side group available for branching or cross-linking. *Cis* or *trans* configurations are not applicable in 1,2-additions of butadiene. See the following reaction diagram for examples of 1,2- and 1,4-addition in a polybutadiene chain.

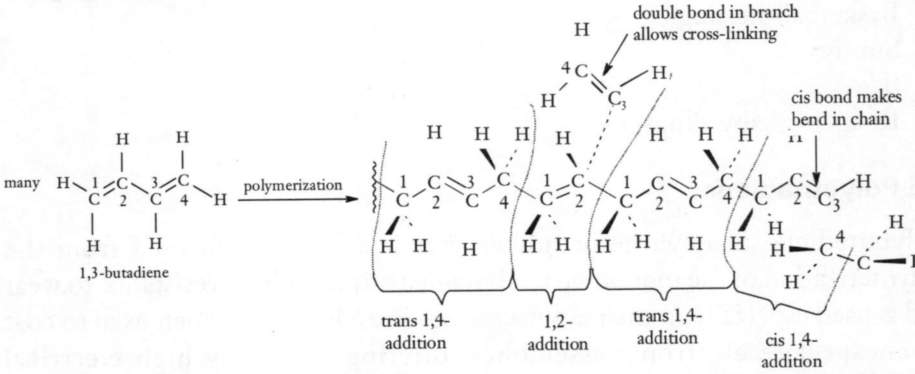

The *trans* double bonds formed during polymerization allow the polymer chain to stay rather straight, allowing sections of polymer chains to line up against each other and effectively form microcrystalline regions in the material. The *cis* double bonds cause a bend in the polymer chain, preventing polymer chains from lining up and forming crystalline regions and resulting in larger regions of amorphous polymer. It has been found that a substantial percentage of *cis* double bond configurations in the polymer will result in a material with flexible elastomer (rubber-like) qualities. In free radical polymerization, both *cis* and *trans* double bonds will form in percentages which depend on temperature. There are different catalysts available which can result in polymerization either in the *cis* or the *trans* configurations.

Properties

Polybutadiene rubber is a highly resilient synthetic rubber. Due to outstanding resilience, it can be used for the manufacturing of golf balls. Heat build up will be less in polybutadiene rubber based on vulcanization subjected to repeated flexing during service. This property leads to the use in the sidewall of the truck tires. Good abrasion resistance of this rubber leads to the use in tread portion of truck tires; however, skidding may be the problem in passenger car tires due to low rolling resistance. For high temperature curing, polybutadiene rubber may be blended with natural rubber and other rubbers, due to resistance in reversion of physical properties. Polybutadiene rubber can be used in water seals for dam due to its low water absorption properties. Rubber bullets and road binders can be also produced by polybutadiene rubber. Processing of Poly Butadiene Rubber will come after Properties

Copolymers

1,3-butadiene is normally copolymerized with other types of monomers such as styrene and acrylonitrile to form rubbers or plastics with various qualities. The most common form is styrene-butadiene copolymer, which is a commodity material for car tires. It is also used in block copolymers and tough thermoplastics such as ABS plastic. This way a copolymer material can be made with good stiffness, hardness, and toughness. Because the chains have a double bond in each and every repeat unit, the material is sensitive to ozone cracking.

Processing of Poly Butadiene Rubber

Seldom PBR is used alone. It is used with other rubbers. Poly Butadiene Rubber is difficult to band in Two roll Mixing Mill. A thin shheet may be prepared and to be kept separate. After proper mastication of Natural Rubber, when it is banded in Two roll mill, Poly butadiene Rubber may be added. Similar practice may be adopted, if PBR has to be mixed with SBR. In Internal Mixer also, Natural Rubber and/or SBR may be put first folloed by Poly Butadiene Rubber.

Plasticity of Poly Butadiene Rubber may not be reduced due to over mastication.

5.6 Polybutylene Terephthalate

Polybutylene terephthalate (PBT) is a plastic that is used as an insulator in the electrical and electronics industries. It is a thermoplastic crystalline polymer, and a type of polyester. PBT is resistant to solvents, shrinks very little during forming, is mechanically strong, heat-resistant up to 150°C (or 200°C with glass-fibre reinforcement) and can be treated with flame retardants to make it noncombustible.

5.7 Polycarbonate

Polycarbonates are a particular group of thermoplastic polymers. They are easily worked, moulded, and thermoformed; as such, these plastics are very widely used in the modern chemical industry. Their interesting features (temperature resistance, impact resistance and optical properties) position them between commodity plastics and engineering plastics. Its plastic identification code is 7.

Physical Properties

Density (ρ)	1200-1220 kg/m³
Abbe number (V)	34.0
Refractive index (n)	1.584-6
Flammability	V0-V2
Limiting oxygen index	25-27%
Water absorption—Equilibrium(ASTM)	0.16-0.35%
Water absorption—over 24 hours	0.1%
Radiation resistance	Fair
Ultraviolet (1-380nm) resistance	Fair

Mechanical Properties

Young's modulus (E)	2-2.4 GPa
Tensile strength (σ_t)	55-75 MPa
Compressive strength (σ_c)	>80 MPa
Elongation (ε) @ break	80-150%
Poisson's ratio (ν)	0.37
Hardness—Rockwell	M70
Izod impact strength	600-850 J/m
Notch test	20-35 kJ/m²
Abrasive resistance—ASTM D1044	10-15 mg/1000 cycles
Coefficient of friction (μ)	0.31

Thermal Properties

Melting temperature (T_m)	267 °C*
Glass transition temperature(T_g)	150 °C
Heat deflection temperature—10 kN (Vicat B)	145 °C
Heat deflection temperature—0.45 MPa	140 °C
Heat deflection temperature—1.8 MPa	128-138 °C
Upper working temperature	115-130 °C

Lower working temperature	-135 °C
Linear thermal expansion coefficient (á)	65-70 × 10^{-6}/K
Specific heat capacity (c)	1.2-1.3 kJ/kg·K
Thermal conductivity (k) @ 23 °C	0.19-0.22 W/(m·K)
Heat transfer coefficient (h)	0.21 W/(m^2·K)

Electrical Properties

Dielectric constant (ε_r) @ 1 MHz	2.9
Permittivity (ε) @ 1 MHz	2.568 × 10^{-11} F/m
Relative permeability (μ_r) @ 1 MHz	0.866(2)
Permeability (μ) @ 1 MHz	1.089(2) μN/A^2
Dielectric strength	15-67 kV/mm
Dissipation factor @ 1 MHz	0.01
Surface resistivity	10^{15} Ω/sq
Volume resistivity (ρ)	10^{12}-10^{14} Ω·m

Near to Short-wave Infrared Transmittance Spectrum

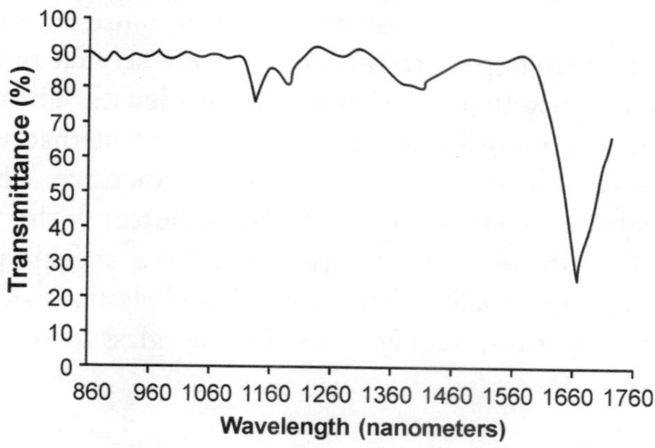

Polycarbonate transmittance in 5/6 of the NIR & 1/5 of the SWIR regions. Also, polycarbonate is almost completely transparent throughout the entire visible region of the spectrum and very sharply cuts off to ~0 per cent transmission at almost exactly 400 nm, blocking all UV light transmission.

Chemical Resistance

Acids—concentrated	Poor	Acids—dilute	Good
Alcohols	Good	Alkalis	Good-Poor
Aromatic hydrocarbons	Poor	Greases & Oils	Good-Fair
Halogenated Hydrocarbons	Good-Poor	Halogens	Poor
Ketones	Poor		

Chemistry

Polycarbonates got their name because they are polymers having functional groups linked together by carbonate groups (-O-(C = O)-O-) in a long molecular chain. Also carbon monoxide was used as a C1-synthon on an industrial scale to produce diphenyl carbonate, being later trans-esterified with a diphenolic derivative affording poly (aromatic carbonate)s.

Taking into consideration the C1-synthon we can divide polycarbonates into poly(aromatic carbonates and poly(aliphatic carbonate)s. The second one, poly(aliphatic carbonate)s are a product of the reaction of carbon dioxide with epoxides, which owing to the thermodynamical stability of carbon dioxide requires the use of a catalyst. The working systems are based on porphyrins, alkoxides, carboxylates, salens and beta-diiminates as organic, chelating ligands and aluminium, zinc, cobalt and chromium as the metal centres. Poly(aliphatic carbonate)s display promising characteristics, have a better biodegradability than the aromatic ones and could be employed to develop other specialty polymers.

One type of polycarbonate plastic is made from bisphenol A (BPA). This polycarbonate is a very durable material, and can be laminated to make bullet-proof "glass", though "bullet-resistant" would be more accurate. Although polycarbonate has high impact-resistance, it has low scratch-resistance and so a hard coating is applied to polycarbonate eyewear lenses. The characteristics of polycarbonate are quite like those of polymethyl methacrylate (*PMMA*; *acrylic*), but polycarbonate is stronger and more expensive. This polymer is highly transparent to visible light and has better light transmission characteristics than many kinds of glass. CR-39 is a specific polycarbonate material—although it is usually referred to as CR-39 plastic—with good optical and mechanical properties, frequently used for eyeglass lenses.

Processing

Polycarbonate has a glass transition temperature of about 150 C, so it softens gradually above this point and flows above about 300 C. Injection moulding is more difficult than other common thermoplastics owing to its non-Newtonian fluid flow behaviour. Tools must be held at high temperatures, generally above 80 C to make strain-and stress-free products. Low molecular mass grades are easier to mould than higher grades, but their strength is lower as a result. The toughest grades have the highest molecular mass, but are much more difficult to process.

Applications

Polycarbonate is becoming more common in housewares as well as laboratories and in industry, especially in applications where any of its main features—

high impact resistance, temperature resistance, optical properties—are required.

Main transformation techniques for polycarbonate resins:

- injection molding into ready articles
- extrusion into tubes, rods and other profiles
- extrusion with cylinders into sheets (0.5-15 mm) and films (below 1 mm), which can be used directly or manufactured into other shapes using thermoforming or secondary fabrication techniques, such as bending, drilling, routing, laser cutting etc.

Typical injected applications:

- drinking glasses
- lighting lenses, sunglass/eyeglass lenses, safety glasses, automotive headlamp lenses
- compact discs, DVDs
- lab equipment, research animal enclosures
- drinking bottles
- MP3/Digital audio player cases

Typical sheet/film application:

- *Industry*: Machined or formed, cases, machine glazing, riot shields, visors, instrument panels.
- *Advertisement*: Signs, displays, poster protection.
- *Building*: Domelights, flat or curved glazing, sound walls.
- *Computers*: Apple, Inc.'s MacBook, and Mac mini.

For use in applications exposed to weathering or UV-radiation, a special surface treatment is needed. This either can be a coating (e.g. for improved abrasion resistance), or a coextrusion for enhanced weathering resistance.

Some polycarbonate grades are used in medical applications and comply with both ISO 10993-1 and USP Class VI standards (occasionally referred to as PC-ISO). Class VI is the most stringent of the six USP ratings. These grades can be sterilized using steam at 120 °C, gamma radiation or the ethylene oxide (EtO) method. See Medical Applications of Polycarbonate for more information. However, scientific research indicates possible problems with biocompatibility. Dow Chemical strictly limits all its plastics with regard to medical applications.

The cockpit canopy of the F-22 Raptor jet fighter is made from a piece of high optical quality polycarbonate, and is the largest piece of its type formed in the world. Being based on bisphenol A (a phenol based on benzene) pricing is largely dependent on phenol and benzene pricing.

Makes

The most common polycarbonate resins are:

- Lexan from SABIC Innovative Plastics (formerly General Electric Plastics).
- Calibre from Dow Chemicals.
- Makrolon from Bayer.
- Panlite from Teijin Chemical Limited.
- Makrolife from Arlaplast.

Potential Hazards in Food Contact Applications

Polycarbonate may be appealing to manufacturers and purchasers of food storage containers due to its clarity and toughness, being described as lightweight and highly break resistant particularly when compared to silica glass. Polycarbonate may be seen in the form of single use and refillable plastic water bottles.

More than 100 studies have explored the bioactivity of bisphenol A leachates from polycarbonates. Bisphenol A appeared to be released from polycarbonate animal cages into water at room temperature and that it may have been responsible for enlargement of the reproductive organs of female mice.

An analysis of the literature on bisphenol A leachate low-dose effects by vom Saal and Hughes published in August 2005 seems to have found a suggestive correlation between the source of funding and the conclusion drawn. Industry funded studies tend to find no significant effects while government funded studies tend to find significant effects.

Research by Ana M. Soto, professor of anatomy and cellular biology at Tufts University School of Medicine, Boston, published Dec. 6 in the online edition of Reproductive Toxicology (DOI: 10.1016/j.reprotox.2006.10.002) describes exposure of pregnant rats to bisphenol A at 2.5 to 1,000 µg per kilogram of body weight per day. At the equivalent of puberty for the pups (50 days old), about 25 per cent of their mammary ducts had precancerous lesions, some three to four times higher than unexposed controls. The study is cited as evidence for the hypothesis that environmental exposure to bisphenol A as a fetus can cause breast cancer in adult women.

An expert panel of 12 scientists has found that there is "some concern that exposure to the chemical bisphenol A in utero causes neural and behavioral effects," according to the draft report prepared by The National Toxicology Program (NTP) Center for the Evaluation of Risks to Human Reproduction.

For the general adult population, the expert panel found a "negligible concern for adverse reproductive effects following exposures." One point of agreement among those studying polycarbonate water and food storage containers may be that using sodium hypochlorite bleach and other alkali cleaners to clean polycarbonate is not recommended, as they catalyze the release of the bisphenol-A. The tendency of polycarbonate to release bisphenol A was discovered after a lab tech used strong cleaners on polycarbonate lab containers. Endocrine disruption later observed on lab rats was traced to exposure from the cleaned containers. On April 18, 2008, Health Canada announced that Bisphenol A is "'toxic' to human health". Canada is the first nation to make this designation. A chemical compatibility chart shows reactivity between chemicals such as polycarbonate and a cleaning agent. Alcohol is one recommended organic solvent for cleaning grease and oils from polycarbonate. For treating mold, borax:H_2O 1:96 to 1:8 may be effective.

Synthesis

Polycarbonate can be synthesized from bisphenol A and phosgene (carbonyl dichloride, $COCl_2$). The first step in the synthesis of polycarbonate from bisphenol A is treatment of bisphenol A with sodium hydroxide. This deprotonates the hydroxyl groups of the bisphenol A molecule.

HO OH + 2 NaOH ⟶ ⁻O O⁻ + 2 Na^+ + H_2C

C, H_3C H_3C

bisphenol A — diphenolate ion of bisphenol A

The deprotonated oxygen reacts with phosgene through carbonyl addition to create a tetrahedral intermediate (not shown here), after which the negatively charged oxygen kicks off a chloride ion (Cl^-) to form a chloroformate.

O⁻ + Cl—C(=O)—Cl ⟶ O—C(=O)—Cl + Cl^-

C, H_3C H_3C

phenolate ion end on bisphenol A — chlorocormate end on bisphenol A

The chloroformate is then attacked by another deprotonated bisphenol A, eliminating the remaining chloride ion and forming a dimer of bisphenol A with a carbonate linkage in between.

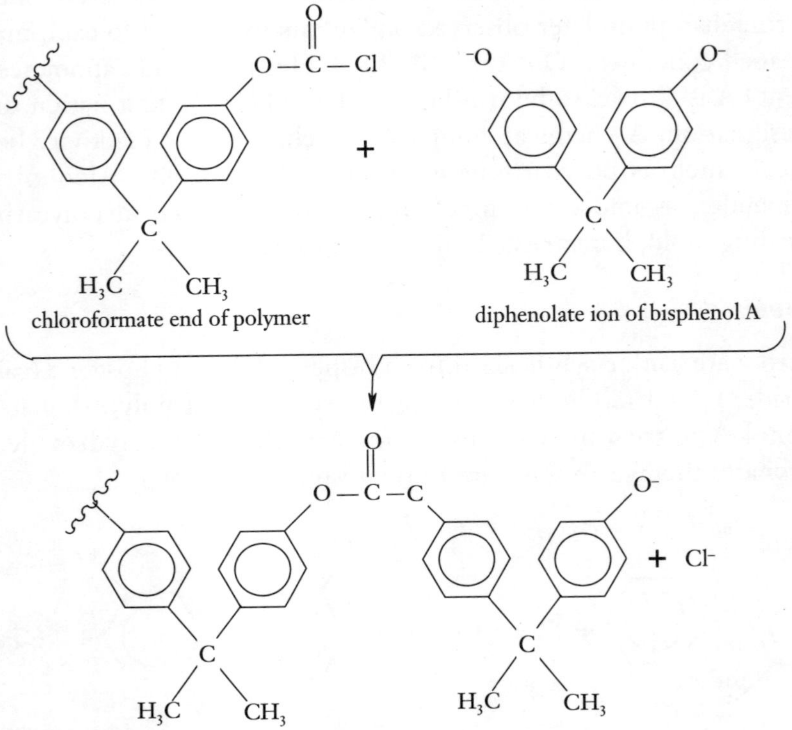

chloroformate end of polymer + diphenolate ion of bisphenol A

polycarbonate polymer being extended + Cl^-

Repetition of this process yields a polycarbonate with alternating carbonate groups and groups from bisphenol A.

Interaction with Other Chemicals

Will damage Polycarbonate	*Require caution*	*Are considered safe*
• Acetone	• Alkali bleaches such as sodium hypochlorite	• Acetic acid
• Acrylonitrile	• Cyclohexanone	• Ammonium chloride
• Ammonia	• Diesel oil	• Antimony trichloride
• Amyl acetate	• Formic acid	• Borax in H_2O
• Benzene	• Gasoline	• Butane
• Bromine	• Glycerine	• Calcium chloride
• Butyl acetate	• Heating oil	• Calcium hypochlorite
• Sodium hydroxide	• Jet fuel	• Carbon dioxide
• Chloroform		• Carbon monoxide

(*Contd.*)

Will damage Polycarbonate	*Require caution*	*Are considered safe*
• Dimethylformamide • Concentrated hydrochloric acid • Concentrated hydrofluoric acid • Iodine • Methanol • Methyl ethyl ketone • Styrene • Tetrachloroethylene • Toluene • Concentrated sulfuric acid • Xylene • Cyanoacrylate monomers	• Concentrated perchloric acid • Sulfur dioxide • Turpentine	• Citric acid 10% • Copper(II) sulfate • Ethyl alcohol, i.e. • ethanol 95% • Ethylene glycol • Formaldehyde 10% • Hydrochloric acid 20% • Hydrofluoric acid 5% • Isopropyl alcohol • Mercury • Methane • Oxygen • Ozone • Sulfur • Urea • Water*

(*Contd.*)

Note: * At room temperature. At temperatures above 60 °C hydrolysis can occur, degrading the plastic. Degradation depends on time and temperature.

Using sodium hypochlorite (bleach) and other alkali cleaners on polycarbonate are not recommended as they cause the release of bisphenol A, a known endocrine disrupter.

5.8 Polysulfone

Polysulfone describes a family of thermoplastic polymers. These polymers are known for their toughness and stability at high temperatures. They contain the subunit aryl-SO_2-aryl, the defining feature of which is the sulfone group. Polysulfones were introduced in 1965 by Union Carbide. Due to the high cost of raw materials and processing, polysulfones are used in specialty applications and often are a superior replacement for polycarbonates.

Production

A typical polysulfone is produced by the reaction of a diphenol and bis(4-chlorophenyl)sulfone, forming a polyether by elimination of sodium chloride:

$$n\,HOC_6H_4OH + n\,(ClC_6H_4)_2SO_2 + n\,Na_2CO_3 \rightarrow [OC_6H_4OC_6H_4SO_2C_6H_4]_n + 2n\,NaCl + n\,H_2O + n\,CO_2$$

The diphenol is typically bisphenol-A or, as shown above, 1,4-dihydroxybenzene. Such step polymerizations require highly pure monomer to ensure high molecular weight products.

Chemical and Physical Properties

These polymers are rigid, high-strength, and transparent, retaining its properties between -100 °C and + 150 °C. It has very high dimensional stability; the size change when exposed to boiling water or + 150 °C air or steam generally falls below 0.1 per cent. Its glass transition temperature is 185 °C. Polysulfone is highly resistant to mineral acids, alkali, and electrolytes, in pH ranging from 2 to 13. It is resistant to oxidizing agents, therefore it can be cleaned by bleaches. It is also resistant to surfactants and hydrocarbon oils. It is not resistant to low-polar organic solvents (eg. ketones and chlorinated hydrocarbons), and aromatic hydrocarbons. Mechanically, polysulfone has high compaction resistance, recommending its use under high pressures. It is also stable in aqueous acids and bases and many non-polar solvents; however it is soluble in dichloromethane and methylpyrrolidone.

Polyethersulfone (PES) is a similar polymer with low protein retention.

Applications

Polysulfone is used as a dielectric in capacitors. Polysulfone allows easy manufacturing of membranes, with reproducible properties and controllable size of pores. Such membranes can be used in applications like hemodialysis, waste water recovery, food and beverage processing, and gas separation. These polymers are also used in the automotive and electronic industries.

Polysulfone can be reinforced with glass fibers. The resulting composite material has twice the tensile strength and three time increase of its modulus.

Polysulfone has the highest service temperature of all melt-processable thermoplastics. Its resistance to high temperatures gives it a role of a flame retardant, without compromising its strength that usually results from addition of flame retardants. Its high hydrolysis stability allows its use in medical applications requiring autoclave and steam sterilization. However, it has low resistance to some solvents and undergoes weathering; this weathering instability can be offset by adding other materials into the polymer.

It is supplied by Solvay Advanced Polymers, BASF, HOS-Technik and PolyOne Corporation. Polysulfone is also used as a copolymer.

5.9 PEEK

Density	1300 kg/m^3
Young's modulus (*E*)	3700 MPa
Tensile strength (σ_t)	90 MPa
Elongation @ break	50%
notch test	55 kJ/m^2
Glass temperature	130-150 °C and 260-290 °C
melting point	~350 °C
Vicat B	—
heat transfer coefficient (λ)	0.25 W/m.K
linear expansion coefficient (α)	1.7 10^{-5}/K
Specific heat (c)	- kJ/kg.K
Water absorption, 24 hours (ASTM D 570)	- 0.1%
Price	40-90 • /kg

Polyetheretherketone (PEEK), also referred to as polyketones, is obtained from aromatic dihalides and bisphenolate salts by nucleophilic substitution. The bisphenolate salt is formed *in situ* from bisphenol and either added sodium or added alkali metal carbonate or hydroxide, by the Williamson ether synthesis.

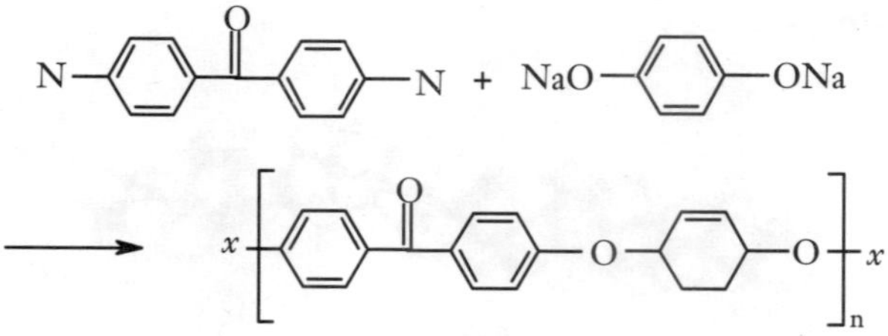

PEEK

The aromatic dihalides are obtained from the 4,4'-Dihydroxy-benzophenone. PEEK is a semicrystalline thermoplastic with extraordinary mechanical properties. The Young's modulus is 3.6 GPa and its tensile strength 90 MPa. PEEK exhibits two glass transition temperatures at around 140°C (284°F) and around 275°C (527°F), depending on cure cycle and precise formulation. PEEK melts at around 350°C (662°F) and is highly resistant to thermal degradation. The material is also resistant to both organic and aqueous environments, and is used in bearings, piston parts, pumps, compressor plate valves, and cable insulation applications. It is one of the few plastics compatible with ultra-high vacuum applications.

PEEK is considered an advanced biomaterial used in medical implants, often in reinforced format using biocompatible fibre fillers such as carbon. Also in carbon fibre reinforced form, PEEK has come under consideration as an aerospace structural material due to its high strength-to-weight ratio. Electronic circuitry also has a high demand for PEEK's large temperature range.

Chemical Resistance

PEEK also exhibits good chemical resistance in many environments, including alkalis (i.e. sodium, potassium and ammonium hydroxides), aromatic hydrocarbons, alcohols (i.e. ethanol, propanol), greases, oils and halogenated hydrocarbons.

However, its performance in acids is very dependent on the type of acid—PEEK shows poor resistance in concentrated sulfuric, nitric, hydrochloric, hydrobromic and other mineral acids (though performance may be adequate for short term use with these acids in very dilute form). Its resistance to hydrofluoric acid and oleum is very poor. PEEK shows good resistance to phosphoric acid and organic acids (acetic, citric, oxalic, tartaric etc.), but varying resistance in the presence of halogens. PEEK is resistant to dissolution by some aldehydes and ketones such as acetone, but not (at higher temperature) methylethyl ketone.

5.10 Polyethylene

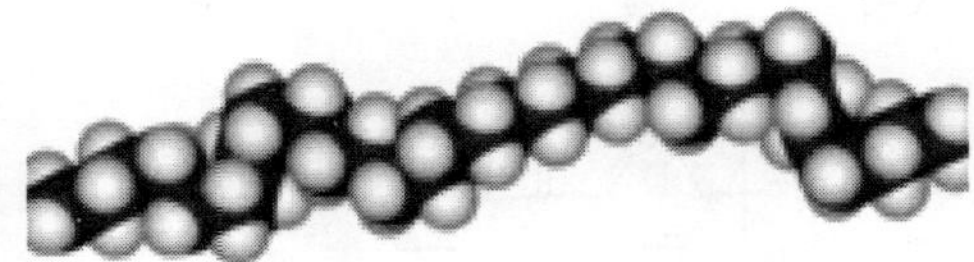

Fig. 5.1: Space-filling model of a polyethylene chain.

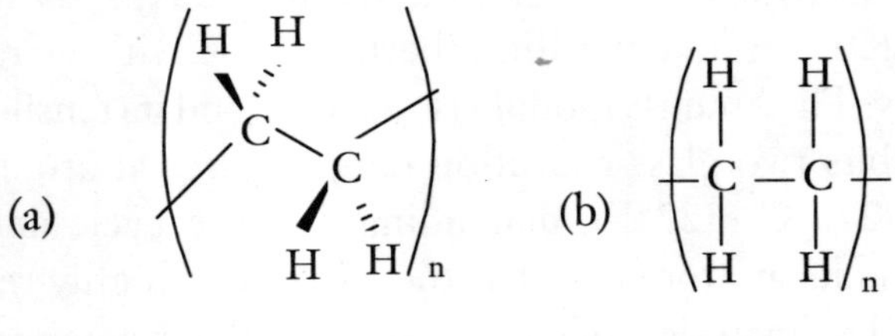

Fig. 5.2: (a) The repeating unit of polyethylene, showing its stereochemistry. (b) A simpler way of representing the repeating unit. Note, however, that the C-H bond angles are not 90° as this diagram would indicate, but are approximately 110°, since each carbon atom is tetrahedral (sp^3).

Polyethylene or polythene (IUPAC name poly(ethene)) is a thermoplastic commodity heavily used in consumer products (notably the plastic shopping bag). Over 60 million tons of the material are produced worldwide every year.

Description

Polyethylene is a polymer consisting of long chains of the monomer ethylene (IUPAC name ethene). The recommended scientific name *polyethene* is systematically derived from the scientific name of the monomer. In certain circumstances it is useful to use a structure–based nomenclature. In such cases IUPAC recommends poly(methylene). The difference is due to the *opening up* of the monomer's double bond upon polymerisation.

In the polymer industry the name is sometimes shortened to PE in a manner similar to that by which other polymers like polypropylene and polystyrene are shortened to PP and PS respectively. In the United Kingdom the polymer is commonly called polythene, although this is not recognized scientifically.

The ethene molecule (known almost universally by its common name ethylene) C_2H_4 is $CH_2 = CH_2$, Two CH_2 groups connected by a double bond, thus:

```
H     H
 \   /
  C=C
 /   \
H     H
```

Polyethylene is created through polymerization of ethene. It can be produced through radical polymerization, anionic addition polymerization, ion coordination polymerization or cationic addition polymerization. This is because ethene does not have any substituent groups that influence the stability of the propagation head of the polymer. Each of these methods results in a different type of polyethylene.

Classification

Polyethylene is classified into several different categories based mostly on its density and branching. The mechanical properties of PE depend significantly on variables such as the extent and type of branching, the crystal structure and the molecular weight.

- ❖ Ultra high molecular weight polyethylene (UHMWPE)
- ❖ Ultra low molecular weight polyethylene (ULMWPE—PE-WAX)
- ❖ High molecular weight polyethylene (HMWPE)

- High density polyethylene (HDPE)
- High density cross-linked polyethylene (HDXLPE)
- Cross-linked polyethylene (PEX)
- Medium density polyethylene (MDPE)
- Low density polyethylene (LDPE)
- Linear low density polyethylene (LLDPE)
- Very low density polyethylene (VLDPE)

UHMWPE is polyethylene with a molecular weight numbering in the millions, usually between 3.1 and 5.67 million. The high molecular weight results in less efficient packing of the chains into the crystal structure as evidenced by densities of less than high density polyethylene (for example, 0.930–0.935 g/cm^3). The high molecular weight results in a very tough material. UHMWPE can be made through any catalyst technology, although Ziegler catalysts are most common. Because of its outstanding toughness and its cut, wear and excellent chemical resistance, UHMWPE is used in a wide diversity of applications. These include can and bottle handling machine parts, moving parts on weaving machines, bearings, gears, artificial joints, edge protection on ice rinks and butchers' chopping boards. It competes with Aramid in bulletproof vests, under the tradenames Spectra and Dyneema, and is commonly used for the construction of articular portions of implants used for hip and knee replacements.

HDPE is defined by a density of greater or equal to 0.941 g/cm^3. HDPE has a low degree of branching and thus stronger intermolecular forces and tensile strength. HDPE can be produced by chromium/silica catalysts, Ziegler-Natta catalysts or metallocene catalysts. The lack of branching is ensured by an appropriate choice of catalyst (for example, chromium catalysts or Ziegler-Natta catalysts) and reaction conditions. HDPE is used in products and packaging such as milk jugs, detergent bottles, margarine tubs, garbage containers and water pipes.

PEX is a medium- to high-density polyethylene containing cross-link bonds introduced into the polymer structure, changing the thermoplast into an elastomer. The high-temperature properties of the polymer are improved, its flow is reduced and its chemical resistance is enhanced. PEX is used in some potable-water plumbing systems because tubes made of the material can be expanded to fit over a metal nipple and it will slowly return to its original shape, forming a permanent, water-tight, connection.

MDPE is defined by a density range of 0.926–0.940 g/cm^3. MDPE can be produced by chromium/silica catalysts, Ziegler-Natta catalysts or metallocene catalysts. MDPE has good shock and drop resistance properties. It also is less notch sensitive than HDPE, stress cracking resistance is better than HDPE.

MDPE is typically used in gas pipes and fittings, sacks, shrink film, packaging film, carrier bags and screw closures.

LLDPE is defined by a density range of 0.915–0.925 g/cm³. LLDPE is a substantially linear polymer with significant numbers of short branches, commonly made by copolymerization of ethylene with short-chain alpha-olefins (for example, 1-butene, 1-hexene and 1-octene). LLDPE has higher tensile strength than LDPE, it exhibits higher impact and puncture resistance than LDPE. Lower thickness (gauge) films can be blown, compared with LDPE, with better environmental stress cracking resistance but is not as easy to process. LLDPE is used in packaging, particularly film for bags and sheets. Lower thickness may be used compared to LDPE. Cable covering, toys, lids, buckets, containers and pipe. While other applications are available, LLDPE is used predominantly in film applications due to its toughness, flexibility and relative transparency.

LDPE is defined by a density range of 0.910–0.940 g/cm³. LDPE has a high degree of short and long chain branching, which means that the chains do not pack into the crystal structure as well. It has, therefore, less strong intermolecular forces as the instantaneous-dipole induced-dipole attraction is less. This results in a lower tensile strength and increased ductility. LDPE is created by free radical polymerization. The high degree of branching with long chains gives molten LDPE unique and desirable flow properties. LDPE is used for both rigid containers and plastic film applications such as plastic bags and film wrap.

VLDPE is defined by a density range of 0.880–0.915 g/cm³. VLDPE is a substantially linear polymer with high levels of short-chain branches, commonly made by copolymerization of ethylene with short-chain alpha-olefins (for example, 1-butene, 1-hexene and 1-octene). VLDPE is most commonly produced using metallocene catalysts due to the greater co-monomer incorporation exhibited by these catalysts. VLDPEs are used for hose and tubing, ice and frozen food bags, food packaging and stretch wrap as well as impact modifiers when blended with other polymers.

Recently much research activity has focused on the nature and distribution of long chain branches in polyethylene. In HDPE a relatively small number of these branches, perhaps 1 in 100 or 1,000 branches per backbone carbon, can significantly affect the rheological properties of the polymer.

Ethylene Copolymers

In addition to copolymerization with alpha-olefins, ethylene can also be copolymerized with a wide range of other monomers and ionic composition that creates ionized free radicals. Common examples include vinyl acetate (the resulting product is ethylene-vinyl acetate copolymer, or EVA, widely used in athletic-shoe sole foams) and a variety of acrylates (applications include packaging and sporting goods).

History

Polyethylene was first synthesized by the German chemist Hans von Pechmann who prepared it by accident in 1898 while heating diazomethane. When his colleagues Eugen Bamberger and Friedrich Tschirner characterized the white, waxy, substance that he had created they recognized that it contained long -CH_2- chains and termed it *polymethylene*.

The first industrially practical polyethylene synthesis was discovered (again by accident) in 1933 by Eric Fawcett and Reginald Gibson at the ICI works in Northwich, England. Upon applying extremely high pressure (several hundred atmospheres) to a mixture of ethylene and benzaldehyde they again produced a white, waxy, material. Because the reaction had been initiated by trace oxygen contamination in their apparatus the experiment was, at first, difficult to reproduce. It was not until 1935 that another ICI chemist, Michael Perrin, developed this accident into a reproducible high-pressure synthesis for polyethylene that became the basis for industrial LDPE production beginning in 1939. Subsequent landmarks in polyethylene synthesis have revolved around the development of several types of catalyst that promote ethylene polymerization at more mild temperatures and pressures. The first of these was a chromium trioxide-based catalyst discovered in 1951 by Robert Banks and J. Paul Hogan at Phillips Petroleum. In 1953 the German chemist Karl Ziegler developed a catalytic system based on titanium halides and organoaluminium compounds that worked at even milder conditions than the Phillips catalyst. The Phillips catalyst is less expensive and easier to work with, however, and both methods are used in industrial practice.

By the end of the 1950s both the Phillips- and Ziegler-type catalysts were being used for HDPE production. Phillips initially had difficulties producing a HDPE product of uniform quality and filled warehouses with off-specification plastic. However, financial ruin was unexpectedly averted in 1957 when the hula hoop, a toy consisting of a circular polyethylene tube, became a fad among youth in the United States. A third type of catalytic system, one based on metallocenes, was discovered in 1976 in Germany by Walter Kaminsky and Hansjörg Sinn. The Ziegler and metallocene catalyst families have since proven to be very flexible at copolymerizing ethylene with other olefins and have become the basis for the wide range of polyethylene resins available today, including very low-density polyethylene and linear low-density polyethylene.

Such resins, in the form of fibers like Dyneema, have (as of 2005) begun to replace aramids in many high-strength applications.

Until recently the metallocenes were the most active single-site catalysts for ethylene polymerisation known—new catalysts are typically compared to zirconocene dichloride. Much effort is currently being exerted on developing new, single-site (so-called post-metallocene) catalysts that may allow greater tuning of the polymer structure than is possible with metallocenes. Recently work by Fujita at the Mitsui corporation (amongst others) has demonstrated that certain salicylaldimine complexes of Group 4 metals show substantially higher activity than the metallocenes.

Physical Properties

Depending on the crystallinity and molecular weight, a melting point and glass transition may or may not be observable. The temperature at which these occur varies strongly with the type of polyethylene. For common commercial grades of medium- and high-density polyethylene the melting point is typically in the range 120 to 130 °C ((250 to 265 °F). The melting point for average, commercial, low-density polyethylene is typically 105 to 115 °C (220 to 240 °F).

Most LDPE, MDPE and HDPE grades have excellent chemical resistance and do not dissolve at room temperature because of their crystallinity. Polyethylene (other than cross-linked polyethylene) usually can be dissolved at elevated temperatures in aromatic hydrocarbons such as toluene or xylene, or in chlorinated solvents such as trichloroethane or trichlorobenzene.

Environmental Issues

The wide use of polyethylene makes it an important environmental issue. Though it can be recycled, most of the commercial polyethylene ends up in landfills and in the oceans (notably the Great Pacific Garbage Patch). Polyethylene is not considered biodegradable, as it takes several centuries until it is efficiently degraded. Recently (May 2008) Daniel Burd, a 16 year old Canadian, won the Canada-Wide Science Fair in Ottawa after discovering that *Sphingomonas*, a type of bacteria, can degrade over 40 per cent of the weight of plastic bags in less than three months. The applicability of this finding is still a matter for the future.

5.11 Polyethylene Glycol

Poly(ethylene glycol) (PEG), also known as poly(ethylene oxide) (PEO) or polyoxyethylene (POE), are the most commercially important polyethers. PEG,

PEO or POE refers to an oligomer or polymer of ethylene oxide. The three names are chemically synonymous, but historically PEG has tended to refer to oligomers and polymers with a molecular mass below 20,000 g/mol, PEO refers to polymers with a molecular mass above 20,000 g/mol, and POE refers to a polymer of any molecular mass. PEG and PEO are liquids or low-melting solids, depending on their molecular weights. PEGs are prepared by polymerization of ethylene oxide and are commercially available over a wide range of molecular weights from 300 g/mol to 10,000,000 g/mol. While PEG and PEO with different molecular weights find use in different applications and have different physical properties (e.g. viscosity) due to chain length effects, their chemical properties are nearly identical. Different forms of PEG are also available dependent on the initiator used for the polymerization process. The most common of which is a monofunctional methyl ether PEG (methoxypoly(ethylene glycol)), abbreviated mPEG. PEGs are also available with different geometries. *Branched* PEGs have 3 to 10 PEG chains emanating from a central core group. *Star* PEGs have 10—100 PEG chains emanating from a central core group. *Comb* PEGs have multiple PEG chains normally grafted to a polymer backbone.

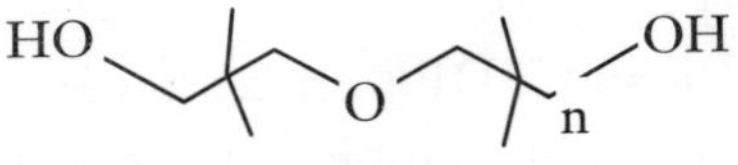

HO ... O ... OH (n)

Polyethylene Glycol

IUPAC name	poly(oxyethylene) {structure-based}, poly(ethylene oxide) {source-based}
Identifiers	
CAS number	[25322-68-3]
Properties	
Molecular formula	$C_{2n+2}H_{4n+6}O_{n+2}$
Molar mass	44n + 62
Hazards	
Flash point	182—287 °C

Their melting points vary depending on the Formula Weight of the polymer. PEG or PEO has the following structure:

$$HO\text{-}(CH_2\text{-}CH_2\text{-}O\text{-})_n\text{-}H$$

The numbers that are often included in the names of PEGs indicate their average molecular weights, e.g. a PEG with n = 80 would have an average molecular weight of approximately 3500 daltons and would be labeled PEG 3500. Most PEGs include molecules with a distribution of molecular weights, i.e. they are polydisperse. The size distribution can be characterized statistically by its weight average molecular weight (Mw) and its number average molecular weight (Mn), the ratio of which is called the polydispersity index (Mw/Mn). Mw and Mn can be measured by mass spectroscopy.

PEGylation is the act of covalently coupling a PEG structure to another larger molecule, for example, a therapeutic protein (which is then referred to as PEGylated). PEGylated interferon alfa-2a or -2b is a commonly used injectable treatment for Hepatitis C infection. PEG is soluble in water, methanol, benzene, dichloromethane and is insoluble in and hexane. It is coupled to hydrophobic molecules to produce non-ionic surfactants.

Production

Poly (ethylene glycol) is produced by the interaction of ethylene oxide with water, ethylene glycol or ethylene glycol oligomers. The reaction is catalyzed by acidic or basic catalysts. Ethylene glycol and its oligomers are preferable as a starting material instead of water, because it allows the creation of polymers with a low polydispersity (narrow molecular weight distribution). Polymer chain length depends on the ratio of reactants.

$$HOCH_2CH_2OH + n(CH_2CH_2O) \rightarrow HO(CH_2CH_2O)_{n+1}H$$

Depending on the catalyst type, the mechanism of polymerization can be cationic or anionic. The anionic mechanism is preferable because it allows one to obtain PEG with a low polydispersity. Polymerization of ethylene oxide is an exothermic process. Overheating or contaminating ethylene oxide with catalysts such as alkalis or metal oxides can lead to runaway polymerization which can end with an explosion after few hours.

Polyethylene oxide or high-molecular polyethylene glycol is synthesized by suspension polymerization. It is necessary to hold the growing polymer chain in solution in the course of the polycondensation process. The reaction is catalyzed by magnesium-, aluminium- or calcium-organoelement compounds. To prevent coagulation of polymer chains from solution, chelating additives such as dimethylglyoxime are used.

Alkali catalysts such as sodium hydroxide NaOH, potassium hydroxide KOH or sodium carbonate Na_2CO_3 are used to prepare low-molecular polyethylene glycol.

Clinical Uses

Polyethylene glycol has a low toxicity and is used in a variety of products. It is the basis of a number of laxatives (e.g. macrogol-containing products such as

Movicol and polyethylene glycol 3350, or MiraLax or GlycoLax). It is the basis of many skin creams, as *cetomacrogol*, and sexual lubricants, frequently combined with glycerin. Whole bowel irrigation (polyethylene glycol with added electrolytes) is used for bowel preparation before surgery or colonoscopy and drug overdoses. It is sold under the brand names GoLYTELY, GlycoLax, Fortrans, TriLyte, and Colyte. When attached to various protein medications, polyethylene glycol allows a slowed clearance of the carried protein from the blood. This makes for a longer acting medicinal effect and reduces toxicity, and it allows longer dosing intervals. Examples include PEG-interferon alpha which is used to treat hepatitis C and PEG-filgrastim (Neulasta) which is used to treat neutropenia. It has been shown that polyethylene glycol can improve healing of spinal injuries in dogs. One of the earlier findings that polyethylene glycol can aid in nerve repair came from the University of Texas (Krause and Bittner). Polyethylene glycol is commonly used to fuse B-cells with myeloma cells in monoclonal antibody production. PEG has recently been proved to give better results in constipation patients than tegaserod.

Research for New Clinical Uses

- High-molecular weight PEG, e.g., PEG 8000, is a strikingly potent dietary preventive agent against colorectal cancer in animal models.
  The Chemoprevention Database shows it is the most effective agent to suppress chemical carcinogenesis in rats. Cancer prevention in humans has not yet been tested in clinical trials.
- The injection of PEG 2000 into the bloodstream of guinea pigs after spinal cord injury leads to rapid recovery through molecular repair of nerve membranes. The effect of this treatment to prevent paraplegia in humans after an accident is not known yet.
- Research is being done in the use of PEG to mask antigens on red blood cells. Various research institutes have reported that using PEG can mask antigens without damaging the functions and shape of the cell.

PEG is being used in the repair of motor neurons damaged in crush or laceration incidence in vivo and in vitro. When coupled with melatonin, 75 per cent of damaged sciatic nerves were rendered viable.

Other Uses

PEG is used in a number of toothpastes as a dispersant; it binds water and helps keep gum uniform throughout the toothpaste. It is also under investigation for use in body armor and tattoos to monitor diabetes.

Polymer segments derived from PEG polyols impart flexibility to polyurethanes for applications such as elastomeric fibers (spandex) and foam cushions.

Since PEG is a flexible, water-soluble polymer, it can be used to create very high osmotic pressures (tens of atmospheres). It also is unlikely to have specific interactions with biological chemicals. These properties make PEG one of the most useful molecules for applying osmotic pressure in biochemistry experiments, particularly when using the osmotic stress technique.

PEO (poly (ethylene oxide)) can serve as the separator and electrolyte solvent in lithium polymer cells. Its low diffusivity often requires high temperatures of operation, but its high viscosity even near its melting point allows very thin electrolyte layers. While crystallization of the polymer can degrade performance, many of the salts used to carry charge can also serve as a kinetic barrier to the formation of crystals. Such batteries carry greater energy for their weight than other lithium ion battery technologies.

When working with phenol in a laboratory situation, PEG 300 can be used on phenol skin burns to deactivate any residual phenol.

Poly (ethylene glycol) is also commonly used as a polar stationary phase for gas chromatography, as well as a heat transfer fluid in electronic testers.

PEG is included in many or all formulations of the soft drink Dr Pepper, purportedly as an anti-foaming agent.

PEG has also been used to preserve objects which have been salvaged from underwater, as was the case with the warship Vasa in Stockholm. It replaces water in wooden objects, which makes the wood dimensionally stable and prevents warping or shrinking of the wood.

PEG is often seen (as a side effect) in mass spectrometry experiments with characteristic fragmentation patterns.

In the field of microbiology, PEG precipitation is used to concentrate viruses and PEG is also used to induce complete fusion (mixing of both inner and outer leaflets) in liposomes reconstituted in vitro.

PEG is also used in lubricant eye drops. PEG derivatives such as narrow range ethoxylates are used as surfactants.

Dimethyl ethers of PEG are the key ingredient of Selexol, a solvent used by coal-burning, integrated gasification combined cycle (IGCC) power plants to remove carbon dioxide and hydrogen sulfide from the gas waste stream.

Polymersomes and gene therapy vectors (such as viruses) can be PEG coated to shield them from inactivation by the immune system and to de-target them away from organs where they may build up and have a toxic effect. The size of the PEG polymer has been shown to be important, with large polymers achieving the best immune protection.

5.12 Polyethylene Terephthalate

	PET
Molecular formula	$(C_{10}H_8O_4)_n$
Density	1370 kg/m^3
Young's modulus(E)	2800–3100 MPa
Tensile strength(σ_t)	55–75 MPa
Elastic limit	50–150%
notch test	3.6 kJ/m^2
Glass temperature	75 °C
melting point	260 °C
Vicat B	170 °C
Thermal conductivity	0.24 W/(m·K)
linear expansion coefficient (α)	$7\times10^{?5}$/K
Specific heat (c)	1.0 kJ/(kg·K)
Water absorption (ASTM)	0.16
Refractive Index	1.5750
Price	0.5–1.25 /kg

Source: A.K. vam der Vegt and L.E. Govaert, Polymeren, van keten tot kunstof, ISBN 90-407-2388-5.

Polyethylene terephthalate (sometimes written poly(ethylene terephthalate), commonly abbreviated PET, PETE, or the obsolete PETP or PET-P), is a thermoplastic polymer resin of the polyester family and is used in synthetic fibers; beverage, food and other liquid containers; thermoforming applications; and engineering resins often in combination with glass fiber. It is one of the most important raw materials used in man-made fibers.

Depending on its processing and thermal history, it may exist both as an amorphous (transparent) and as a semi-crystalline (opaque and white) material. Its monomer can be synthesized by the esterification reaction between terephthalic acid and ethylene glycol with water as a byproduct, or the transesterification reaction between ethylene glycol and dimethyl terephthalate with methanol as a byproduct. Polymerization is through a polycondensation reaction of the monomers (done immediately after esterification/ transesterification) with ethylene glycol as the byproduct (the ethylene glycol is recycled in production).

The majority of the world's PET production is for synthetic fibers (in excess of 60%) with bottle production accounting for around 30 per cent of global demand. In discussing textile applications, PET is generally referred to as simply "polyester" while "PET" is used most often to refer to packaging applications. Some of the trade names of PET products are Dacron, Diolen, Terylene, and Trevira fibers, Cleartuf, Eastman PET and Polyclear bottle resins,

Hostaphan, Melinex, and Mylar films, and Arnite, Ertalyte, Impet, Rynite and Valox injection molding resins. The polyester Industry makes up about 18 per cent of world polymer production and is third after polyethylene (PE) and polypropylene (PP).

Fig. 5.3: Chemical structure of polyethylene terephthalate.

Uses

PET can be semi-rigid to rigid, depending on its thickness, and is very lightweight. It makes a good gas and fair moisture barrier, as well as a good barrier to alcohol (requires additional "Barrier" treatment) and solvents. It is strong and impact-resistant. It is naturally colourless with high transparency.

When produced as a thin film (often known by the tradename Mylar), PET is often coated with aluminium to reduce its permeability, and to make it reflective and opaque. PET bottles are excellent barrier materials and are widely used for soft drinks, (see carbonation). PET or Dacron is also used as a thermal insulation layer on the outside of the International Space Station as seen in an episode of *Modern Marvels* "Sub Zero". For certain specialty bottles, PET sandwiches an additional polyvinyl alcohol to further reduce its oxygen permeability.

When filled with glass particles or fibers, it becomes significantly stiffer and more durable. This glass-filled plastic, in a semi-crystalline formulation, is sold under the tradename Rynite, Arnite, Hostadur, and Crastin.

While most thermoplastics can, in principle, be recycled, PET bottle recycling is more practical than many other plastic applications. The primary reason is that plastic carbonated soft drink bottles and water bottles are almost exclusively PET which makes them more easily identifiable in a recycle stream. PET has a resin identification code of 1. PET, as with many plastics, is also an excellent candidate for thermal recycling (incineration) as it is composed of carbon, hydrogen and oxygen with only trace amounts of catalyst elements (no sulphur) and has the energy content of soft coal.

One of the uses for a recycled PET bottle is for the manufacture of polar fleece material. It can also make fiber for polyester products.

PET was patented in 1941 by the Calico Printers' Association of Manchester. The PET bottle was patented in 1973.

Intrinsic Viscosity

One of the most important characteristics of PET is referred to as I.V. (intrinsic viscosity). The I.V. of the material, measured in deciliters per gram (dl/g) is dependent upon the length of its polymer chains. The longer the chains, the stiffer the material, and therefore the higher the I.V. The average chain length of a particular batch of resin can be controlled during polymerization.

An I.V. of about:

- 0.60 dl/g: Would be appropriate for fibre
- 0.65 dl/g: Film
- 0.76-0.84 dl/g: Bottles
- 0.85 dl/g: Tire cord

Drying

PET is hygroscopic, meaning that it naturally absorbs water from its surroundings. However, when this 'damp' PET is then heated a chemical reaction known as hydrolysis takes place between the water and the PET which reduces its molecular weight (IV) and its physical properties. This means that before the resin can be processed in a molding machine, as much moisture as possible must be removed from the resin. This is achieved through the use of a desiccant or dryers before the PET is fed into the processing equipment.

Inside the dryer, hot dry air is pumped into the bottom of the hopper containing the resin so that it flows up through the pellets, removing moisture on its way. The hot wet air leaves the top of the hopper and is first run through an after-cooler, because it is easier to remove moisture from cold air than hot air. The resulting cool wet air is then passed through a desiccant bed. Finally the cool dry air leaving the desiccant bed is re-heated in a process heater and sent back through the same processes in a closed loop. Typically residual moisture levels in the resin must be less than 40 parts per million (parts of water per million parts of resin, by weight) before processing. Dryer residence time should not be shorter than about four hours. This is because drying the material in less than 4 hours would require a temperature above 160 °C, at which level hydrolysis would begin inside the pellets before they could be dried out.

Copolymers

In addition to pure (homopolymer) PET, PET modified by copolymerization is also available. In some cases, the modified properties of copolymer are more desirable for a particular application. For example, cyclohexane dimethanol (CHDM) can be added to the polymer backbone in place of ethylene glycol. Since this building block is much larger (6 additional carbon atoms) than the ethylene glycol unit it replaces, it does not fit in with the neighbouring chains the way an ethylene glycol unit would. This interferes with crystallization and lowers the polymer's melting temperature. Such PET is generally known as PETG (Eastman Chemical and SK Chemicals are the only two manufacturers).

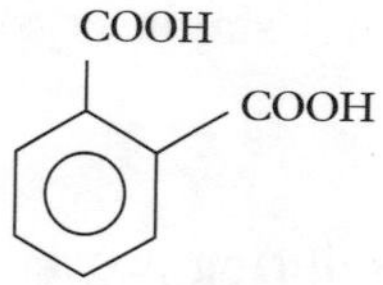

phthalic acid 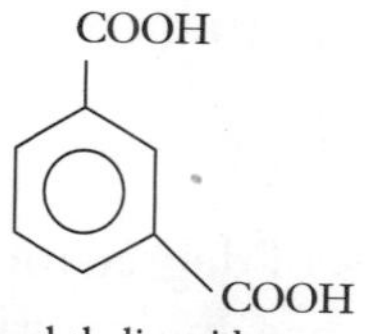　　isophthalic acid 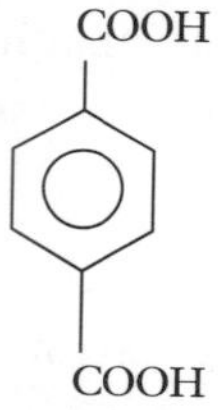　　isophthalic acid

Replacing terephthalic acid (right) with isophthalic acid (center) creates a kink in the PET chain, interfering with crystallization and lowering the polymer's melting point. Another common modifier is isophthalic acid, replacing some of the 1,4-(*para-*) linked terephthalate units. The 1,2-(*ortho-*) or 1,3-(*meta-*) linkage produces an angle in the chain, which also disturbs crystallinity. Such copolymers are advantageous for certain moulding applications, such as thermoforming, which is used for example to make tray or blister packagings from PETG film, or PETG sheet. On the other hand, crystallization is important in other applications where mechanical and dimensional stability are important, such as seat belts. For PET bottles, the use of small amounts of CHDM or other comonomers can be useful: if only small amounts of comonomers are used, crystallization is slowed but not prevented entirely. As a result, bottles are obtainable via stretch blow molding ("SBM"), which are both clear and crystalline enough to be an adequate barrier to aromas and even gases, such as carbon dioxide in carbonated beverages.

Crystals

Crystallization occurs when polymer chains fold up on themselves in a repeating, symmetrical pattern. Long polymer chains tend to become entangled on themselves, which prevents full crystallization in all but the most carefully controlled circumstances. PET is no exception to this rule; 60 per cent crystallization is the upper limit for commercial products, with the exception of polyester fibers.

PET in its natural state is a crystalline resin. Clear products can be produced by rapidly cooling molten polymer to form an amorphous solid. Like glass, amorphous PET forms when its molecules are not given enough time to arrange themselves in an orderly fashion as the melt is cooled. At room temperature the molecules are frozen in place, but if enough heat energy is put back into them, they begin to move again, allowing crystals to nucleate and grow. This procedure is known as solid-state crystallization.

Like most materials, PET tends to produce many small crystallites when crystallized from an amorphous solid, rather than forming one large single crystal. Light tends to scatter as it crosses the boundaries between crystallites and the amorphous regions between them. This scattering means that crystalline PET is opaque and white in most cases. Fiber drawing is among the few industrial processes that produces a nearly single-crystal product.

Degradation

PET is subject to various types of degradations during processing. The main degradations that can occur are hydrolytic, thermal and probably most important thermal oxidation. When PET degrades, several things happen: discoloration, chain scissions resulting in reduced molecular weight, formation of acetaldehyde and cross-links ("gel" or "fish-eye" formation). Discoloration is due to the formation of various chromophoric systems following prolonged thermal treatment at elevated temperatures. This becomes a problem when the optical requirements of the polymer are very high like for example in packaging applications. Acetaldehyde is normally a colorless gas with a fruity smell. It forms naturally in fruit, but it can cause an off-taste in bottled water. Acetaldehyde forms in PET through the "abuse" of the material. High temperatures (PET decomposes above 300 °C or 570 °F), high pressures, extruder speeds (excessive shear flow raises temperature) and long barrel residence times all contribute to the production of acetaldehyde. When acetaldehyde is produced, some of it remains dissolved in the walls of a container and then diffuses into the product stored inside, altering the taste and aroma. This is not such a problem for non-consumables such as shampoo, for fruit juices, which already contain acetaldehyde or for strong-tasting drinks, such as soft drinks. For bottled water, low acetaldehyde content is quite important, because if nothing masks the aroma, even extremely low concentrations (10-20 parts per billion parts of resin, by weight) of acetaldehyde can produce an off-taste. The thermal and thermooxidative degradation results in poor processability characteristics and performance of the material.

One way to alleviate this is to use a copolymer. Comonomers such as CHDM or isophthalic acid lower the melting temperature and reduces the degree of crystallinity of PET (especially important when the material is used for bottle manufacturing). Thus the resin can be plastically formed at lower temperatures and/or with lower force. This helps to prevent degradation, reducing the acetaldehyde content of the finished product to an acceptable (that is, unnoticeable) level. See copolymers, above. Other ways to improve the stability of the polymer is by using stabilizers, mainly antioxidants such as phosphites. Recently, molecular level stabilization of the material using nanostructured chemicals has also been considered.

Antimony

Antimony trioxide (Sb_2O_3) is a catalyst that is often used in the production of PET. It remains in the material and can thus in principle migrate out into food and drinks. Although antimony trioxide is of low toxicity, its presence is still of concern. The Swiss Federal Office of Public Health investigated the amount of antimony migration, comparing waters bottled in PET and glass: the antimony concentrations of the water in PET bottles was higher, but still well below the allowed maximal concentrations. (report available in German and French only) The Swiss Federal Office of Public Health concluded that small amounts of antimony migrate from the PET into bottled water, but that the health risk of the resulting low concentrations is negligible (1% of the "tolerable daily intake" determined by the WHO). A later (2006) but more widely publicized study by a group of geochemists at the University of Heidelberg headed by William Shotyk found similar amounts of antimony in water in PET bottles. The most recent WHO risk assessment for antimony in drinking water can be found here:

Re-crystallization

PET can be used to explore the crystallization of amorphous solids. The resin identification code can be used to verify the type of plastic it is made of: many plastic beverage bottles have the letters PET or PETE and a code of 1 on the bottom, near the center. When a flame is held several inches below the bottle and slowly brought closer, part of the material will visibly change. This happens because high temperatures melt the PET. This releases the tension that was frozen in during the blow molding process and the polymer chains will shift to a more relaxed and disordered state, which results in shrinkage of the softened area. Because of the decreased order of the polymer chains, there are now fewer crystal nuclei. Consequently, when the crystallites re-form upon cooling they grow larger than the original crystallites in the bottle wall. Because the new crystallites are larger than the wave length of light, they will now cause light to scatter, giving the material an opaque white appearance.

PETE has SPI resin ID code 1

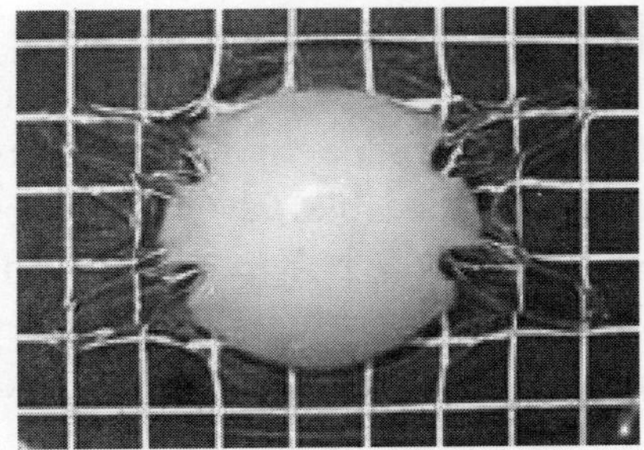

Recrystallized PET

Processing Equipment

There are two basic molding methods, one-step and two-step. In two-step molding, two separate machines are used. The first machine injection molds the preform. The preform looks like a test tube. The bottle-cap threads are already molded into place, and the body of the tube is significantly thicker, as it will be inflated into its final shape in the second step using stretch blow molding. In the second process, the preforms are heated rapidly and then inflated against a two-part mold to form them into the final shape of the bottle. Preforms (uninflated bottles) are now also used as containers for candy.

In one-step machines, the entire process from raw material to finished container is conducted within one machine, making it especially suitable for molding non-standard shapes (custom molding), including jars, flat oval, flask shapes etc. Its greatest merit is the reduction in space, product handling and energy, and far higher visual quality than can be achieved by the two-step system.

5.13 Polyimide

Polyimide (sometimes abbreviated PI) is a polymer of imide monomers. The structure of imide is as shown. Thermosetting polyimides are commercially available as uncured resins, stock shapes, thin sheets, laminates and machines parts. Thermoplastic polyimides are very often called *pseudothermoplastic*. There are two general types of polyimides. One type, so-called linear polyimides, are made by combining imides into long chains. Aromatic heterocyclic polyimides are the other usual kind, where R' and R? are two carbon atoms of an aromatic ring. Examples of polyimide films include Apical, Kapton and Kaptrex. Polyimide parts and shapes include Meldin, Vespel and Plavis. Polyimides have been in mass production since 1955.

$$R'-\overset{\displaystyle O}{\overset{\|}{C}}-\underset{\displaystyle R}{\underset{|}{N}}-\overset{\displaystyle O}{\overset{\|}{C}}-R''$$

Density	1430 kg/m^3
Young's modulus(E)	3200 MPa
Tensile strength($_t$)	75-90 MPa
Elongation @ break	4-8%
notch test	4-8 kJ/m
Glass temperature	>400 °C
melting point	none
Vicat B	220(?) °C
Thermal conductivity (k)	0.52 W/m.K
linear expansion coefficient ()	5.5 10^{-5}/K
Specific heat (c)	1.15 kJ/kg.K
Water absorption (ASTM)	0.32
Dielectric constant (Dk) at 1MHz	3.5

Properties

Thermosetting polyimides are known for thermal stability, good chemical resistance, excellent mechanical properties, and characteristic orange/yellow color. Polyimides compounded with graphite or glass fiber reinforements have flexural strengths of up to 50,000 p.s.i. and flexural moduli of 3 million p.s.i. Thermoset polyimides exhibit very low creep and high tensile strength. These properties are maintained during continuous use to temperatures of 450°F (232°C) and for short excursions, as high as 900°F (482°C). Molded polyimide parts and laminates have very good heat resistance. Normal operating temperatures for such parts and laminates range from cryogenic to those exceeding 500°F (260°C). Polyimides are also inherently resistant to flame combustion and do not usually need to be mixed with flame retardants. Most carry a UL rating of VTM-0. Polyimide laminates have a flexural strength half life at 480°F (249°C) of 400 hours.

Typical polyimide parts are not affected by commonly used solvents and oils—including hydrocarbons, esters, ethers, alcohols and freons. They also resist weak acids but are not recommended for use in environments that contains alkalis or inorganic acids. Some polyimides, such as CP1 and CORIN XLS, are solvent-soluble and exhibit high optical clarity. The solubility properties lend them towards spray and low temperature cure applications.

Application

Polyimide is often used in the electronics industry for flexible cables, as an insulating film on magnet wire and for medical tubing. For example, in a laptop computer, the cable that connects the main logic board to the display (which must flex every time the laptop is opened or closed) is often a polyimide base with copper conductors. The semiconductor industry uses polyimide as a high-temperature adhesive; it is also used as a mechanical stress buffer. Some polyimide can be used like a photoresist; both "positive" and "negative" types of photoresist-like polyimide exist in the market.

5.14 Polypropylene

Polypropylene or polypropene (PP) is a thermoplastic polymer, made by the chemical industry and used in a wide variety of applications, including packaging, textiles (e.g., ropes, thermal underwear and carpets), stationery, plastic parts and reusable containers of various types, laboratory equipment, loudspeakers, automotive components, and polymer banknotes. An addition polymer made from the monomer propylene, it is rugged and unusually resistant to many chemical solvents, bases and acids. Its resin identification code is ♳ PETE. Melt processing of polypropylene can be achieved via extrusion and molding.

Common extrusion methods include production of melt blown and spun bond fibers to form long rolls for future conversion into a wide range of useful products such as face masks, filters, nappies and wipes.

The most common shaping technique is injection molding, which is used for parts such as cups, cutlery, vials, caps, containers, housewares and automotive parts such as batteries. The related techniques of blow molding and injection-stretch blow molding are also used, which involve both extrusion and molding.

IUPAC name	poly(1-methylethylene)	
Other names	Polypropylene; Polypropene; Polipropene 25 [USAN]; Propene polymers; Propylene polymers; 1-Propene homopolymer	
Identifiers		
	CAS number	[9003-07-0]
Properties		
Molecular formula	$(C_3H_6)_x$	
Density	0.855 g/cm^3, amorphous 0.946 g/cm^3, crystalline	
Melting point	~ 160 °C	

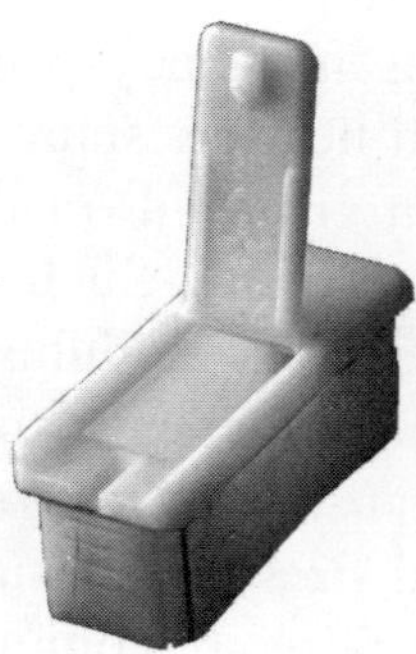

Polypropylene lid of a Tic Tacs box, with a living hinge and the resin identification code under its flap

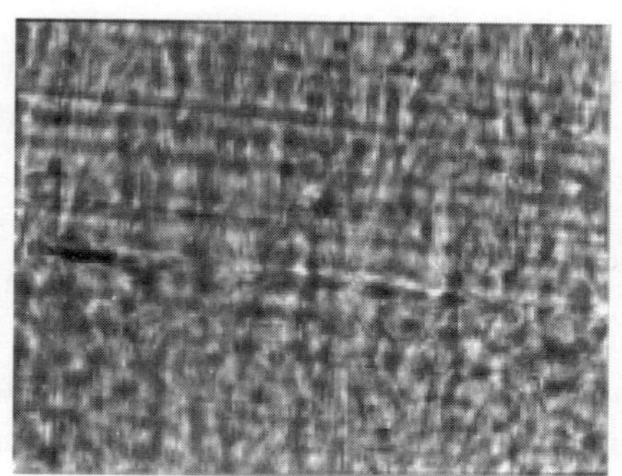

Micrograph of polypropylene

The large number of end use applications for PP are often possible because of the ability to tailor grades with specific molecular properties and additives during its manufacture. For example, antistatic additives can be added to help PP surfaces resist dust and dirt. Many physical finishing techniques can also be used on PP, such as machining. Surface treatments can be applied to PP parts in order to promote adhesion of printing ink and paints.

Chemical and Physical Properties

Most commercial polypropylene is isotactic and has an intermediate level of crystallinity between that of low density polyethylene (LDPE) and high density polyethylene (HDPE); its Young's modulus is also intermediate. Through the incorporation of rubber particles, PP can be made both tough and flexible, even at low temperatures. This allows polypropylene to be used as a replacement for engineering plastics, such as ABS. Polypropylene is rugged, often somewhat stiffer than some other plastics, reasonably economical, and can be made translucent when uncolored but is not as readily made transparent as polystyrene, acrylic or certain other plastics. It can also be made opaque and/or have many kinds of colors through the use of pigments. Polypropylene has very good resistance to fatigue, so that most plastic living hinges, such as those on flip-top bottles, are made from this material. Very thin sheets of polypropylene are used as a dielectric within certain high performance pulse and low loss RF capacitors.

Polypropylene has a melting point of ~160°C (320°F), as determined by Differential Scanning Calorimetry (DSC). Many plastic items for medical or laboratory use can be made from polypropylene because it can withstand the heat in an autoclave. Food containers made from it will not melt in the dishwasher, and do not melt during industrial hot filling processes. For this reason, most plastic tubs for dairy products are polypropylene sealed with aluminium foil (both heat-resistant materials). After the product has cooled, the tubs are often given lids made of a less heat-resistant material, such as LDPE or polystyrene. Such containers provide a good hands-on example of the difference in modulus, since the rubbery (softer, more flexible) feeling of LDPE with respect to PP of the same thickness is readily apparent. Rugged, translucent, reusable plastic containers made in a wide variety of shapes and sizes for consumers from various companies such as Rubbermaid and Sterilite are commonly made of polypropylene, although the lids are often made of somewhat more flexible LDPE so they can snap on to the container to close it. Polypropylene can also be made into disposable bottles to contain liquid, powdered or similar consumer products, although HDPE and polyethylene terephthalate are commonly also used to make bottles. Plastic pails, car batteries, wastebaskets, cooler containers, dishes and pitchers are often made of polypropylene or HDPE, both of which commonly have rather similar appearance, feel, and properties at ambient temperature.

The MFR (Melt Flow Rate) or MFI (Melt Flow Index) is an indication of PP's molecular weight. This helps to determine how easily the melted raw material will flow during processing. Higher MFR PPs fill the plastic mold more easily during the injection or blow molding production process. As the melt flow increases, however, some physical properties, like impact strength, will decrease.

There are three general types of PP: homopolymer, random copolymer and impact or block copolymer. The comonomer used is typically ethylene. Ethylene-propylene rubber added to PP homopolymer increases its low temperature impact strength. Randomly polymerized ethylene monomer added to PP homopolymer decreases the polymer crystallinity and makes the polymer more transparent.

Degradation

The polymer is however, liable to chain degradation from exposure to UV radiation such as that present in sunlight. This is one main reason for not using it transparent instead of glass. For external applications, UV-absorbing additives must be used. Carbon black also provides some protection from UV attack. The polymer can also be oxidised at high temperatures, a common problem during moulding operations. Anti-oxidants are normally added to prevent polymer degradation.

Synthesis

```
    H   H   H   H   H   H   H   H   H   H   H   H   H   H   H   H
    |   |   |   |   |   |   |   |   |   |   |   |   |   |   |   |
——C——C——C——C——C——C——C——C——C——C——C——C——C——C——C——C——
    |   |   |   |   |   |   |   |   |   |   |   |   |   |   |   |
   CH3  H  CH3  H  CH3  H  CH3  H  CH3  H  CH3  H  CH3  H  CH3  H

   CH3  H   H   H  CH3  H   H   H  CH3  H   H   H  CH3  H   H   H
    |   |   |   |   |   |   |   |   |   |   |   |   |   |   |   |
——C——C——C——C——C——C——C——C——C——C——C——C——C——C——C——C——
    |   |   |   |   |   |   |   |   |   |   |   |   |   |   |   |
   CH3  H  CH3  H  CH3  H  CH3  H  CH3  H  CH3  H  CH3  H  CH3  H
```

Fig. 5.4: Short segments of polypropylene, showing examples of isotactic (above) and syndiotactic (below) tacticity.

An important concept in understanding the link between the structure of polypropylene and its properties is tacticity. The relative orientation of each methyl group (CH_3 in the figure at left) relative to the methyl groups on neighboring monomers has a strong effect on the finished polymer's ability to form crystals, because each methyl group takes up space and constrains backbone bending.

Like most other vinyl polymers, useful polypropylene cannot be made by radical polymerization due to the higher reactivity of the allylic hydrogen (leading to dimerization) during polymerization. Moreover, the material that would result from such a process would have methyl groups arranged randomly, so called *atactic* PP. The lack of long-range order prevents any crystallinity in such a material, giving an amorphous material with very little strength and only specialized qualities suitable for niche end uses.

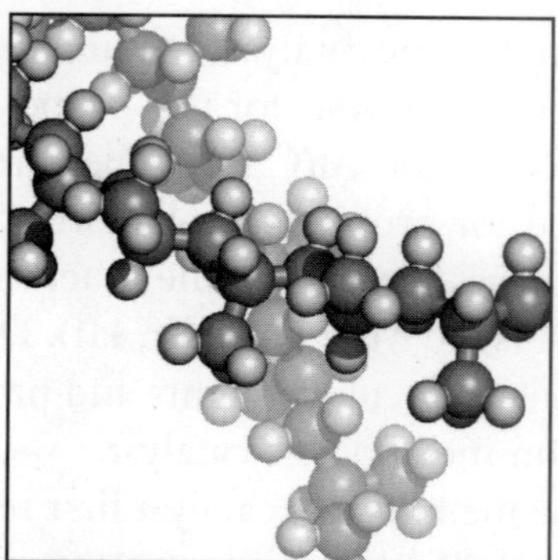

Fig. 5.5: A ball-and-stick model of syndiotactic polypropylene.

A Ziegler-Natta catalyst is able to limit incoming monomers to a specific orientation, only adding them to the polymer chain if they face the right direction. Most commercially available polypropylene is made with such Ziegler-Natta catalysts, which produce mostly isotactic polypropylene (the upper chain in the figure above). With the methyl group consistently on one side, such molecules tend to coil into a helical shape; these helices then line up next to one another to form the crystals that give commercial polypropylene many of its desirable properties.

More precisely engineered Kaminsky catalysts have been made, which offer a much greater level of control. Based on metallocene molecules, these catalysts use organic groups to control the monomers being added, so that a proper choice of catalyst can produce isotactic, syndiotactic, or atactic polypropylene, or even a combination of these. Aside from this qualitative control, they allow better quantitative control, with a much greater ratio of the desired tacticity than previous Ziegler-Natta techniques. They also produce narrower molecular weight distributions than traditional Ziegler-Natta catalysts, which can further improve properties.

To produce a rubbery polypropylene, a catalyst can be made which yields isotactic polypropylene, but with the organic groups that influence tacticity held in place by a relatively weak bond. After the catalyst has produced a short length of polymer which is capable of crystallization, light of the proper frequency is used to break this weak bond, and remove the selectivity of the catalyst so that the remaining length of the chain is atactic. The result is a mostly amorphous material with small crystals embedded in it. Since each chain has one end in a crystal but most of its length in the soft, amorphous bulk, the crystalline regions serve the same purpose as vulcanization.

Mechanism of Metallocene Catalysts

The reaction of many metallocene catalysts requires a co catalyst for activation. One of the most common co catalysts for this purpose is Methylalmuinoxane (MAO). Other co catalysts include, $Al(C_2H_5)_3$. There are numerous metallocene catalysts that can be used for propylene polymerization. (Some metallocene catalysts are used for industrial process, while others are not, due to their high cost.) One of the simplest is Cp_2MCl_2 (M = Zr, Hf). Different catalyst can lead to polymers with different molecular weights and properties. Active research is still being conducted on metallocene catalyst.

In the mechanism the metallocene catalyst first reacts with the co catalyst. If MAO is the co catalyst, the first step is to replace one of the Cl atoms on the catalyst with a methyl group from the MAO. The methyl group on the MAO is replaced by the Cl from the catalyst. The MAO then removes another Cl from the catalyst. This makes the catalyst positively charged and susceptible to attack from propylene.

Once the catalyst is activated, the double bond on the propene coordinates with the metal of the catalyst. The methyl group on the catalyst then migrates to the propene, and the double bond is broken. This starts the polymerization. Once the methyl migrates the positively charged catalyst is reformed and another propene can coordinate to the metal. The second propene coordinates and the carbon chain that was formed migrates to the propene. The process of coordination and migration continues and a polymer chain is grown off of the metallocene catalyst.

History

Polypropylene was first polymerized by Dr. Karl Rehn at Hoechst AG in Germany in 1951, who didn't recognize the importance of his discovery. It was then rediscovered on March 11 1954 by Giulio Natta. At first it was thought that it would be cheaper than polyethylene.

Practical Applications

A common application for polypropylene is as Biaxially Oriented polypropylene (BOPP). These BOPP sheets are used to make a wide variety of materials including clear bags. When polypropylene is biaxially oriented, it becomes crystal clear and serves as an excellent packaging material for artistic and retail products. In India, polypropylene is used in manufacturing mats to be used at home. In New Zealand and also in the US military, polypropylene, or 'polypro' (New Zealand 'polyprops'), is the material used for the fabrication of cold-weather gear, such as a long-sleeve shirt or long underwear, in addition to warm-weather gear such as Under Armour clothing, which can easily wick away sweat. These polypro clothes are not easily flammable, however, they can melt, which may result in severe burns if the service member is involved in an explosion or fire of any kind. Polypropylene is also used as an alternative to polyvinyl chloride (PVC) as insulation for electrical cables for LSZH cable in low-ventilation environments, primarily tunnels. This is because it emits less smoke and no toxic halogens, which may lead to production of acid in high temperature conditions. Polypropylene is also used in particular roofing membranes as the waterproofing top layer of single ply systems as opposed to modified bit systems. Its most common medical use is in the synthetic, nonabsorbable suture Prolene, manufactured by Ethicon Inc.

Polypropylene is most commonly used for plastic moldings where it is injected into a mold while molten, forming complex shapes at relatively low cost and high volume, examples include bottle tops, bottles and fittings.

Recently it has been produced in sheet form and this has been widely used for the production of stationary folders, packaging and storage boxes. The wide colour range, durability and resistance to dirt make it ideal as a protective cover for papers and other materials. It is used in Rubik's cube stickers because of these characteristics.

The availability of sheet polypropylene has provided an opportunity for the use of the material by designers. The light weight, durable and colourful plastic makes an ideal medium for the creation of light shades and a number of designs have been developed using interlocking sections to create elaborate designs. Polypropylene sheets are a popular choice for trading card collectors; these come with pockets (nine for standard size cards) for the cards to be inserted and are used to protect their condition and are meant to be stored in a binder. Polypropylene has been used in hernia repair operations to protect the body from new hernias in the same location. A small patch of the material is placed over the spot of the hernia, below the skin, and is painless and is rarely, if ever, rejected by the body.

The material has recently been introduced into the fashion industry through the work of designers such as Anoush Waddington who have developed specialized techniques to create jewellery and wearable items from polypropylene. Expanded Polypropylene (EPP) is a foam form of polypropylene. EPP has very good impact characteristics due to its low stiffness, this allows EPP to resume its shape after impacts. EPP is extensively used in model aircraft and other radio controlled vehicles by hobbyists. This is mainly due to its ability to absorb impacts, making this an ideal material for RC aircraft for beginners and amateurs.

Future Applications

Recent development of a specialty extruded polypropylene called Innegra S has produced a high modulus thread that is 66 per cent lighter than comparably deniered fiberglass. Dr. Brian Morin patented a process that creates a high modulus fiber that has potential in specialty applications ranging from marine ropes, marine hulls, kayak paddles, aircraft pallets, ballistic applications and other lightweight structural applications.

5.15 Polytetrafluoroethylene

In chemistry, poly(tetrafluoroethene) or poly(tetrafluoroethylene) (PTFE) is a synthetic fluoropolymer which finds numerous applications. PTFE's most well known trademark in the industry is the DuPont brand name Teflon. Water and water-containing substances like most foods do not wet PTFE, therefore adhesion to PTFE surfaces is inhibited. Due to this property PTFE is used as a non-stick coating for pans and other cookware. It is very non-reactive, and so is often used in containers and pipework for reactive and corrosive chemicals. Where used as a lubricant, PTFE reduces friction, wear and energy consumption of machinery.

$$\left(-\overset{\displaystyle F}{\underset{\displaystyle F}{\overset{|}{\underset{|}{C}}}} - \overset{\displaystyle F}{\underset{\displaystyle F}{\overset{|}{\underset{|}{C}}}} - \right)_n$$

IUPAC name	*Poly(tetrafluoroethene)*
Systematic name	Poly(tetrafluoroethylene)
Other names	Teflon
Identifiers	
Abbreviations	PTFE
CAS number	[9002-84-0]
Properties	
Molecular formula	C_nF_{2n}
Density	2200 kg m^{-3}
Melting point	327 °C
Supplementary data page	
Structure and properties	n, ε_r, etc.
Thermodynamic data	Phase behaviour: Solid, liquid, gas
Spectral data	UV, IR, NMR, MS
Except where noted otherwise, data are given for materials in their standard state (at 25 °C, 100 kPa) Infobox references	

History

PTFE was accidentally invented by Roy Plunkett of Kinetic Chemicals in 1938. While Plunkett was attempting to make a new CFC refrigerant, the perfluorethylene polymerized in its pressurized storage container. (In this original chemical reaction, iron from the inside of the container acted as a catalyst.) Kinetic Chemicals patented it in 1941 and registered the Teflon trademark in 1944. The original patent number is US 2,230,654.

Teflon was first sold commercially in 1946. By 1950, DuPont had acquired full interest in Kinetic Chemicals and was producing over a million pounds (450 t) per year in Parkersburg, West Virginia. In 1954, French engineer Marc Grégoire created the first pan coated with Teflon non-stick resin under the brandname of Tefal after his wife urged him to try the material that he'd been using on fishing tackle on her cooking pans. In the United States, Kansas

City, Missouri resident Marion A. Trozzolo, who had been using the substance on scientific utensils, marketed the first frying pan, "The Happy Pan," in 1961.

An early advanced use was in the Manhattan Project as a material to coat valves and seals in the pipes holding highly reactive uranium hexafluoride in the vast uranium enrichment plant at Oak Ridge, Tennessee, when it was known as K-25.

Properties

PTFE is a white solid at room temperature, with a density of about 2.2 g/cm^3. According to DuPont its melting point is 327 °C (620.6 °F), but its properties degrade above 260 °C (500 °F). The coefficient of friction of plastics is usually measured against polished steel. PTFE's coefficient of friction is 0.1 or less, which is the second lowest of any known solid material (Diamond-like carbon being the first). PTFE's resistance to van der Waals forces means that it is the only known surface to which a gecko cannot stick.

PTFE has excellent dielectric properties. This is especially true at high radio frequencies, making it suitable for use as an insulator in cables and connector assemblies and as a material for printed circuit boards used at microwave frequencies. Combined with its high melting temperature, this makes it the material of choice as a high-performance substitute for the weaker and lower melting point polyethylene that is commonly used in low-cost applications. Its extremely high bulk resistivity makes it an ideal material for fabricating long life electrets, useful devices that are the electrostatic analogues of magnets.

Because of its chemical inertness, PTFE cannot be cross-linked like an elastomer. Therefore it has no "memory," and is subject to creep (also known as "cold flow" and "compression set"). This can be both good and bad. A little bit of creep allows PTFE seals to conform to mating surfaces better than most other plastic seals. Too much creep, however, and the seal is compromised. Compounding fillers control unwanted creep and improve wear, friction, and other properties. Sometimes metal springs apply continuous force to PTFE seals to give good contact, while permitting some creep.

Applications

Due to its low friction, it is used for applications where sliding action of parts is needed: bearings, bushings, gears, slide plates, etc. In these applications it performs significantly better than nylon and acetal; it is comparable to ultra high-molecular weight polyethylene (UHMWPE), although UHMWPE is more resistant to wear than Teflon. For these applications, versions of teflon with mineral oil or molybdenum disulfide embedded as additional lubricants in its matrix are being manufactured.

Fig. 5.6: The roof of the Hubert H. Humphrey Metrodome is made of 20 acres of teflon-coated fiberglass.

Gore-Tex is a material incorporating fluoropolymer membrane with micropores. The roof of the Hubert H. Humphrey Metrodome in Minneapolis is one of the largest applications of Teflon PTFE coatings on Earth, using 20 acres (about 8 hectares) of the material in a double-layered, white dome, made with PTFE-coated fiberglass, that gives the stadium its distinctive appearance. The Millennium Dome in London is also substantially made of PTFE.

Powdered PTFE is used in pyrotechnic compositions as oxidizer together with powdered metals such as aluminium and magnesium. Upon ignition these mixtures form carbonaceous soot and the corresponding metal fluoride and release large amounts of heat. Hence they are used as infrared decoy flares and igniters for solid-fuel rocket propellants.

PTFE is also used in body piercings, such as a sub-clavicle piercing, due to its flexibility and bio-compatibility.

In optical radiometry, sheets made from PTFE are used as measuring heads in spectroradiometers and broadband radiometers (e.g. illuminance meter and UV radiometer) due to its capability to diffuse a transmitting light nearly perfectly. Moreover, optical properties of PTFE stay constant over a wide range of wavelengths, from UV up to near infrared. In this region, the relation of its regular transmittance to diffuse transmittance is negligibly small so light transmitted through a diffuser (PTFE sheet) radiates like Lambert's cosine law. Thus, PTFE enables cosinusoidal angular response for a detector measuring the power of optical radiation at a surface, e.g., in solar irradiance measurements.

PTFE is also used to coat certain types of hardened, armor-piercing bullets, so as to reduce the amount of wear on the firearm's rifling. These are often referred to as "cop-killer" bullets by virtue of PTFE's supposed ability to ease a bullet's passage through body armor. However, this is simply an urban myth as PTFE has no effect in the bullet's ability to penetrate soft body armor.

PTFE's low frictional properties have also been utilized as computer mice feet such as the Logitech G5 and Logitech G7 computer mice series from Logitech or most Razer gaming mice (e.g the Deathadder, Lachesis). The low-friction provided by PTFE allows the mice to be moved and glide across surfaces smoothly and with less effort.

PTFE's high corrosion resistance makes it ideal for laboratory environments as containers, magnetic stirrers and tubing for highly corrosive chemicals such as hydrofluoric acid, which will dissolve glass containers.

PTFE can be used as a thread seal tape in plumbing applications.

PTFE grafts can be used to bypass stenotic arteries in peripheral vascular disease, if a suitable autologous vein graft is not available.

PTFE can be used to prevent insects climbing up surfaces painted with the material. PTFE is so slippery that insects cannot get a grip and tend to fall off. For example PTFE is used to prevent ants climbing out of formicariums.

Production

PTFE is either synthesized by the emulsion polymerization of tetrafluoroethylene monomer under pressure, using free-radical catalysts, or it may be produced by the direct substitution of hydrogen atoms on polyethylene with fluorine, using polyethylene and fluorine gas at 20 °C.

Safety

While PTFE itself is chemically inert and non-toxic, it begins to deteriorate after the temperature of cookware reaches about 500 °F (260 °C), and decompose above 660 °F (350 °C). These degradation products can be lethal to birds, and can cause flu-like symptoms in humans.

By comparison, cooking fats, oils, and butter will begin to scorch and smoke at about 392 °F (200 °C), and meat is usually fried between 400–450 °F (200–230 °C), but empty cookware can exceed this temperature if left unattended on a hot burner. A 1959 study, (conducted before the U.S. Food and Drug Administration approved the material for use in food processing equipment) showed that the toxicity of fumes given off by the coated pan on dry heating was less than that of fumes given off by ordinary cooking oils.

Carcinogens in Production

The United States Environmental Protection Agency's scientific advisory board found in 2005 that perfluorooctanoic acid (PFOA), a chemical compound used to make Teflon, is a "likely carcinogen." This finding was part of a draft report that has yet to be made final. DuPont settled for $300 million in a 2004 lawsuit filed by residents near its manufacturing plant in Ohio and West Virginia based on groundwater pollution from this chemical. Currently this chemical is not regulated by the EPA.

In January 2006, DuPont, the only company that manufactures PFOA in the US, agreed to eliminate releases of the chemical from its manufacturing plants by 2015, but did not commit to completely phasing out its use of the chemical. This agreement is said to apply to not only PTFE used in cookware but also other products such as food packaging, clothing, and carpeting. DuPont also stated that it cannot produce PTFE without the use of the chemical PFOA, although it is looking for a substitute.

PFOA is used only during the manufacture of the product—only a trace amount of PFOA remains after the curing process. DuPont maintains that there should be no measurable amount of PFOA on a finished pan, provided that it has been properly cured.

Similar Polymers

Teflon is also used as the trade name for a polymer with similar properties, perfluoroalkoxy polymer resin (PFA).

```
 / F   F \ / F   F \
|  |   |  ||  |   |  |
 - C — C --- C — C -
|  |   |  ||  |   |  |
 \ F   F /n\ F   O /m
                 |
             F   C   F

                 F
```

Other polymers with similar composition are also known by the Teflon name:

- ❖ PFA (perfluoroalkoxy polymer resin).
- ❖ FEP (fluorinated ethylene-propylene)

They retain the useful properties of PTFE of low friction and non-reactivity, but are more easily formable. FEP is softer than PTFE and melts at 260°C; it is highly transparent and resistant to sunlight.

5.16 Polystyrene

Polystyrene: (IUPAC Polyphenylethene) is an aromatic polymer made from the aromatic monomer styrene, a liquid hydrocarbon that is commercially manufactured from petroleum by the chemical industry. Polystyrene is a thermoplastic substance, normally existing in solid state at room temperature, but melting if heated (for molding or extrusion), and becoming solid again when cooling off.

Polystyrene	
Density	1050 kg/m^3
Density of EPS	25-200 kg/m^3
Specific Gravity	1.05
Electrical conductivity (s)	10^{-16} S/m
Thermal conductivity (k)	0.08 W/(m·K)
Young's modulus (E)	3000-3600 MPa
Tensile strength (s_t)	46–60 MPa
Elongation at break	3–4%
Notch test	2–5 kJ/m^2
Glass temperature	95 °C
Melting point	240 °C
Vicat B	90 °C
Heat transfer coefficient (Q)	0.17 W/(m^{2}K)
Linear expansion coefficient (a)	8 10^{-5}/K
Specific heat (c)	1.3 kJ/(kg·K)
Water absorption (ASTM)	0.03–0.1
Decomposition	X years, still decaying

Pure solid polystyrene is a colorless, hard plastic with limited flexibility. It can be cast into molds with fine detail. Polystyrene can be transparent or can be made to take on various colours. It is economical and is used for producing plastic model assembly kits, license plate frames, plastic cutlery, CD "jewel" cases, and many other objects where a fairly rigid, economical plastic is desired.

History

Polystyrene was discovered in 1839 by Eduard Simon, an apothecary in Berlin. From storax, the resin of *Liquidambar orientalis*, he distilled an oily substance, a monomer which he named styrol. Several days later Simon found that the styrol had thickened, presumably from oxidation, into a jelly he dubbed styrol oxide ("Styroloxyd"). By 1845 English chemist John Blyth and German chemist August Wilhelm von Hofmann showed that the same transformation of styrol took place in the absence of oxygen. They called their substance metastyrol. Analysis later showed that it was chemically identical to Styroloxyd. In 1866 Marcelin Berthelot correctly identified the formation of metastyrol from styrol as a polymerization process. About 80 years went by before it was realized that heating of styrol starts a chain reaction which produces macromolecules, following the thesis of German organic chemist Hermann Staudinger (1881–

1965). This eventually led to the substance receiving its present name, polystyrene. The I. G. Farben company began manufacturing polystyrene in Ludwigshafen, Germany, about 1931, hoping it would be a suitable replacement for die cast zinc in many applications. Success was achieved when they developed a reactor vessel that extruded polystyrene through a heated tube and cutter, producing polystyrene in pellet form. Polystyrene is about as strong as unalloyed aluminium, but much more flexible.

Structure

The chemical makeup of polystyrene is a long chain hydrocarbon with every other carbon connected to a Phenyl group (the name given to the aromatic ring benzene, when bonded to complex carbon substituents).

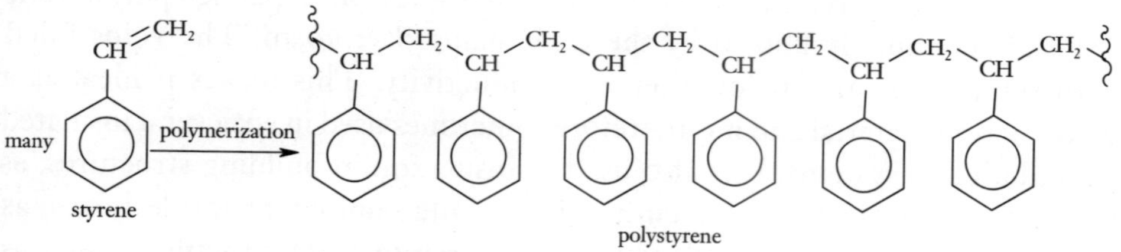

many styrene —polymerization→ polystyrene

A 3-D model would show that each of the chiral backbone carbons lies at the center of a tetrahedron, with its 4 bonds pointing toward the vertices. Say the -C-C- bonds are rotated so that the backbone chain lies entirely in the plane of the diagram. From this flat schematic, it is not evident which of the phenyl (benzene) groups are angled toward us from the plane of the diagram, and which ones are angled away. The isomer where all of them are on the same side is called *isotactic* polystyrene, which is not produced commercially. Ordinary *atactic* polystyrene has these large phenyl groups randomly distributed on both sides of the chain. This random positioning prevents the chains from ever aligning with sufficient regularity to achieve any crystallinity, so the plastic has no melting temperature, T_m. But metallocene-catalyzed polymerization can produce an ordered *syndiotactic* polystyrene with the phenyl groups on alternating sides. This form is highly crystalline with a T_m of 270 °C.

Solid Foam

Polystyrene's most common use is as expanded polystyrene (EPS). Expanded polystyrene is produced from a mixture of about 90-95 per cent polystyrene and 5-10 per cent gaseous blowing agent, most commonly pentane or carbon dioxide. The solid plastic is expanded into a foam through the use of heat, usually steam.

Fig. 5.7: Expanded polysterene tray with tomato seedlings.

Extruded polystyrene (XPS), which is different from expanded polystyrene (EPS), is commonly known by the trade name Styrofoam. The voids filled with trapped air give it low thermal conductivity. This makes it ideal as a construction material and it is therefore sometimes used in structural insulated panel building systems. It is also used as insulation in building structures, as molded packing material for cushioning fragile equipment inside boxes, as packing "peanuts", as non-weight-bearing architectural structures (such as pillars), and also in crafts and model building, particularly architectural models. Foamed between two sheets of paper, it makes a more-uniform substitute for corrugated cardboard, tradenamed Foamcore. A more unexpected use for the material is as a lightweight fill for embankments in the civil engineering industry. Expanded polystyrene used to contain CFCs, but other, more environmentally-safe blowing agents are now used. Because it is an aromatic hydrocarbon, it burns with an orange-yellow flame, giving off soot, as opposed to non-aromatic hydrocarbon polymers such as polyethylene, which burn with a light yellow flame (often with a blue tinge) and no soot. Production methods include sheet stamping (PS) and injection molding (both PS and HIPS).

The density of expanded polystyrene varies greatly from around 25 kg/m^3 to 200 kg/m^3 depending on how much gas was admixed to create the foam. A density of 200 kg/m^3 is typical for the expanded polystyrene used in surfboards.

Standard Markings

The resin identification code symbol for polystyrene, developed by the Society of the Plastics Industry so that items can be labeled for easy recycling. However, the majority of polystyrene products are currently not recycled because of a lack of suitable recycling facilities. Furthermore, when it is "recycled," it is not a closed loop—polystyrene cups and other packaging materials are usually recycled into fillers in other plastics, or other items that cannot themselves be recycled and are thrown away.

Copolymers

Pure polystyrene is brittle, but hard enough that a fairly high-performance product can be made by giving it some of the properties of a stretchier material, such as polybutadiene rubber. The two such materials can never normally be mixed because of the amplified effect of intermolecular forces on polymer insolubility (*see* plastic recycling), but if polybutadiene is added during polymerization it can become chemically bonded to the polystyrene, forming a graft copolymer which helps to incorporate normal polybutadiene into the final mix, resulting in high-impact polystyrene or HIPS, often called "high-impact plastic" in advertisements. One commercial name for HIPS is Bextrene. Common applications include use in toys and product casings. HIPS is usually injection molded in production. Autoclaving polystyrene can compress and harden the material.

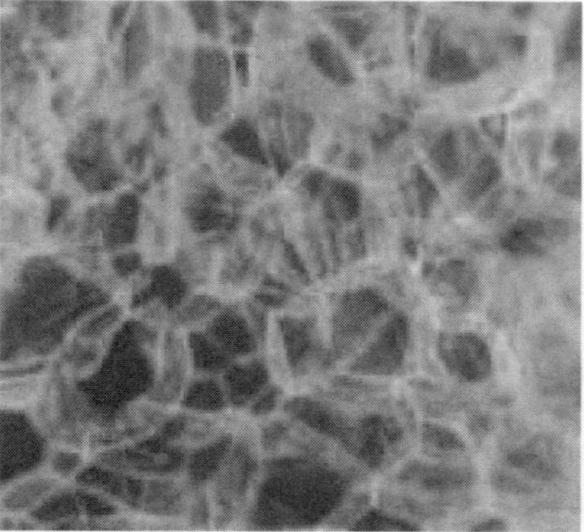

Fig. 5.8: Structure of expanded polystyrene (microscope).

Acrylonitrile butadiene styrene or ABS plastic is similar to HIPS: a copolymer of acrylonitrile and styrene, toughened with polybutadiene. Most electronics cases are made of this form of polystyrene, as are many sewer pipes. ABS pipes may become brittle over time. SAN is a copolymer of styrene with acrylonitrile and SMA one with maleic anhydride. Styrene can be copolymerized with other monomers; for example, divinylbenzene for cross-linking the polystyrene chains.

Cutting and Shaping

Expanded polystyrene is very easily cut with a hot-wire foam cutter, which is easily made by a heated taut length of wire, usually nichrome because of nichrome's resistance to oxidation at high temperatures and its suitable electrical conductivity. The hot wire foam cutter works by heating the wire to the point where it can vaporize foam immediately adjacent to it. The foam gets vaporized before actually touching the heated wire, which yields exceptionally smooth cuts.

Polystyrene, shaped and cut with hot wire foam cutters, is used in architecture models, actual signage, amusement parks, movie sets, airplane construction, and much more. Such cutters may cost just a few dollars (for a completely manual cutter) to tens of thousands of dollars for large CNC machines that can be used in high-volume industrial production.

Polystyrene can also be cut with a traditional cutter. In order to do this without ruining the sides of the blade one must first dip the blade in water and cut with the blade at an angle of about 30°. The procedure has to be repeated multiple times for best results.

Polystyrene can also be cut on 3 and 5-axis routers, enabling large-scale prototyping and model-making. Special polystyrene cutters are available that look more like large cylindrical rasps.

Use in Biology

Petri dishes and other containers such as test tubes, made of polystyrene, play an important role in biomedical research and science. For these uses, articles are almost always made by injection molding, and often sterilized post molding, either by irradiation or treatment with ethylene oxide. Post mold surface modification, usually with oxygen rich plasmas, is often done to introduce polar groups. Much of modern biomedical research relies on the use of such products; they therefore play a critical role in pharmaceutical research.

Finishing

In the United States, environmental protection regulations prohibit the use of solvents on polystyrene (which would dissolve the polystyrene and de-foam most of foams anyway).

Some acceptable finishing materials are:

- Water-based paint (artists have created paintings on polystyrene with gouache)
- Mortar or acrylic/cement render, often used in the building industry as a weather-hard overcoat that hides the foam completely after finishing the objects.
- Cotton wool or other fabrics used in conjunction with a stapling implement.

Dangers and Fire Hazard

Benzene, a material used in the production of polystyrene, is a known human carcinogen. Moreover, butadiene and styrene (in ABS), when combined, become benzene-like in both form and function.

The EPA Claims

"Styrene is primarily used in the production of polystyrene plastics and resins. Acute (short-term) exposure to styrene in humans results in mucous membrane and eye irritation, and gastrointestinal effects. Chronic (long-term) exposure to styrene in humans results in effects on the central nervous system (CNS), such as headache, fatigue, weakness, and depression, CSN dysfunction, hearing loss, and peripheral neuropathy. Human studies are inconclusive on the reproductive and developmental effects of styrene; several studies did not report an increase in developmental effects in women who worked in the plastics industry, while an increased frequency of spontaneous abortions and decreased frequency of births were reported in another study. Several epidemiologic studies suggest there may be an association between styrene exposure and an increased risk of leukemia and lymphoma. However, the evidence is inconclusive due to confounding factors. EPA has not given a formal carcinogen classification to styrene."

Polystyrene is classified according to DIN4102 as a "B3" product, meaning highly flammable or "easily ignited". Consequently, though it is an efficient insulator at low temperatures, it is prohibited from being used in any exposed installations in building construction as long the material is not flame retarded e.g. with hexabromocyclododecane. It must be concealed behind drywall, sheet metal or concrete. Foamed plastic materials have been accidentally ignited and caused huge fires and losses. Examples include the Düsseldorf International Airport, the Channel tunnel, where it was inside a railcar and caught on fire, and the Browns Ferry Nuclear Power Plant, where fire reached through a fire retardant, reached the foamed plastic underneath, inside a firestop that had not been tested and certified in accordance with the final installation. In addition to fire hazard, substances that contain acetone (such as most aerosol paint sprays), and cyanoacrylate glues can dissolve polystyrene.

Environmental Concerns and Bans

Expanded polystyrene is not easily recyclable because of its light weight and low scrap value. It is generally not accepted in curbside programs. Expanded polystyrene foam takes 900 years to decompose in the environment and has been documented to cause starvation in birds and other marine wildlife. According to the California Coastal Commission, it is a principal component of marine debris. Restricting the use of foamed polystyrene takeout food packaging is a priority of many solid waste environmentalist organizations, like Californians Against Waste.

The city of Berkeley, California was one of the first cities in the world to ban polystyrene food packaging (called Styrofoam in the media announcements). It was also banned in Portland, OR, and Suffolk County, NY in 1990. Now, over 20 US cities have banned polystyrene food packaging, including Oakland, CA on Jan 1st 2007. San Francisco introduced a ban on the packaging on June 1 2007:

> "This is a long time coming. Polystyrene foam products rely on nonrenewable sources for production, are nearly indestructible and leave a legacy of pollution on our urban and natural environments. If McDonald's could see the light and phase out polystyrene foam more than a decade ago, it's about time San Francisco got with the program."
>
> —Board of Supervisors President, Aaron Peskin

The overall benefits of the ban in Portland have been questioned, as have the general environmental concepts of the use of paper versus polystyrene.

A campaign to achieve the first ban of polystyrene foam from the food & beverage industry in Canada has been launched in Toronto as of January 2007, by local non-profit organization NaturoPack. The California and New York legislatures are currently considering bills which would effectively ban expanded polystyrene in all takeout food packaging state-wide.

Explosives

Polystyrene is used in some polymer-bonded explosives:

Name	*Explosive Ingredients*	*Binder Ingredients*	*Usage*
PBX-9205	RDX 92%	Polystyrene 6%; DOP 2%	
PBX-9007	RDX 90%	Polystyrene 9.1%; DOP 0.5%; resin 0.4%	

It is also a component of Napalm and a component of most designs of hydrogen bombs.

Cleaning

Polystyrene can be dishwashed at 70 °C without deformation since it has a glass transition temperature of 95 °C

5.17 Polytrimethylene Terephthalate

Polytrimethylene terephthalate, or PTT, has been a commercially available polymer for nearly 50 years. It is produced by a method called condensation polymerization or transesterification. The two monomer units used in producing this polymer are: 1,3-propanediol and terephthalic acid. Similar to polyethylene terephthalate, PTT is used in the production of polyesters.

PTT's value as a commercial polymer has improved only lately, with the development of more economical and efficient methods to produce propane diol (PDO), the main raw material used for PTT synthesis. DuPont is currently experimenting with the synthesis of PDO via bioplastic routes, i.e. using corn as the base material for PDO production. These developments will allow PTT to effectively compete against PBT and PET, two polyesters that have been far more successful than PTT to date.

5.18 Polyurethane

A polyurethane, commonly abbreviated PU, is any polymer consisting of a chain of organic units joined by urethane links. Polyurethane polymers are formed by reacting a monomer containing at least two isocyanate functional groups with another monomer containing at least two alcohol groups in the presence of a catalyst.

$$O{=}C{=}N{-}R^1{-}N{=}C{=}O + HO{-}R^2{-}OH + O{=}C{=}N{-}R^1{-}N{=}C{=}O + HO{-}R^2{-}OH + \cdots \longrightarrow$$

$$\cdots -\overset{O}{\overset{\|}{C}}-\underset{H}{N}-R^1-\underset{H}{N}-\overset{O}{\overset{\|}{C}}-O-R^2-O-\overset{O}{\overset{\|}{C}}-\underset{H}{N}-R^1-\underset{H}{N}-\overset{O}{\overset{\|}{C}}-O-R^2-O- \cdots$$

PU polymer formed by reacting a diisocyanate with a polyol.

Polyurethane formulations cover an extremely wide range of stiffness, hardness, and densities. These materials include:

- low density flexible foam used in upholstery and bedding,
- low density rigid foam used for thermal insulation and e.g. automobile dashboards,
- soft solid elastomers used for gel pads and print rollers, and
- hard solid plastics used as electronic instrument bezels and structural parts.

Polyurethanes are widely used in high resiliency flexible foam seating, rigid foam insulation panels, microcellular foam seals and gaskets, durable elastomeric wheels and tires, electrical potting compounds, high performance adhesives and sealants, Spandex fibres, seals, gaskets, carpet underlay, and hard plastic parts.

Polyurethane products are often called "urethanes". They should not be confused with the specific substance urethane, also known as ethyl carbamate. Polyurethanes are not produced from ethyl carbamate, nor do they contain it.

History

The pioneering work on polyurethane polymers was conducted by Otto Bayer and his coworkers in 1937 at the laboratories of I.G. Farben in Leverkusen, Germany. They recognized that using the polyaddition principle to produce polyurethanes from liquid diisocyanates and liquid polyether or polyester diols seemed to point to special opportunities, especially when compared to already existing plastics that were made by polymerizing olefins, or by polycondensation. The new monomer combination also circumvented existing patents obtained by Wallace Carothers on polyesters. Initially, work focused on the production of fibres and flexible foams. With development constrained by World War II (when PUs were applied on a limited scale as aircraft coating), it was not until 1952 that polyisocyanates became commercially available. Commercial production of flexible polyurethane foam began in 1954, based on toluene diisocyanate (TDI) and polyester polyols. The invention of these foams (initially called *imitation swiss cheese* by the inventors) was thanks to water accidentally introduced in the reaction mix. These materials were also used to produce rigid foams, gum rubber, and elastomers. Linear fibres were produced from hexamethylene diisocyanate (HDI) and 1,4-butanediol (BDO).

The first commercially available polyether polyol, poly(tetramethylene ether) glycol, was introduced by DuPont in 1956 by polymerizing tetrahydrofuran. Less expensive polyalkylene glycols were introduced by BASF and Dow Chemical the following year, 1957. These polyether polyols offered technical and commercial advantages such as low cost, ease of handling, and better hydrolytic stability; and quickly supplanted polyester polyols in the manufacture of polyurethane goods. Another early pioneer in PUs was the Mobay corporation. In 1960 more than 45,000 tons of flexible polyurethane foams were produced. As the decade progressed, the availability of chlorofluoroalkane blowing agents, inexpensive polyether polyols, and methylene diphenyl diisocyanate (MDI) heralded the development and use of polyurethane rigid foams as high performance insulation materials. Rigid foams based on polymeric MDI (PMDI) offered better thermal stability and combustion characteristics than those based on TDI. In 1967, urethane modified polyisocyanurate rigid foams were introduced, offering even better thermal stability and flammability resistance to low density insulation products. Also during the 1960s, automotive interior safety components such as instrument and door panels were produced by back-filling thermoplastic skins with semi-rigid foam.

In 1969, Bayer AG exhibited an all plastic car in Dusseldorf, Germany. Parts of this car were manufactured using a new process called RIM, Reaction

Injection Molding. RIM technology uses high-pressure impingement of liquid components followed by the rapid flow of the reaction mixture into a mold cavity. Large parts, such as automotive fascia and body panels, can be molded in this manner. Polyurethane RIM evolved into a number of different products and processes. Using diamine chain extenders and trimerization technology gave poly(urethane urea), poly(urethane isocyanurate), and polyurea RIM. The addition of fillers, such as milled glass, mica, and processed mineral fibres gave arise to RRIM, reinforced RIM, which provided improvements in flexural modulus (stiffness) and thermal stability. This technology allowed production of the first plastic-body automobile in the United Sates, the Pontiac Fiero, in 1983. Further improvements in flexural modulus were obtained by incorporating preplaced glass mats into the RIM mold cavity, also known as SRIM, or structural RIM.

Starting in the early 1980s, water-blown microcellular flexible foam was used to mold gaskets for panel and radial seal air filters in the automotive industry. Since then, increasing energy prices and the desire to eliminate PVC plastisol from automotive applications have greatly increased market share. Costlier raw materials are offset by a significant decrease in part weight and in some cases, the elimination of metal end caps and filter housings. Highly filled polyurethane elastomers, and more recently unfilled polyurethane foams are now used in high-temperature oil filter applications.

Polyurethane foam (including foam rubber) is often made by adding small amounts of volatile materials, so-called blowing agents, to the reaction mixture. These simple volatile chemicals yield important performance characteristics, primarily thermal insulation. In the early 1990s, because of their impact on ozone depletion, the Montreal Protocol led to the greatly reduced use of many chlorine-containing blowing agents, such as trichlorofluoromethane (CFC-11). Other haloalkanes, such as the hydrochlorofluorocarbon 1,1-dichloro-1-fluoroethane (HCFC-141b), were used as interim replacements until their phase out under the IPPC directive on greenhouse gases in 1994 and by the Volatile Organic Compounds (VOC) directive of the EU in 1997 (*see*: Haloalkanes). By the late 1990s, the use of blowing agents such as carbon dioxide, pentane, 1,1,1,2-tetrafluoroethane (HFC-134a) and 1,1,1,3,3-pentafluoropropane (HFC-245fa) became more widespread in North America and the EU, although chlorinated blowing agents remained in use in many developing countries.

Building on existing polyurethane spray coating technology and polyetheramine chemistry, extensive development of two-component polyurea spray elastomers took place in the 1990s. Their fast reactivity and relative insensitivity to moisture make them useful coatings for large surface area projects, such as secondary containment, manhole and tunnel coatings, and tank liners. Excellent adhesion to concrete and steel is obtained with the proper primer and surface treatment. During the same period, new two-component polyurethane and hybrid polyurethane-polyurea elastomer technology was used to enter the marketplace of spray-in-place load bed liners. This technique for coating pickup truck beds and other cargo bays creates a durable, abrasion resistant composite with the metal substrate, and eliminates corrosion and brittleness associated with drop-in thermoplastic bed liners.

The use of polyols derived from vegetable oils to make polyurethane products began garnering attention beginning around 2004, partly due to the rising costs of petrochemical feedstocks and partially due to an enhanced public desire for environmentally friendly green products. One of the most vocal supporters of these polyurethanes made using natural oil polyols is the Ford Motor Company.

Chemistry

$$R^1 — N = C = O + R^2 — O — H \quad R1 — \overset{H}{\underset{}{N}} — \overset{O}{\overset{\|}{C}} — O — R^2$$

generalized polyurethane reaction

Polyurethanes are in the class of compounds called reaction polymers, which include epoxies, unsaturated polyesters, and phenolics. A urethane linkage is produced by reacting an isocyanate group, -N = C = O with a hydroxyl (alcohol) group, -OH. Polyurethanes are produced by the polyaddition reaction of a polyisocyanate with a polyalcohol (polyol) in the presence of a catalyst and other additives. In this case, a polyisocyanate is a molecule with two or more isocyanate functional groups, $R\text{-}(N = C = O)_{n=2}$ and a polyol is a molecule with two or more hydroxyl functional groups, $R'\text{-}(OH)_{n=2}$. The reaction product is a polymer containing the urethane linkage, -RNHCOOR'-. Isocyanates will react with any molecule that contains an active hydrogen. Importantly, isocyanates react with water to form a urea linkage and carbon dioxide gas; they also react with polyetheramines to form polyureas. Commercially, polyurethanes are produced by reacting a liquid isocyanate with a liquid blend of polyols, catalyst, and other additives. These two components are referred to as a polyurethane system, or simply a system. The isocyanate is commonly referred to in North America as the 'A-side' or just the 'iso'. The blend of polyols and other additives is commonly referred to as the 'B-side' or as the 'poly'. This mixture might also be called a 'resin' or 'resin blend'. In Europe the meanings for 'A-side' and 'B-side' are reversed. Resin blend additives may include chain extenders, cross linkers, surfactants, flame retardants, blowing agents, pigments, and fillers.

The first essential component of a polyurethane polymer is the isocyanate. Molecules that contain two isocyanate groups are called diisocyanates. These molecules are also referred to as monomers or monomer units, since they themselves are used to produce polymeric isocyanates that contain three or more isocyanate functional groups. Isocyanates can be classed as aromatic, such as diphenylmethane diisocyanate (MDI) or toluene diisocyanate (TDI); or aliphatic, such as hexamethylene diisocyanate (HDI) or isophorone diisocyanate (IPDI). An example of a polymeric isocyanate is polymeric diphenylmethane diisocyanate, which is a blend of molecules with two-, three-, and four- or more isocyanate groups, with an average functionality of 2.7. Isocyanates can be further modified by partially reacting them with a polyol to form a prepolymer. A quasi-prepolymer is formed when the stoichiometric ratio of isocyanate to hydroxyl groups is greater than 2:1. A true prepolymer is formed when the stoichiometric ratio is equal to 2:1. Important characteristics of isocyanates are their molecular backbone, per cent NCO content, functionality, and viscosity.

The second essential component of a polyurethane polymer is the polyol. Molecules that contain two hydroxyl groups are called diols, those with three hydroxyl groups are called triols, et cetera. In practice, polyols are distinguished from short chain or low-molecular weight glycol chain extenders and cross linkers such as ethylene glycol (EG), 1,4-butanediol (BDO), diethylene glycol (DEG), glycerine, and trimethylol propane (TMP). Polyols are polymers in their own right. They are formed by base-catalyzed addition of propylene oxide (PO), ethylene oxide (EO) onto a hydroxyl or amine containing initiator, or by polyesterification of a di-acid, such as adipic acid, with glycols, such as ethylene glycol or dipropylene glycol (DPG). Polyols extended with PO or EO are polyether polyols. Polyols formed by polyesterification are polyester polyols. The choice of initiator, extender, and molecular weight of the polyol greatly affect its physical state, and the physical properties of the polyurethane polymer. Important characteristics of polyols are their molecular backbone, initiator, molecular weight, per cent primary hydroxyl groups, functionality, and viscosity.

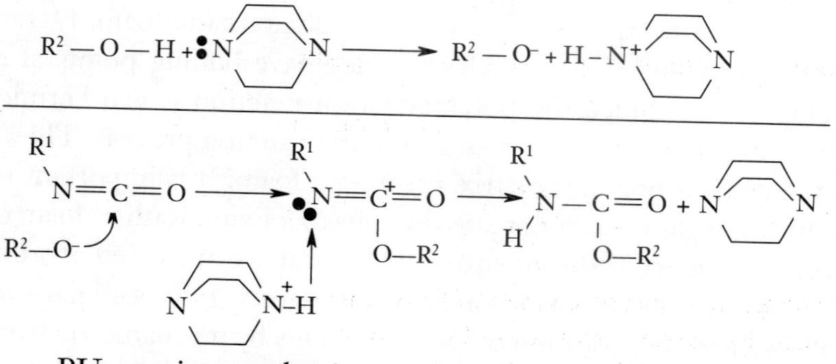

PU reaction mechanism catalyzed by a tertiary amine

$$R-N=C=O+H_2O \xrightarrow{\text{step 1}} R-\underset{\underset{H}{|}}{N}-\overset{\overset{O}{||}}{C}-O-H \xrightarrow[\text{decomposes}]{\text{step 2}} R-NH_2 + CO_2^{gas}$$

$$R-N=C=O+R-NH_2 \xrightarrow{\text{step 3}} R-\underset{\underset{H}{|}}{N}-\overset{\overset{O}{||}}{C}-\underset{\underset{H}{|}}{N}-R-$$

carbon dioxide gas formed by reacting water and isocyanate

The polymerization reaction is catalyzed by tertiary amines, such as dimethylcyclohexylamine, and organometallic compounds, such as dibutyltin dilaurate or bismuth octanoate. Furthermore, catalysts can be chosen based on whether they favor the urethane (gel) reaction, such as 1,4-diazabicyclo[2.2.2]octane (also called DABCO or TEDA), or the urea (blow) reaction, such as bis-(2-dimethylaminoethyl)ether, or specifically drive the isocyanate trimerization reaction, such as potassium octoate.

One of the most desirable attributes of polyurethanes is their ability to be turned into foam. Blowing agents such as water, certain halocarbons such as HFC-245fa (1,1,1,3,3-pentafluoropropane) and HFC-134a (1,1,1,2-tetrafluoroethane), and hydrocarbons such as n-pentane, can be incorporated into the poly side or added as an auxiliary stream. Water reacts with the isocyanate to create carbon dioxide gas, which fills and expands cells created during the mixing process. The reaction is a three step process. A water molecule reacts with an isocyanate group to form a carbamic acid. Carbamic acids are unstable, and decompose forming carbon dioxide and an amine. The amine reacts with more isocyanate to give a substituted urea. Water has a very low molecular weight, so even though the weight per cent of water may be small, the molar proportion of water may be high and considerable amounts of urea produced. The urea is not very soluble in the reaction mixture and tends to form separate "hard segment" phases consisting mostly of polyurea. The concentration and organization of these polyurea phases can have a significant impact on the properties of the polyurethane foam. Halocarbons and hydrocarbons are chosen such that they have boiling points at or near room temperature. Since the polymerization reaction is exothermic, these blowing agents volatilize into a gas during the reaction process. They fill and expand the cellular polymer matrix, creating a foam. It is important to know that the blowing gas does not create the cells of a foam. Rather, foam cells are a result of blowing gas diffusing into bubbles that are nucleated or stirred into the system at the time of mixing. In fact, high density microcellular foams can be formed without the addition of blowing agents by mechanically frothing or nucleating the polyol component prior to use.

Surfactants are used to modify the characteristics of the polymer during the foaming process. They are used to emulsify the liquid components, regulate cell size, and stabilize the cell structure to prevent collapse and surface defects. Rigid foam surfactants are designed to produce very fine cells and a very high closed cell content. Flexible foam surfactants are designed to stabilize the reaction mass while at the same time maximizing open cell content to prevent the foam from shrinking. The need for surfactant can be affected by choice of isocyanate, polyol, component compatibility, system reactivity, process conditions and equipment, tooling, part shape, and shot weight.

Raw Materials

For the manufacture of polyurethane polymers, two groups of at least bifunctional substances are needed as reactants; compounds with isocyanate groups, and compounds with active hydrogen atoms. The physical and chemical character, structure, and molecular size of these compounds influence the polymerization reaction, as well as ease of processing and final physical properties of the finished polyurethane. In addition, additive such as catalysts, surfactants, blowing agents, cross linkers, flame retardants, light stabilizers, and fillers are used to control and modify the reaction process and performance characteristics of the polymer.

Isocyanates

Isocyanates with two or more functional groups are required for the formation of polyurethane polymers. Volume wise, aromatic isocyanates account for the vast majority of global diisocyanate production. Aliphatic and cycloaliphatic isocyanates are also important building blocks for polyurethane materials, but in much smaller volumes. There are a number of reasons for this. First, the aromatically linked isocyanate group is much more reactive than the aliphatic one. Second, aromatic isocyanates are more economical to use. Aliphatic isocyanates are used only if special properties are required for the final product. For example, light stable coatings and elastomers can only be obtained with aliphatic isocyanates. Even within the same class of isocyanates, there is a significant difference in reactivity of the functional groups based on steric hindrance. In the case of 2,4-toluene diisocyanate, the isocyanate group in the para position to the methyl group is much more reactive than the isocyanate group in the ortho position.

Phosgenation of corresponding amines is the main technical process for the manufacture of isocyanates. The amine raw materials are generally manufactured by the hydrogenation of corresponding nitro compounds. For example, toluenediamine (TDA) is manufactured from dinitrotoluene, which then converted to toluene diisocyanate (TDI). Diamino diphenylmethane or methylenedianiline (MDA) is manufactured from nitrobenzene via aniline, which is then converted to diphenylmethane diisocyanate (MDI).

The two most important aromatic isocyanates are toluene diisocyanate (TDI) and diphenylmethane diisocyanate (MDI). TDI consists of a mixture of the 2,4- and 2,6-diisocyanatotoluene isomers. The most important product is TDI-80 (TD-80), consisting of 80 per cent of the 2,4-isomer and 20 per cent of the 2,6-isomer. This blend is used extensively in the manufacture of polyurethane flexible slabstock and molded foam. TDI, and especially crude TDI and TDI/MDI blends can be used in rigid foam applications, but have been supplanted by polymeric MDI. TDI-polyether and TDI-polyester prepolymers are used in high performance coating and elastomer applications. Prepolymers are available that have been vacuum stripped of TDI monomer, which greatly reduces their toxicity. Diphenylmethane diisocyanate (MDI) has three isomers, 4,4'-MDI, 2,4'-MDI, and 2,2'-MDI, and is also polymerized to provide oligomers of functionality three and higher.

4,4' OCN–C6H4–CH_2–C6H4–NCO

2,4' (NCO)C6H4–CH_2–C6H4–NCO

2,2' (NCO)C6H4–CH_2–C6H4(OCN)

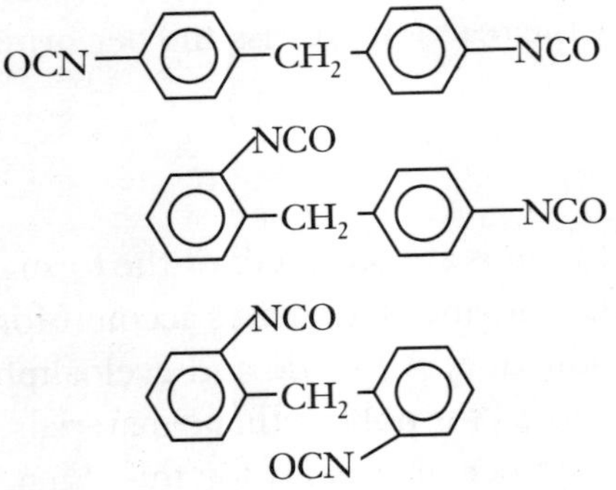

Pure MDI's

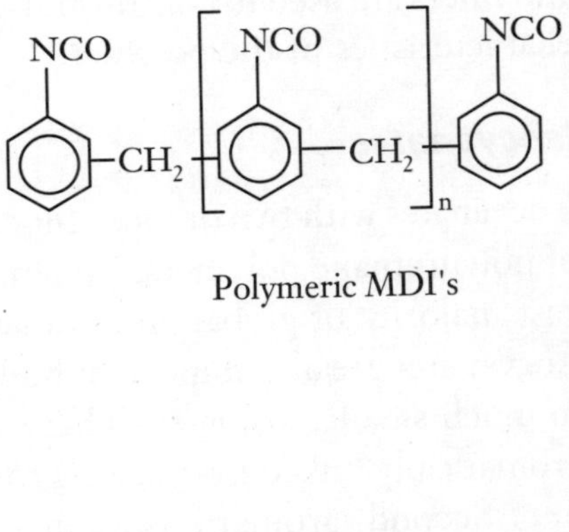

Polymeric MDI's

Only the 4,4'-MDI monomer is sold commercially as a single isomer. It is provided either as a frozen solid or flake, or in molten form, and is used to manufacture high performance prepolymers. Monomer blends, consisting of approximately 50 per cent of the 4,4'-isomer and 50 per cent of the 2,4'-isomer, are liquid at room temperature and are used to manufacture prepolymers for polyurea spray elastomer applications. 4,4'-MDI blends containing MDI uretonimine, carbodiimide, and allophonate moieties are also liquid at room temperature, and are used in the manufacture of integral skin and microcellular foams. 4,4'-MDI-glycol prepolymers offer increased mechanical properties in the same applications, but are prone to freezing at temperatures below 20°C.

Polymeric MDI (PMDI) is used in rigid pour-in-place, spray foam, and molded foam applications. Polymeric MDI that contains a very high portion of high-functionality oligomers is used to manufacture polyurethane and polyisocyanurate rigid insulation boardstock. Modified PMDI, which contains high levels of MDI monomer, is used in the production of polyurethane flexible molded and microcellular foam. The relative percentage of the 4,4'- and 2,4'-isomers is adjusted to change the reactivity and storage stability of the isocyanate blend, as well as the firmness and other physical properties of the finished goods. Other aromatic isocyanate include p-phenylene diisocyante (PPDI), naphthalene diisocyanate (NDI), and o-tolidine diisocyanate (TODI).

The most important aliphatic and cycloaliphatic isocyanates are 1,6-hexamethylene diisocyanate (HDI), 1-isocyanato-3-isocyanatomethyl-3,5,5-trimethyl-cyclohexane (isophorone diisocyanate, IPDI), and 4,4'-diisocyanato dicyclohexylmethane (H_{12}MDI). They are used to produce light stable, non-yellowing polyurethane coatings and elastomers. Because of their toxicity, aliphatic isocyanate monomers are converted into prepolymers, biurets, dimers, and trimers for commercial use. HDI adducts are used extensively for weather and abrasion resistant coatings and lacquers. IPDI is used in the manufacture of coatings, elastomeric adhesives and sealants. H_{12}MDI prepolymers are used to produce high performance coatings and elastomers with optical clarity and hydrolysis resistance. Other aliphatic isocyanates include cyclohexane diisocyanate (CHDI), tetramethylxylene diisocyanate (TMXDI), and 1,3-bis(isocyanatomethyl)cyclohexane (H_6XDI).

Polyols

Polyols are higher molecular weight materials manufactured from an initiator and monomeric building blocks. They are most easily classified as polyether polyols, which are made by the reaction of epoxides (oxiranes) with an active hydrogen containing starter compounds, or polyester polyols, which are made by the polycondensation of multifunctional carboxylic acids and hydroxyl compounds. They can be further classified according to their end use as flexible or rigid polyols, depending on the functionality of the initiator and their molecular weight. Taking into account functionality, flexible polyols have molecular weights from 2,000 to 10,000 (OH# from 18 to 56). Rigid polyols have molecular weights from 250 to 700 (OH# from 300 to 700). Polyols with molecular weights from 700 to 2,000 (OH# 60 to 280) are used to add stiffness or flexibility to base systems, as well as increase solubility of low molecular weight glycols in high molecular weight polyols.

Polyether polyols come in a wide variety of grades based on their end use, but are all constructed in a similar manner. Polyols for flexible applications use low functionality initiators such as dipropylene glycol (f = 2) or glycerine (f = 3). Polyols for rigid applications use high functionality initiators such sucrose (f = 8), sorbitol (f = 6), toluenediamine (f = 4), and Mannich bases (f = 4). Propylene oxide is then added to the initiators until the desired molecular weight is achieved. Polyols extended with propylene oxide are terminated with secondary hydroxyl groups. In order to change the compatibility, rheological properties, and reactivity of a polyol, ethylene oxide is used as a co-reactant to create random or mixed block heteropolymers. Polyols capped with ethylene oxide contain a high percentage of primary hydroxyl groups, which are more reactive than secondary hydroxyl groups. Because of their high viscosity (470 OH# sucrose polyol, 33,000 cps at 25°C), carbohydrate initiated polyols often use glycerine or diethylene glycol as a co-initiate in order to lower the viscosity to ease handling and processing (490 OH# sucrose-glycerine polyol, 5,500 cps at 25°C). Graft polyols (also called filled polyols or polymer polyols) contain finely dispersed styrene-acrylonitrile, acrylonitrile, or polyurea (PHD) polymer solids chemically grafted to a high molecular weight polyether backbone. They are used to increase the load bearing properties of low density high-resiliency (HR) foam, as well as add toughness to microcellular foams and cast elastomers. PHD polyols are also used to modify the combustion properties of HR flexible foam. Solids content ranges from 14 per cent to 50 per cent, with 22 per cent and 43 per cent being typical. Initiators such as ethylenediamine and triethanolamine are used to make low molecular weight rigid foam polyols that have built-in catalytic activity due to the presence of nitrogen atoms in the backbone. They are used to increase system reactivity and physical property build, and to reduce the friability of rigid foam molded parts. A special class of polyether polyols, poly(tetramethylene ether) glycols are made by polymerizing tetrahydrofuran. They are used in high performance coating and elastomer applications.

Polyester polyols fall into two distinct categories according to composition and application. Conventional polyester polyols are based on virgin raw materials and are manufactured by the direct polyesterification of high-purity diacids and glycols, such as adipic acid and 1,4-butanediol. They are distinguished by the choice of monomers, molecular weight, and degree of branching. While costly and difficult to handle because of their high viscosity, they offer physical properties not obtainable with polyether polyols, including superior solvent, abrasion, and cut resistance. Other polyester polyols are based on reclaimed raw materials. They are manufactured by transesterification (glycolysis) of recycled poly(ethyleneterephthalate) (PET) or dimethylterephthalate (DMT) distillation bottoms with glycols such as diethylene glycol. These low molecular weight, aromatic polyester polyols are used in the manufacture of rigid foam, and bring low cost and excellent flammability characteristics to polyisocyanurate (PIR) boardstock and polyurethane spray foam insulation.

Specialty polyols include polycarbonate polyols, polycaprolactone polyols, polybutadiene polyols, and polysulfide polyols. The materials are used in elastomer, sealant, and adhesive applications that require superior weatherability, and resistance to chemical and environmental attack. Natural oil polyols derived from castor oil and other vegetable oils are used to make elastomers, flexible bunstock, and flexible molded foam.

Chain Extenders and Cross Linkers

Chain extenders (f = 2) and cross linkers (f = 3 or greater) are low molecular weight hydroxyl and amine terminated compounds that play an important role in the polymer morphology of polyurethane fibers, elastomers, adhesives, and certain integral skin and microcellular foams. The elastomeric properties of these materials are derived from the phase separation of the hard and soft copolymer segments of the polymer, such that the urethane hard segment domains serve as cross-links between the amorphous polyether (or polyester) soft segment domains. This phase separation occurs because the mainly non-polar, low melting soft segments are incompatible with the polar, high melting hard segments. The soft segments, which are formed from high molecular weight polyols, are mobile and are normally present in coiled formation, while the hard segments, which are formed from the isocyanate and chain extenders, are stiff and immobile. Because the hard segments are covalently coupled to the soft segments, they inhibit plastic flow of the polymer chains, thus creating elastomeric resiliency. Upon mechanical deformation, a portion of the soft segments are stressed by uncoiling, and the hard segments become aligned in the stress direction. This reorientation of the hard segments and consequent powerful hydrogen bonding contributes to high tensile strength, elongation, and tear resistance values. The choice of chain extender also determines flexural, heat, and chemical resistance properties. The most important chain extenders are ethylene glycol, 1,4-butanediol (1,4-BDO or BDO), 1,6-hexanediol, cyclohexane dimethanol and hydroquinone bis(2-hydroxyethyl) ether (HQEE). All of these glycols form polyurethanes that phase separate well and form well defined hard segment domains, and are melt processable. They are all suitable for thermoplastic polyurethanes with the exception of ethylene glycol, since the its derived bis-phenyl urethane undergoes unfavorable degradation at high hard segment levels. Diethanolamine and triethanolamine are used in flex molded foams to build firmness and add catalytic activity. Diethyltoluenediamine is used extensively in RIM, and in polyurethane and polyurea elastomer formulations.

Table 5.3: Table of chain extenders and cross linkers.

	MW	s.g.	f.p. °C	b.p. °C
hydroxyl compounds—difunctional molecules				
ethylene glycol	62.1	1.110	-13.4	197.4
diethylene glycol	106.1	1.111	-8.7	245.5
triethylene glycol	150.2	1.120	-7.2	287.8
tetraethylene glycol	194.2	1.123	-9.4	325.6
propylene glycol	76.1	1.032	supercools	187.4
dipropylene glycol	134.2	1.022	supercools	232.2
tripropylene glycol	192.3	1.110	supercools	265.1
1,3-propanediol	76.1	1.060	-28	210
1,3-butanediol	92.1	1.005	—	207.5
1,4-butanediol	92.1	1.017	20.1	235
neopentyl glycol	104.2	—	130	206
1,6-hexanediol	118.2	1.017	43	250
1,4-cyclohexanedimethanol	—	—	—	—
HQEE	—	—	—	—
ethanolamine	61.1	1.018	10.3	170
diethanolamine	105.1	1.097	28	271
methyldiethanolamine	119.1	1.043	-21	242
phenyldiethanolamine	181.2	—	58	228
hydroxyl compounds—trifunctional molecules				
glycerol	92.1	1.261	18.0	290
trimethylolpropane	—	—	—	—
1,2,6-hexanetriol	—	—	—	—
triethanolamine	149.2	1.124	21	—
hydroxyl compounds—tetrafunctional molecules				
pentaerythritol	136.2	—	260.5	—
N, N, N', N'-tetrakis (2-hydroxypropyl) ethylenediamine	—	—	—	—
amine compounds—difunctional molecules				
diethyltoluenediamine	178.3	1.022	—	308
dimethylthiotoluenediamine	214.0	1.208	—	—

Catalysts

Polyurethane catalysts can be classified into two broad categories, amine compounds and organometallic complexes. They can be further classified as to their specificity, balance, and relative power or efficiency. Traditional amine catalysts have been tertiary amines such as triethylenediamine (TEDA, also known as 1,4-diazabicyclo[2.2.2]octane or DABCO), dimethylcyclohexylamine (DMCHA), and dimethylethanolamine (DMEA). Tertiary amine catalysts are selected based on whether they drive the urethane (polyol + isocyanate, or gel) reaction, the urea (water + isocyanate, or blow) reaction, or the isocyanate trimerization reaction. Since most tertiary amine catalysts will drive all three reactions to some extent, they are also selected based on how much they favor one reaction over another. For example, tetramethylbutanediamine (TMBDA) preferentially drives the gel reaction over the blow reaction. On the other hand, both pentamethyldipropylenetriamine and N-(3-dimethylaminopropyl)-N, N-diisopropanolamine balance the blow and gel reactions, although the former is more potent than the later on a weight basis. 1,3,5-(tris(3-dimethylamino)propyl)-hexahydro-s-triazine is a trimerization catalyst that also strongly drives the blow reaction. Molecular structure gives some clue to the strength and selectivity of the catalyst. Blow catalysts generally have an ether linkage two carbons away from a tertiary nitrogen. Examples include bis-(2-dimethylaminoethyl)ether and N-ethylmorpholine. Strong gel catalysts contain alkyl-substituted nitrogens, such as triethylamine (TEA), 1,8-diazabicyclo[5.4.0]undecene-7 (DBU), and pentamethyldiethylenetriamine (PMDETA). Weaker gel catalysts contain ring-substituted nitrogens, such as benzyldimethylamine (BDMA). Trimerization catalysts contain the triazine structure, or are quaternary ammonium salts. Two trends have emerged since the late 1980s. The requirement to fill large, complex tooling with increasing production rates has led to the use of blocked catalysts to delay front end reactivity while maintaining back end cure. In the United States, acid- and quaternary ammonium salt-blocked TEDA and bis-(2-dimethylaminoethyl)ether are common blocked catalysts used in molded flexible foam and microcellular integral skin foam applications. Increasing aesthetic and environmental awareness has led to the use of non-fugitive catalysts for vehicle interior and furnishing applications in order to reduce odor, fogging, and the staining of vinyl coverings. Catalysts that contain a hydroxyl group or an active amino hydrogen, such as N, N, N'-trimethyl-N'-hydroxyethyl-bis(aminoethyl)ether and N'-(3-(dimethylamino)propyl)-N, N-dimethyl-1,3-propanediamine that react into the polymer matrix can replace traditional catalysts in these applications.

Organometallic compounds based on mercury, lead, tin (dibutyltin dilaurate), bismuth (bismuth octanoate), and zinc are used as polyurethane catalysts. Mercury carboxylates, such as phenylmercuric neodeconate, are particularly effective catalysts for polyurethane elastomer, coating and sealant applications, since they are very highly selective towards the polyol + isocyanate reaction. Mercury catalysts can be used at low levels to give systems a long pot life while still giving excellent back-end cure. Lead catalysts are used in highly reactive rigid spray foam insulation applications, since they maintain their potency in low-temperature and high-humidity conditions. Due to their toxicity and the necessity to dispose of mercury and lead catalysts and catalyzed material as hazardous waste in the United States, formulators have been searching for suitable replacements. Since the 1990s, bismuth and zinc carboxylates have been used as alternatives but have short comings of their own. In elastomer applications, long pot life systems do not build green strength as fast as mercury catalyzed systems. In spray foam applications, bismuth and zinc do not drive the front end fast enough in cold weather conditions and must be otherwise augmented to replace lead. Alkyl tin carboxylates, oxides and mercaptides oxides are used in all types of polyurethane applications. For example, dibutyltin dilaurate is a standard catalyst for polyurethane adhesives and sealants, dioctyltin mercaptide is used in microcellular elastomer applications, and dibutyltin oxide is used in polyurethane paint and coating applications. Tin mercaptides are used in formulations that contain water, as tin carboxylates are susceptible to degradation from hydrolysis.

Surfactants

Surfactants are used to modify the characteristics of both foam and non-foam polyurethane polymers. They take the form of polydimethylsiloxane-polyoxyalkylene block copolymers, silicone oils, nonylphenol ethoxylates, and other organic compounds. In foams, they are used to emulsify the liquid components, regulate cell size, and stabilize the cell structure to prevent collapse and sub-surface voids. In non-foam applications they are used as air release and anti-foaming agents, as wetting agents, and are used to eliminate surface defects such as pin holes, orange peel, and sink marks.

Production

The main polyurethane producing reaction is between a diisocyanate (aromatic and aliphatic types are available) and a polyol, typically a polypropylene glycol or polyester polyol, in the presence of catalysts and materials for controlling the cell structure, (surfactants) in the case of foams. Polyurethane can be made in a variety of densities and hardnesses by varying the type of monomer(s) used and adding other substances to modify their characteristics, notably density, or enhance their performance. Other additives can be used to improve the fire performance, stability in difficult chemical environments and other properties of the polyurethane products.

Though the properties of the polyurethane are determined mainly by the choice of polyol, the diisocyanate exerts some influence, and must be suited to the application. The cure rate is influenced by the functional group reactivity and the number of functional isocyanate groups. The mechanical properties are influenced by the functionality and the molecular shape. The choice of diisocyanate also affects the stability of the polyurethane upon exposure to light. Polyurethanes made with aromatic diisocyanates yellow with exposure to light, whereas those made with aliphatic diisocyanates are stable.

Softer, elastic, and more flexible polyurethanes result when linear difunctional polyethylene glycol segments, commonly called polyether polyols, are used to create the urethane links. This strategy is used to make spandex elastomeric fibers and soft rubber parts, as well as foam rubber. More rigid products result if polyfunctional polyols are used, as these create a three-dimensional cross-linked structure which, again, can be in the form of a low-density foam. An even more rigid foam can be made with the use of specialty trimerization catalysts which create cyclic structures within the foam matrix, giving a harder, more thermally stable structure, designated as polyisocyanurate foams. Such properties are desired in rigid foam products used in the construction sector.

Careful control of viscoelastic properties—by modifying the catalysts and polyols used—can lead to memory foam, which is much softer at skin temperature than at room temperature.

There are then two main foam variants: one in which most of the foam bubbles (cells) remain closed, and the gas(es) remains trapped, the other being systems which have mostly open cells, resulting after a critical stage in the foam-making process (if cells did not form, or became open too soon, foam would not be created). This is a vitally important process: if the flexible foams have closed cells, their softness is severely compromised, they become pneumatic in feel, rather than soft; so, generally speaking, flexible foams are required to be open-celled.

The opposite is the case with most rigid foams. Here, retention of the cell gas is desired since this gas (especially the fluorocarbons referred to above) gives the foams their key characteristic: high thermal insulation performance.

A third foam variant, called microcellular foam, yields the tough elastomeric materials typically experienced in the coverings of car steering wheels and other interior automotive components.

Health and Safety

Fully reacted polyurethane polymer, CAS # 9009-54-5 (CAS registry number), is chemically inert. In the United States, no exposure limits have been established by OSHA (Occupational Safety and Health Administration) or ACGIH (American Conference of Governmental Industrial Hygienists). It is not regulated by OSHA for carcinogenicity. Polyurethane polymer is a combustible solid and will ignite if exposed to an open flame for a sufficient period of time. Decomposition products include carbon monoxide, oxides of nitrogen, and hydrogen cyanide. Firefighters should wear self-contained breathing apparatus in enclosed areas. When heated above about 200°C the PU polymer will thermally degrade and emit not only the isocyanates it was made from but also a number of mono isocyanates like methyl isocyanate (MIC) and isocyanic acid (ICA), depending on the type of PU being heated. Heating of any PU material (e. g. soft foam, paint dust after sanding, textiles, PU painted flooring etc.) should be avoided at any cost. Polyurethane polymer dust can cause mechanical irritation to the eyes and lungs. Proper hygiene controls and personal protective equipment (PPE), such as gloves, dust masks, respirators, mechanical ventilation, and protective clothing and eye wear should be used. Clothes should be changed and hands, hair and face should be cleaned before smoking.

Liquid resin blends and isocyanates may contain hazardous or regulated components. They should be handled in accordance with manufacturer recommendations found on product labels, and in MSDS (Material Safety Data Sheet) and product technical literature. Isocyanates are known skin and respiratory sensitizers, and proper engineering controls should be in place to prevent exposure to isocyanate liquid and vapor.

In the United States, additional health and safety information can be found through organizations such as the Polyurethane Manufacturers Association (PMA) and the Center for the Polyurethanes Industry (CPI), as well as from polyurethane system and raw material manufacturers. In Europe, health and safety information is available from ISOPA, the European Diisocyanate and Polyol Producers Association. Regulatory information can be found in the Code of Federal Regulations Title 21 (Food and Drugs) and Title 40 (Protection of the Environment).

Uses

Polyurethane products have many uses. Over three quarters of the global consumption of polyurethane products is in the form of foams, with flexible and rigid types being roughly equal in market size. In both cases, the foam is usually behind other materials: flexible foams are behind upholstery fabrics in commercial and domestic furniture; rigid foams are inside the metal and plastic walls of most refrigerators and freezers, or behind paper, metals and other surface materials in the case of thermal insulation panels in the construction sector. Its use in garments is growing: for example, in lining the cups of brassieres. Polyurethane is also used for moldings which include door frames, columns, balusters, window headers, pediments, medallions and rosettes.

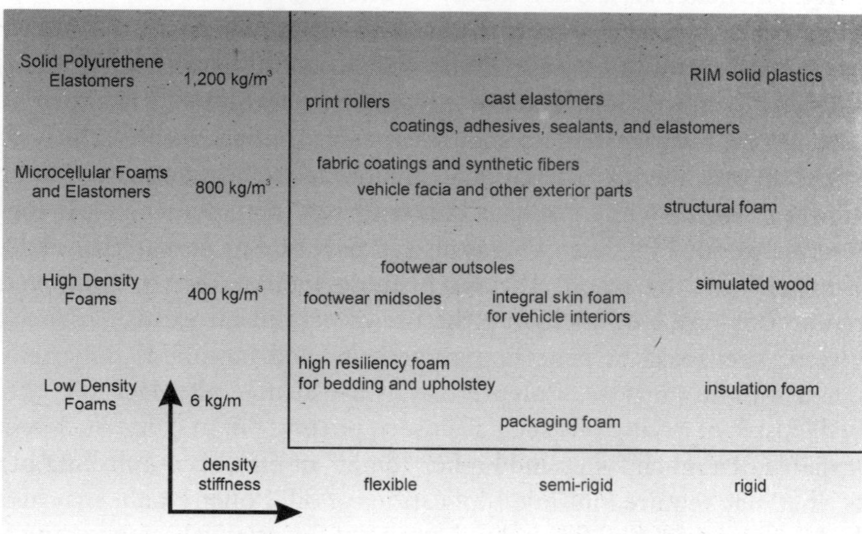

Fig. 5.9: Characteristics of polyurethane materials.

The precursors of expanding polyurethane foam are available in many forms, for use in insulation, sound deadening, flotation, industrial coatings, packing material, and even cast-in-place upholstery padding. Since they adhere to most surfaces and automatically fill voids, they have become quite popular in these applications. The following table shows how polyurethanes are used (US data from 2004):

Table 5.4: Table shows how polyurethanes are used.

Application	*Amount of polyurethane used (millions of pounds)*	*Percentage of total*
Building & Construction	1,459	26.8%
Transportation	1,298	23.8%
Furniture & Bedding	1,127	20.7%
Appliances	278	5.1%
Packaging	251	4.6%
Textiles, Fibers & Apparel	181	3.3%
Machinery & Foundry	178	3.3%
Electronics	75	1.4%
Footwear	39	0.7%
Other uses	558	10.2%
Total	5,444	100.0%

Varnish

Polyurethane materials are commonly formulated as paints and varnishes for finishing coats to protect or seal wood. This use results in a hard, abrasion-resistant, and durable coating that is popular for hardwood floors, but considered by some to be difficult or unsuitable for finishing furniture or other detailed pieces. Relative to oil or shellac varnishes, polyurethane varnish forms a harder film which tends to de-laminate if subjected to heat or shock, fracturing the film and leaving white patches. This tendency increases when it is applied over softer woods like pine. This is also in part due to polyurethane's lesser penetration into the wood. Various priming techniques are employed to overcome this problem, including the use of certain oil varnishes, specified "dewaxed" shellac, clear penetrating epoxy, or "oil-modified" polyurethane designed for the purpose. Polyurethane varnish may also lack the "hand-rubbed" lustre of drying oils such as linseed or tung oil; in contrast, however, it is capable of a much faster and higher "build" of film, accomplishing in two coats what may require multiple applications of oil. Polyurethane may also be applied over a straight oil finish, but because of the relatively slow curing time of oils, the presence of volatile byproducts of curing, and the need for extended exposure of the oil to oxygen, care must be taken that the oils are sufficiently cured to accept the polyurethane.

Unlike drying oils and alkyds which cure, after evaporation of the solvent, upon reaction with oxygen from the air, polyurethane coatings cure after evaporation of the solvent by a variety of reactions of chemicals within the original mix, or by reaction with moisture from the air. Certain products are "hybrids" and combine different aspects of their parent components. "Oil-modified" polyurethanes, whether water-borne or solvent-borne, are currently the most widely used wood floor finishes.

Exterior use of polyurethane varnish may be problematic due to its susceptibility to deterioration through ultra-violet light exposure. It must be noted, however, that all clear or transluscent varnishes, and indeed all film-polymer coatings (i.e. paint, stain, epoxy, synthetic plastic, etc.) are susceptible to this damage in varying degrees. Pigments in paints and stains protect against UV damage, while UV-absorbers are added to polyurethane and other varnishes (in particular "spar" varnish) to work against UV damage. Polyurethanes are typically the most resistant to water exposure, high humidity, temperature extremes, and fungus or mildew, which also adversely affect varnish and paint performance.

Wheels

Polyurethane is also used in making solid tires. Modern roller blading and skateboarding became economical only with the introduction of tough, abrasion-resistant polyurethane parts, helping to usher in the permanent popularity of what had once been an obscure 60s craze. The durability of Polyurethane wheel allowed the range of tricks and stunts performed on skateboards to expand considerably. Other constructions have been developed for pneumatic tires, and microcellular foam variants are widely used in tires on wheelchairs, bicycles and other such uses. These latter foam types are also widely encountered in car steering wheels and other interior and exterior automotive parts, including bumpers and fenders.

Furniture

Polyurethane is also used in furniture manufacture for casting soft edges around table tops and panel that are stylish, very durable and prevent injury. These are used in school tables, hospital and bank furniture as well as shop counters and displays. It is used on the bottom of some mouse pads.

Much of the foam used in chairs (including beanbag chairs), couches, mattresses is polyurethane foam. This type of foam is made by mixing polyols, diisocyanates, catalysts, blowing agents and other additives and allowing the resulting foam to rise freely. This can be done in a batch process where relatively small blocks of foam are made in an open-topped mold, or continuously where the components are poured onto an inclined moving belt. The foam is then cut to the desired shape and size for use in making furniture. Safety concerns about the flammability of polyurethane foam, particularly in upholstered furniture, sometimes requires the addition of flame retardants to this foam.

Polyurethane is used for flooring in houses, offices and public areas.

Automobile Seats

Flexible and semi-flexible polyurethane foams are used extensively for interior components of automobiles, in seats, headrests, armrests, roof liners, dashboards and instrument panels.

Polyurethane foam in the lower half of the mold in which it was made. When assembled into a car seat, this foam makes up the seat back. The forward-facing part of the seat back is the surface of the foam which face-down in the mold. The two holes in the foam at the top of the picture are for the headrest posts. Polyurethanes are used to make automobile seats in a remarkable manner. The seat manufacturer has a mold for each seat model. The mold is a closeable "clamshell" sort of structure that will allow quick casting of the seat cushion, so-called molded flexible foam, which is then upholstered after removal from the mold. It is possible to combine these two steps, so-called in-situ, foam-in-fabric or direct moulding. A complete, fully-assembled seat cover is placed in the mold and held in place by vacuum drawn through small holes in the mold. Sometimes a thin pliable plastic film backing on the fabric is used to help the vacuum work more effectively. The metal seat frame is placed into the mold and the mold closed. At this point the mold contains what could be visualized as a "hollow seat", a seat fabric held in the correct position by the vacuum and containing a space with the metal frame in place.

Polyurethane chemicals are injected by a mixing head into the mold cavity. Then the mold is held at a preset reaction temperature until the chemical mixture has foamed, filled the mold, and formed a stable soft foam. The time required is two to three minutes, depending on the size of the seat and the precise formulation and operating conditions. Then the mold is usually opened slightly for a minute or two for an additional cure time, before the fully upholstered seat is removed.

Houses, Sculptures, and Decorations

The walls and ceiling (not just the insulation) of the futuristic Xanadu House were built out of polyurethane foam. Domed ceilings and other odd shapes are easier to make with foam than with wood. Foam was used to build oddly-shaped buildings, statues, and decorations in the Seuss Landing section of the Islands of Adventure theme park. Speciality rigid foam manufactures sell foam that replace wood in carved sign and 3D topography industries.

Construction Sealants and Firestopping

Head-of-Wall Firestop Joint: the presence of penetrants demonstrates the need to have both operational and fire-tested compatibility between the joint sealant and mechanical/electrical through-penetrations. In other words, it is easier to insist on the use of joint firestops that can also be used for penetration seals, as otherwise penetrants may be run by mechanical and electrical subtrades that unintentionally void the fire-resistance rating of the wall, which jeopardises the entire fire safety plan in place for a building.

Head-of-Wall Firestop Joint penetrated by both electrical and mechanical services, demonstrating the need for operational and fire-tested compatibility between the joint firestop system and penetrants, be they electrical, mechanical or structural.

Polyurethane sealants are available in 1, 2 and even 3 part systems, either in cartridge, bucket or drum format. Polyurethane sealants are also sold for firestopping applications. Obviously, the sealant by itself provides no serious hindrance to fire, as its hydrocarbon bonds readily support combustion.

However, when backed by inorganic insulation, such as rockwool or ceramic fibres, it can act as an effective seal to thwart smoke and hose-stream passage, particularly in inorganic joints. It is, however, advisable to avoid direct contact with metallic penetrants and through-penetrating cables, as the heat carried by the penetrants may jeopardise the sealant. This, however, requires a lot of vigilance. In concrete to concrete, or concrete to masonry joints, however, that are free of mechanical or electrical penetrants, it works well and dependably.

Watercraft

Some surfboards are made with a solid polyurethane core. A rigid foam blank is molded, shaped to specification, then covered with fiberglass cloth and polyester resin.

The hull of the *Boston Whaler* motorboat is polyurethane foam sandwiched in a fiberglass skin. The foam provides strength, buoyancy, and sound deadening.

Tennis Grips

Polyurethane has been used to make several Tennis Overgrips such as Yonex Supergrap, Wilson Pro Overgrip and many other grips. These grips are highly stretchable to ensure the grip wraps neatly around the racquet's handle.

Electronic Components

Often electronic components are protected from environmental influence and mechanical shock by enclosing them in polyurethane. Typically polyurethanes are selected for the excellent abrasion resistances, good electrical properties, excellent adhesion, impact strength, and low temperature flexibility. The disadvantage of polyurethanes is the limited upper service temperature (typically 250 °F (121 °C)). In production the electronic manufacture would purchase a two part urethane (resin and catalyst) that would be mixed and poured onto the circuit assembly (see Resin casting). In most cases, the final circuit board assembly would be unrepairable after the urethane has cured. Because of its physical properties and low cost, polyurethane encapsulation (potting) is a popular option in the automotive manufacturing sector for automotive circuits and sensors.

Adhesives

Polyurethane is used as an adhesive, especially as a woodworking glue. Its main advantage over more traditional wood glues is its water resistance. It was introduced in the general North American market in the 1990s as *Gorilla Glue* and *Excel*, but has been used much longer in Europe.

On the way to a new and better glue for bookbinders, a new adhesive system was introduced for the first time in 1985. The base for this system is polyether or polyester, whereas polyurethane (PUR) is used as prepolymer. Its special feature is the coagulation at room temperature and the reacting to moisture.

1st Generation (1988 at the drupa)

- Low starting solidity,
- High viscosity,
- Cure time of more than 3 days.

2nd Generation (1996 at the drupa)

- Low starting solidity,
- High viscosity,
- Cure time of less than 3 days.

3rd Generation (2000 at the drupa)

- Good starting solidity,
- Low viscosity,
- Cure time between 6 and 16 hours.

4th Generation (present)

- Good starting solidity,
- Very low viscosity,
- Cure reached within a few seconds due to dual-core systems.

Advantages of polyurethane glue in the bookbinding industry: PUR is real wonder compared to hotmelt and cold glue. Because of the missing moisture in the glue, papers with wrong grain direction can be processed without problems. Even printed and supercalandered paper can be bound without problems. It is the most economical glue with an application thickness of theoretical 0.01 mm. But in reality it is not possible to apply less than 0.03 mm. The PUR glue is very weather-proof and stable at temperatures from -40 °C to 100 °C.

Watch Band Wrapping

Polyurethane is used as a black wrapping for timepiece bracelets over the main material which is generally stainless steel. It is used for comfort, style, and durability.

Abrasion Resistance

Thermoset polyurethanes are also used as a protective coating against abrasion. Cast polyurethane over materials such as steel will absorb particle impact more effieciently. Polyurethanes have been proven to last in excess of 25 years in abrasisive environments where non-coated steel would erode in less than 8 years. Polyurethanes are used in industries such as:

- Mining & Mineral Processisng
- Aggregate
- Transportation
- Concrete
- Paper Processing
- Power

Filling of Spaces and Cavities

Two binary chemicals, one of which is a polyurethane (either T6 or 16), when mixed and aerated, expand into a hard, space-filling aerosolid. Underground caves and voids can be filled quickly this way.

Testing

Effects of Visible Light

Polyurethanes, especially those made using aromatic isocyanates, contain chromophores which interact with light. This is of particular interest in the area of polyurethane coatings, where light stability is a critical factor and is the main reason that aliphatic isocyanates are used in making polyurethane coatings. When PU foam, which is made using aromatic isocyanates, is exposed to visible light it discolors, turning from off-white to yellow to reddish brown. It has been generally accepted that apart from yellowing, visible light has little effect on foam properties. This is especially the case if the yellowing happens on the outer portions of a large foam, as the deterioration of properties in the outer portion has little effect on the overall bulk properties of the foam itself.

It has been reported that exposure to visible light can affect the variability of some physical property test results. Increasing exposure time and/or light intensity during the storage of foam samples under ambient laboratory conditions increased the amount of permanent set induced in some compression set tests (the samples did not fully return to their original size and/or shape). Variability resulted from uncontrolled light exposure of cut samples prior to being compressed. Other foam properties were not substantively affected. It was recommended that specimen preparation and testing be done rapidly to minimize variation in results or if specimens are prepared but not tested for a week or more, that the samples should be protected from light exposure.

Higher-energy UV radiation promotes chemical reactions in foam, some of which are detrimental to the foam structure.

5.19 Polyvinyl Chloride

Polyvinyl chloride, (IUPAC Polychloroethene) commonly abbreviated PVC, is a widely used thermoplastic polymer. In terms of revenue generated, it is one of the most valuable products of the chemical industry. Around the world, over 50 per cent of PVC manufactured is used in construction. As a building material, PVC is cheap, durable, and easy to assemble. In recent years, PVC has been replacing traditional building materials such as wood, concrete and clay in many areas.

Polyvinyl chloride	
Density	1380 kg/m^3
Young's modulus (E)	2900-3300 MPa
Tensile strength(σ_t)	50-80 MPa
Elongation at break	20-40%
Notch test	2-5 kJ/m^2
Glass temperature	87 °C
Melting point	80 °C
Vicat B[1]	85 °C
Heat transfer coefficient (λ)	0.16 W/(m·K)
Effective heat of combustion	17.95 MJ/kg
Linear expansion coefficient (α)	8 10^{-5}/K
Specific heat (c)	0.9 kJ/(kg·K)
Water absorption (ASTM)	0.04-0.4
Price	0.5-1.25 /kg

[1] Deformation temperature at 10 kN needle load.

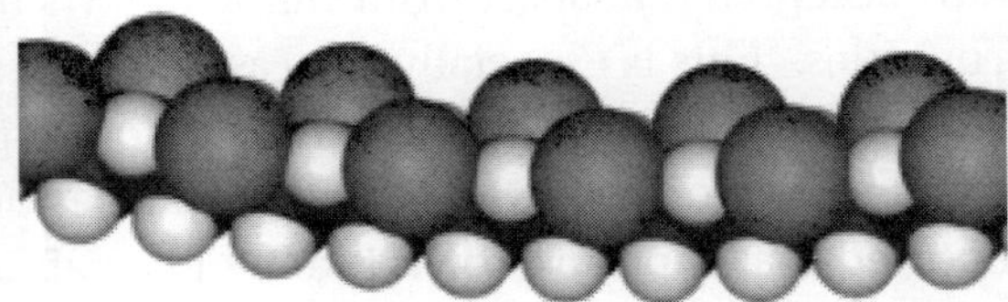

Fig. 5.10: Polyvinyl chloride.

Polyvinyl chloride is used in a variety of applications. As a hard plastic, it is used as vinyl siding, magnetic stripe cards, window profiles, gramophone records (which is the source of the term *vinyl records*), pipe, plumbing and conduit fixtures. The material is often used in Plastic Pressure Pipe Systems for pipelines in the water and sewer industries because of its inexpensive nature and flexibility. PVC pipe plumbing is typically white, as opposed to ABS, which is commonly available in grey and black, as well as white.

It can be made softer and more flexible by the addition of plasticizers, the most widely-used being phthalates. In this form, it is used in clothing and upholstery, and to make flexible hoses and tubing, flooring, to roofing membranes, and electrical cable insulation. It is also commonly used in figurines.

Preparation

Polyvinyl chloride is produced by polymerization of the monomer vinyl chloride, as shown. Since about 57 per cent of its mass is chlorine, creating a given mass of PVC requires less petroleum than many other polymers.

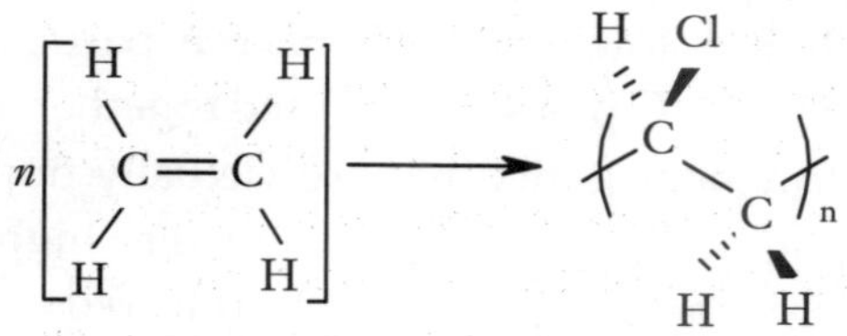

$$n\,[\,H_2C{=}CH_2\text{-type monomer: } CH_2{=}CHCl\,] \longrightarrow -(CH_2{-}CHCl)_n-$$

History

Polyvinyl chloride was accidentally discovered on at least two different occasions in the 19th century, first in 1835 by Henri Victor Regnault and in 1872 by Eugen Baumann. On both occasions, the polymer appeared as a white solid inside flasks of vinyl chloride that had been left exposed to sunlight. In the early 20th century, the Russian chemist Ivan Ostromislensky and Fritz Klatte of the German chemical company Griesheim-Elektron both attempted to use PVC (polyvinyl chloride) in commercial products, but difficulties in processing the rigid, sometimes brittle polymer blocked their efforts. In 1926, Waldo Semon and the B.F. Goodrich Company developed a method to plasticize PVC by blending it with various additives. The result was a more flexible and more easily-processed material that soon achieved widespread commercial use.

Applications

Electric Wires

PVC is commonly used as the insulation on electric wires; the plastic used for this purpose needs to be plasticized. In a fire, PVC-coated wires can form HCl fumes; the chlorine serves to scavenge free radicals and is the source of the material's fire retardance. While HCl fumes can also pose a health hazard in their own right, HCl dissolves in moisture and breaks down onto surfaces, particularly in areas where the air is cool enough to breathe, and is not available for inhalation. Frequently in applications where smoke is a major hazard (notably in tunnels) PVC-free LSOH (low-smoke, zero-halogen) cable insulation is preferred. PVC insulation may be eaten or degraded by mice, termites, antechinus and cockatoos. The applicable building code should be consulted to determine the type of electrical wires approved for the intended use.

Pipes

PVC pipes in use with intumescent firestops at Nortown Casitas, North York, Ontario. Polyvinyl chloride is also widely used for producing pipes. In the water distribution market it accounts for 66 per cent of the market in the US, and in sanitary sewer pipe applications, it accounts for 75 per cent. Its light weight, high strength, and low reactivity make it particularly well-suited to this purpose. In addition, PVC pipes can be fused together using various solvent cements, creating permanent joints that are virtually impervious to leakage. Despite PVC's many advantages, in cases where very high strength or ease of disassembly is necessary, metal pipes are still preferred.

In February 2007, the California Building Standards Code was updated to approve the use of chlorinated polyvinyl chloride (CPVC) pipe for use in residential water supply piping systems. CPVC has been a nationally-accepted material in the US since 1982; however, California has only permitted its use on a limited basis since 2001. The Department of Housing and Community Development prepared and certified an Environmental Impact Report resulting in a recommendation that the Commission adopt and approve the use of CPVC. The Commission's vote was unanimous and CPVC has been placed in the 2007 California Plumbing Code.

Portable Electronic Accessories

PVC is finding increased use as a composite for the production of accessories or housings for portable electronics. Through a fusing process, it can adopt cleaning properties possessed by materials such as wool or cotton which can absorb dust particles and bacteria. Its inherent ability to absorb particles from the LCD screen and its form fitting characteristics make it effective.

Signs

In flat sheet form, polyvinyl chloride is formed in a variety of thicknesses and colors. As flat sheets, PVC is often expanded to create voids in the interior of the material, providing additional thickness without additional weight and cost. Sheets are cut using saw and rotary cutting equipment (*see* CNC). Plasticized

PVC is also used to produce thin, colored, or clear, adhesive-backed films referred to simply as vinyl. These films are typically cut on a computer-controlled plotter or printed in a wide-format printer. These sheets and films are used to produce a wide variety of commercial signage products and markings on vehicles.

Unplasticized Polyvinyl Chloride (uPVC)

PVC or Rigid PVC is often used in the building industry as a low-maintenance material, particularly in the UK, and in the USA where it is known as vinyl, or vinyl siding. The material comes in a range of colors and finishes, including a photo-effect wood finish, and is used as a substitute for painted wood, mostly for window frames and sills when installing double glazing in new buildings, or to replace older single glazed windows. It has many other uses including fascia, and siding or weatherboarding. The same material has almost entirely replaced the use of cast iron for plumbing and drainage, being used for waste pipes, drainpipes, gutters and downpipes.

Due to environmental concerns use of PVC is discouraged by some local authorities in countries such as Germany and The Netherlands. This concerns both flexible PVC and rigid uPVC as not only the plasticizers in PVC are seen as a problem but also the emissions from manufacturing and disposal. The use of modern impact modifiers offer great stability. The issues of migration and brittleness of the PVC compound are overcome.

Health and Safety

Phthalate Plasticizers

Many vinyl products contain additional chemicals to change the chemical consistency of the product. Some of these additional chemicals called additives can leach out of vinyl products. Plasticizers that must be added to make PVC flexible have been an additive of particular concern.

Because soft PVC toys have been made for babies for years, there are concerns that these additives leach out of soft toys into the mouths of the children chewing on them. Additionally, adult sex toys have been demonstrated to leach significant additives. In January 2006, the European Union placed a ban on six types of phthalate softeners, including DEHP (diethylhexyl phthalate), used in toys. In the USA most companies have voluntarily stopped manufacturing PVC toys with DEHP and in 2003 the US Consumer Product Safety Commission (CPSC) denied a petition for a ban on PVC toys made with an alternative plasticizer, DINP (diisononyl phthalate). In April 2006, the European Chemicals Bureau of the European Commission published an assessment of DINP which found risk "unlikely" for children and newborns.

Vinyl IV bags used in neo-natal intensive care units have also been shown to leach DEHP. In a draft guidance paper published in September 2002, the US FDA recognizes that many medical devices with PVC containing DEHP are not used in ways that result in significant human exposure to the chemical. However, FDA is suggesting that manufacturers consider eliminating the use of DEHP in certain devices that can result in high aggregate exposures for sensitive patient populations such as neonates.

Other vinyl products, including car interiors, shower curtains, flooring, initially release chemical gases into the air. Some studies indicate that this outgassing of additives may contribute to health complications, and have resulted in a call for banning the use of DEHP on shower curtains, among other uses.

In 2004, a joint Swedish-Danish research team found a statistical association between allergies in children and indoor air levels of DEHP and BBzP (butyl benzyl phthalate), which is used in vinyl flooring. In December 2006, the European Chemicals Bureau of the European Commission released a final draft risk assessment of BBzP which found "no concern" for consumer exposure including exposure to children.

In November 2005, one of the largest hospital networks in the U.S., Catholic Healthcare West, signed a contract with B.Braun for vinyl-free intravenous bags and tubing. According to the Center for Health, Environment & Justice in Falls Church, VA, which helps to coordinate a "precautionary" "PVC Campaign", several major corporations including Microsoft, Wal-Mart, and Kaiser Permanente announced efforts to eliminate PVC from products and packaging in 2005. Even Target is reducing its sale of items with PVC. (http://besafenet.com/pvc/newsreleases/target_to_reduce_use.htm)

The FDA Paper titled "Safety Assessment of Di(2-ethylhexyl)phthalate (DEHP) Released from PVC Medical Devices" states that [3.2.1.3] Critically ill or injured patients may be at increased risk of developing adverse health effects from DEHP, not only by virtue of increased exposure, relative to the general population, but also because of the physiological and pharmacodynamic changes that occur in these patients, compared to healthy individuals.

In 2008, The European Union's Scientific Committee on Emerging and Newly Identified Health Risks (SCENIHR) reviewed the safety of DEHP in medical devices. The SCENIHR report states that certain medical procedures used in high risk patients result in a significant exposure to DEHP and concludes there is still a reason for having some concerns about the exposure of prematurely born male babies to medical devices containing DEHP. The Committee said there are some alternative plasticisers available for which there is sufficient toxicological data to indicate a lower hazard compared to DEHP but added that the functionality of these plasticisers should be assessed before they can be used as an alternative for DEHP in PVC medical devices.

Vinyl Chloride Monomer

In the early 1970s, Dr. John Creech and Dr. Maurice Johnson were the first to clearly link and recognize the carcinogenicity of vinyl chloride monomer to humans when workers in the polyvinyl chloride polymerization section of a B.F. Goodrich plant near Louisville, Kentucky, were diagnosed with liver angiosarcoma also known as hemangiosarcoma, a rare disease. Since that time, studies of PVC workers in Australia, Italy, Germany, and the UK have all associated certain types of occupational cancers with exposure to vinyl chloride. The link between angiosarcoma of the liver and long-term exposure to vinyl chloride is the only one that has been confirmed by the International Agency for Research on Cancer. All the cases of angiosarcoma developed from exposure to vinyl chloride monomer, were in workers who were exposed to very high VCM levels, routinely, for many years. These workers cleaned accretions in reactors, a practice that has now been replaced by automated high pressure water jets.

A 1997 U.S. Centers for Disease Control and Prevention (CDC) report concluded that the development and acceptance by the PVC industry of a closed loop polymerization process in the late 1970s "almost completely eliminated worker exposures" and that "new cases of hepatic angiosarcoma in vinyl chloride polymerization workers have been virtually eliminated."

According to the EPA, "vinyl chloride emissions from polyvinyl chloride (PVC), ethylene dichloride (EDC), and vinyl chloride monomer (VCM) plants cause or contribute to air pollution that may reasonably be anticipated to result in an increase in mortality or an increase in serious irreversible, or incapacitating reversible illness. Vinyl chloride is a known human carcinogen that causes a rare cancer of the liver." EPA's 2001 updated Toxicological Profile and Summary Health Assessment for VCM in its Integrated Risk Information System (IRIS) database lowers EPA's previous risk factor estimate by a factor of 20 and concludes that "because of the consistent evidence for liver cancer in all the studies ... and the weaker association for other sites, it is concluded that the liver is the most sensitive site, and protection against liver cancer will protect against possible cancer induction in other tissues."

A 1998 front-page series in the Houston Chronicle claimed the vinyl industry has manipulated vinyl chloride studies to avoid liability for worker exposure and to hide extensive and severe chemical spills into local communities. Retesting of community residents in 2001 by the U.S. Agency for Toxic Substances and Disease Registry (ATSDR) found dioxin levels similar to those in a comparison community in Louisiana and to the U.S. population. Cancer rates in the community were similar to Louisiana and US averages.

Dioxins

The environmentalist group Greenpeace has advocated the global phase-out of PVC because they claim dioxin is produced as a byproduct of vinyl chloride manufacture and from incineration of waste PVC in domestic garbage. The European Industry, however, asserts that it has improved production processes to minimize dioxin emissions.

Also, scientific tests wherein municipal refuse containing several known concentrations of PVC was burned in a commercial-scale incinerator showed no relationship between the PVC content of the waste and dioxin emissions.

PVC produces HCl upon combustion (almost quantitatively related to its chlorine content). Extensive studies in Europe indicate that the chlorine found in emitted dioxins is not derived from HCl in the flue gases. Instead, most dioxins arise in the condensed solid phase by the reaction of inorganic chlorides with graphitic structures in char-containing ash particles. Copper acts as a catalyst for these reactions.

Dioxins are a global health threat because they persist in the environment and can travel long distances. At very low levels, near those to which the general population is exposed, dioxins have been linked to immune system suppression, reproductive disorders, a variety of cancers, and endometriosis. According to a 1994 report by the British firm, ICI Chemicals & Polymers Ltd., "It has been known since the publication of a paper in 1989 that these oxychlorination reactions [used to make vinyl chloride and some chlorinated solvents] generate polychlorinated dibenzodioxins (PCDDs) and dibenzofurans (PCDFs). The reactions include all of the ingredients and conditions necessary to form PCDD/PCDFs.. It is difficult to see how any of these conditions could be modified so as to prevent PCDD/PCDF formation without seriously impairing the reaction for which the process is designed." In other words, dioxins are an undesirable byproduct of producing vinyl chloride and eliminating the production of dioxins while maintaining the oxychlorination reaction may be difficult. Dioxins created by vinyl chloride production are released by on-site incinerators, flares, boilers, wastewater treatment systems and even in trace quantities in vinyl resins. The US EPA estimate of dioxin releases from the PVC industry was 13 grams TEQ in 1995, or less than 0.5 per cent of the total dioxin emissions in the US; by 2002, PVC industry dioxin emissions had been further reduced by 23 per cent.

The largest well-quantified source of dioxin in the US EPA inventory of dioxin sources is barrel burning of household waste. Studies of household waste burning indicate consistent increases in dioxin generation with increasing PVC concentrations. According to the EPA dioxin inventory, landfill fires are likely to represent an even larger source of dioxin to the environment. A survey of international studies consistently identifies high dioxin concentrations in areas affected by open waste burning and a study that looked at the homologue pattern found the sample with the highest dioxin concentration was "typical for the pyrolysis of PVC". Other EU studies indicate that PVC likely "accounts for the overwhelming majority of chlorine that is available for dioxin formation during landfill fires."

The next largest sources of dioxin in the EPA inventory are medical and municipal waste incinerators. Studies have shown a clear correlation between dioxin formation and chloride content and indicate that PVC is a significant contributor to the formation of both dioxin and PCB in incinerators.

In February 2007, the Technical and Scientific Advisory Committee of the US Green Building Council (USGBC) released its report on a PVC avoidance related materials credit for the LEED Green Building Rating system. The report concludes that "no single material shows up as the best across all the human health and environmental impact categories, nor as the worst" but that the "risk of dioxin emissions puts PVC consistently among the worst materials for human health impacts."

Bans

The State of California is currently considering a bill that would ban the use of PVC in consumer packaging due to the threats it poses to human and environmental health and its effect on the recycling stream.

Recycling

The symbol, or 'SPI code', for polyvinyl chloride developed by the Society of the Plastics Industry so that items can be labeled for easy recycling is: ♵ The Unicode character for this symbol is U + 2675 (HTML character reference & # 9845;).

Post-consumer PVC is not typically recycled due to the prohibitive cost of regrinding and recompounding the resin compared to the cost of virgin (unrecycled) resin.

Some PVC manufacturers have placed vinyl recycling programs into action, recycling both manufacturing waste back into their products, as well as post consumer PVC construction materials to reduce the load on landfills.

The thermal depolymerization process can safely and efficiently convert PVC into fuel and minerals, according to the company that developed it. It is not yet in widespread use.

A new process of PVC Recycling is being developed in Europe called Texiloop. This process is based on a technology already applied industrially in Europe and Japan, called Vinyloop, which consists of recovering PVC plastic from composite materials through dissolution and precipitation. It strives to be a closed loop system, recycling its key solvent and hopefully making PVC a future technical nutrient.

5.20 Polyvinylidene Fluoride

Polyvinylidene Fluoride, or PVDF is a highly non-reactive and pure thermoplastic fluoropolymer. It is also known as KYNAR, HYLAR or SYGEF. PVDF is a specialty plastic material in the fluoropolymer family; it is used generally in applications requiring the highest purity, strength, and resistance to solvents, acids, bases and heat and low smoke generation during a fire event. Compared to other fluoropolymers, it has an easier melt process because of its relatively low melting point.

It has a relatively low density (1.78) and low cost compared to the other fluoropolymers. It is available as piping products, sheet, tubing, films, plate and an insulator for premium wire. It can be injected, molded or welded and is commonly used in the chemical, semiconductor, medical and defense industries, as well as in lithium ion batteries.

A fine powder grade, KYNAR 500 PVDF or HYLAR 5000 PVDF, is also used as the principal ingredient of high-end paints for metals. These PVDF paints have extremely good gloss and color retention, and they are in use on many prominent buildings around the world, e.g. the Petronas Towers in Malaysia and Taipei 101 in Taiwan, as well as on commercial and residential metal roofing. PPG Industries, Inc. is a well-known supplier of such PVDF-containing coatings.

IUPAC name	*poly-1,1-difluoroethene*
Other names	polyvinylidene difluoride
Identifiers	
CAS number	[24937-79-9]
Properties	
Molecular formula	$-(CH_2CF_2)_n-$
Appearance	whitish or translucent solid
Solubility in water	not soluble in water
Structure	
Crystal structure	mm2[(Kawai, 1969)]
Dipole moment	2.1 D[(Zhang, 2002)]
Related compounds	
Related compounds	PVC, PTFE, P(VDF-TrFE)

$$\left[\begin{array}{c} \text{H} \quad \text{F} \\ | \quad\;\; | \\ -\text{C}-\text{C}- \\ | \quad\;\; | \\ \text{H} \quad \text{F} \end{array}\right]_n$$

PVDF membranes are used for western blots for immobilization of proteins, due to its non-specific affinity for amino acids.

Properties

In 1969, the strong piezoelectricity of PVDF was observed by Kawai *et al.* The piezoelectric coefficient of poled thin films of the material were reported to be as large as 6-7 pCN^{-1}: 10 times larger than that observed in any other polymer.

PVDF has a glass transition temperature (Tg) of about -35°C and is typically 50-60 per cent crystalline. To give the material its piezoelectric properties, it is mechanically stretched to orient the molecular chains and then poled under tension. PVDF exists in several forms: alpha (TGTG'), beta (TTTT), and gamma (TTTGTTTG') phases, depending on the chain conformations as trans (T) or gauche (G) linkages. When poled, PVDF is a ferroelectric polymer, exhibiting efficient piezoelectric and pyroelectric properties. These characteristics make it useful in sensor and battery applications. Thin films of PVDF are used in some newer thermal camera sensors.

Unlike other popular piezoelectric materials, such as PZT, PVDF has a negative d_{33} value. Physically, this means that PVDF will compress instead of expand or vice versa when exposed to the same electric field.

Processing

PVDF may be synthesized from the gaseous VDF monomer via a free radical (or controlled radical) polymerization process. This may be followed by processes such as melt casting, or processing from a solution (e.g. solution casting, spin coating, and film casting). Langmuir-Blodgett (LB) films have also been made. In the case of solution-based processing, typical solvents used include DMF as well as the more volatile MEK. For characterization of the molecular weight via gel permeation chromatography (also called SEC), solvents such as DMSO or THF may be used.

Processed materials are typically in the non-piezoelectric alpha phase. The material must either be stretched or annealed to obtain the piezoelectric beta phase. The exception to this is for PVDF thin films (thickness in the order of micrometres). Residual stresses between thin films and the substrates on which they are processed are great enough to cause the Beta phase to form.

In order to obtain a piezoelectric response, the material must first be poled in a large electric field. Poling of the material typically requires an external field of > = 30 MV/m. Thick films (typically >100 μm) must be heated during the poling process in order to achieve a large piezoelectric response. Thick films are usually heated to 70-100 °C during the poling process.

A quantitative defluorination process was described by mechanochemistry (Zhang, 2001), for safe eco-friendly PVDF waste processing.

Copolymers

Copolymers of PVDF are also used in piezoelectric and electrostrictive applications. One of the most commonly-used copolymers is P(VDF-TrFE), usually available in ratios of about 50:50 wt per cent and 65:35 wt per cent (equivalent to about 56:44 mol% and 70:30 mol%). Another one is P(VDF-TFE). They improve the piezoelectric response by improving the crystallinity of the material.

While the copolymers' unit structures are less polar than that of pure PVDF, the copolymers typically have a much higher crystallinity. This results in a larger piezoelectric response. d_{33} values for P(VDF-TrFE) have been recorded to be as high as -38 pC/N[7] versus -33 pC/N in pure PVDF[5].

5.21 Polyvinylpyrrolidone

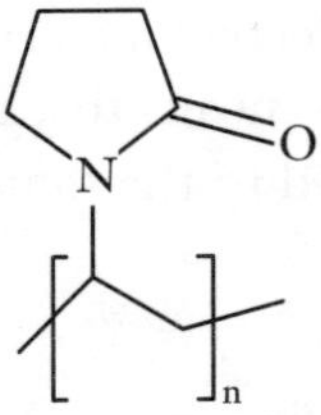

Polyvinylpyrrolidone (PVP, povidone, polyvidone) is a water-soluble polymer made from the monomer *N*-Vinylpyrrolidone:

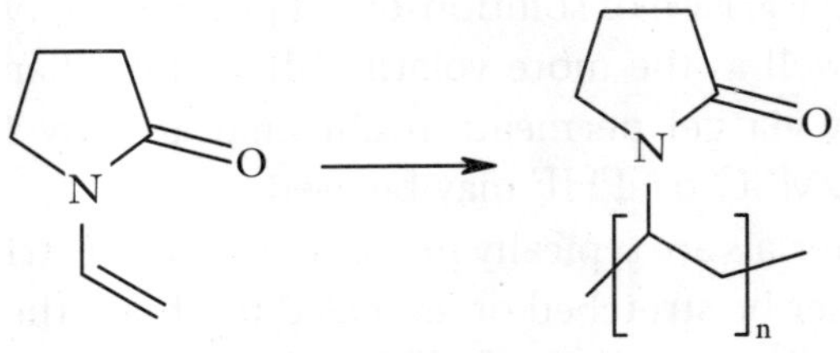

Fig. 5.11: Polyvinylpyrrolidone.

IUPAC name	*Polyvinylpyrrolidone*
Other names	• 1-Ethenyl-2-pyrrolidon homopolymer • Poly[1-(2-oxo-1-pyrrolidinyl)ethylen] • Polyvidone • 1-Vinyl-2-pyrrolidinon-Polymere • PVP
Identifiers	
CAS number	[9003-39-8]
Properties	
Molecular formula	$(C_6H_9NO)_n$
Molar mass	111,14 g·mol? 1
Appearance	white to light yellow, hygroscopic, amorphous powder
Density	1,2 g/cm³
Melting point	130-180 °C (glass temperature)
Hazards	
MSDS	MSDS for PVP from EMD Chemicals

Except where noted otherwise, data are given for materials in their standard state (at 25 °C, 100 kPa) Infobox references

Properties

PVP is soluble in water and other polar solvents. In water it has the useful property of Newtonian viscosity. When dry it is a light flaky powder, which readily absorbs up to 40 per cent of its weight in atmospheric water. In solution, it has excellent wetting properties and readily forms films. This makes it good as a coating or an additive to coatings.

History

PVP was invented by Prof. Walter Reppe and a patent is filed in 1939 for one of the most interesting derivatves of acetylene chemistry. PVP is initially used as a blood plasma substitute and later in a wide variety of applications in medicine, pharmacy, cosmetics and industrial production.

Uses

The monomer is carcinogenic and is extremely toxic to aquatic life. However the polymer PVP in its pure form is so safe that not only is it edible by humans, but it was used as a blood plasma expander for trauma victims after the first half of 20th century.

It is used as a binder in many pharmaceutical tablets; being completely inert to humans, it simply passes through the body when taken orally. PVP added to Iodine forms a complex (Povidone-iodine [5]) that possesses disinfectant properties. This complex is contained in various products like solutions, ointment, pessaries, liquid soaps and surgical scrubs. It is known for instance under the trade name Betadine.

PVP binds to polar molecules exceptionally well, owing to its polarity. This has led to its application in coatings for photo-quality ink-jet papers and transparencies, as well as in inks for inkjet printers.

PVP is also used in personal care products, such as shampoos and toothpastes, in paints, and adhesives that you have to moisten, such as old-style postage stamps and envelopes. It has also been used in contact lens solutions and in steel-quenching solutions. PVP is the basis of the early formulas for hair sprays and hair gels, and still continues to be a component of some.

As a food additive, PVP is a stabilizer and has E number E1201. PVPP is E1202. It is also used in the wine industry as a fining agent for white wine. Other references state that as polyvinyl pyrrolidone and its derivatives are fully from mineral synthetic origin. Therefore, its use in the production should not be a problem for vegans.

In molecular biology, PVP can be used as a blocking agent during Southern blot analysis as a component of Denhardt's buffer. It is also exceptionally good at adsorbing polyphenols during DNA purification. Polyphenols are common in many plant tissues and can deactivate proteins if not removed and therefore inhibit many downstream reactions like PCR.

PVP is also used in many technical applications:

- as adhesive in glue stick and hot melts;
- as special additive for batteries, ceramics, fiberglass, inks and inkjet paper;
- as emulsifier and disintegrant for solution polymerization;
- as photoresist for cathode ray tubes (CRT);
- use in aqueous metal quenching;
- for production of membranes, such as dialysis filter;
- as binder and complexation agent in agro applications such as crop protection, seed treatment and coating;
- as a thickening agent in tooth whitening gels[6]; and
- as an aid for increasing the solubility of drugs in liquid and semi-liquid dosage forms (syrups, soft gelatine capsules) and as an inhibitor of recrystallisation.

Cross-linked Derivatives

A cross-linked form of PVP is also used as a disintegrant in pharmaceutical tablets [1,4]. Is is also known as cross-linked polyvinyl pyrrolidone, Polyvinyl Polypyrrolidone (PVPP), crospovidone, crospolividone. Basically, PVPP is a highly cross-linked version of PVP, which makes it insoluble in water but it still absorbs water and swells very rapidly and generate a swelling force. That is why it can be used a disintegrant in tablets. It is also used to bind impurities to remove them from solutions.It is also used as a fining to extract impurities (via agglomeration followed by filtration). Using the same principle it is uses to remove polyphenols in beer production and thus clear beers with stable foam are produced. PVPP can be used as well as a drug taken as a tablet or suspension and it absorbs compounds (so called Endotoxins) causing diarrhoea. (*Cf.* bone char, charcoal.)

5.22 Kevlar

Kevlar's molecular structure; BOLD: monomer unit; DASHED: hydrogen bonds. Kevlar is the registered trademark for a light, strong para-aramid synthetic fiber, related to other aramids such as Nomex and Technora. Developed at DuPont in 1965 by Stephanie Kwolek and Roberto Berendt, it was first commercially used in the early 1970s as a replacement for steel in racing tires. Typically it is spun into ropes or fabric sheets that can be used as such or as an ingredient in composite material components. Currently, Kevlar has many applications, ranging from bicycle tires and racing sails to body armor because of its high strength-to-weight ratio—famously: *"..5 times stronger than steel on an equal weight basis.."* A similar fiber called Twaron with roughly the same chemical structure was introduced by Akzo in 1978, and now manufactured by Teijin.

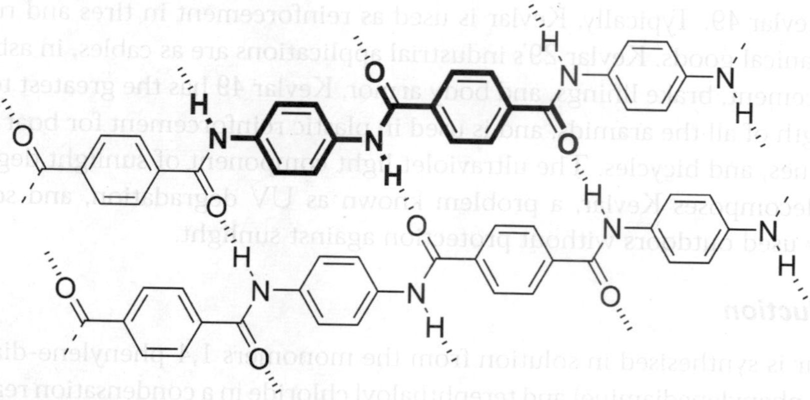

Fig. 5.12: Kevlar.

History

In the 1970s, one of its most significant achievements in the development of body armor was the invention of DuPont's Kevlar ballistic fabric. The fabric was originally intended to replace steel belting in vehicle tires. The development of Kevlar body armor by NIJ was a four-phase effort that took place over several years. The first phase involved testing Kevlar fabric to determine whether it could stop a lead bullet. The second phase involved determining the number of layers of material necessary to prevent penetration by bullets of varying speeds and calibers and developing a prototype vest that would protect officers against the most common threats: the 38 Special and the 22 Long Rifle bullets. By 1973, researchers at the Army's Edgewood Arsenal responsible for the bullet proof vest design had developed a garment made of seven layers of Kevlar fabric for use in field trials. It was determined that the penetration resistance of Kevlar was degraded when wet. The bullet resistant properties of the fabric also diminished upon exposure to ultraviolet light, including sunlight. Dry-cleaning agents and bleach also had a negative effect on the antiballistic properties of the fabric, as did repeated washing. To protect against these problems, the vest was designed with waterproofing, as well as with fabric coverings to prevent exposure to sunlight and other degrading agents.

Properties

When Kevlar is spun, the resulting fiber has great tensile strength (ca. 3 000 MPa), and a relative density of 1.44. When used as a woven material, it is suitable for mooring lines and other underwater application objects.

There are three grades of Kevlar: (i) Kevlar, (ii) Kevlar 29, and (iii) Kevlar 49. Typically, Kevlar is used as reinforcement in tires and rubber mechanical goods. Kevlar 29's industrial applications are as cables, in asbestos replacement, brake linings, and body armor. Kevlar 49 has the greatest tensile strength of all the aramids, and is used in plastic reinforcement for boat hulls, airplanes, and bicycles. The ultraviolet light component of sunlight degrades and decomposes Kevlar, a problem known as UV degradation, and so it is rarely used outdoors without protection against sunlight.

Production

Kevlar is synthesised in solution from the monomers 1,4-phenylene-diamine (*para*-phenylenediamine) and terephthaloyl chloride in a condensation reaction yielding hydrochloric acid as a byproduct. The result has liquid-crystalline behaviour, and mechanical drawing orients the polymer chains in the fiber's direction. Hexamethylphosphoramide (HMPA) was the polymerization solvent first used, but toxicology tests demonstrated it provoked tumors in the noses of rats, so DuPont replaced it by a *N*-methyl-pyrrolidone and calcium chloride as the solvent. As this process was patented by Akzo in the production of Twaron, a patent war ensued.

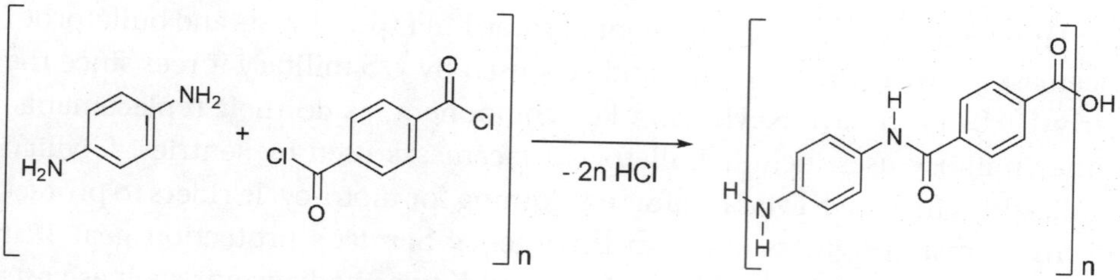

Kevlar (poly paraphenylene terephthalamide) production is expensive because of the difficulties arising from using corrosive concentrated sulfuric acid, needed to keep the water-insoluble polymer in solution during its synthesis and spinning.

Chemical Properties

Fibers of Kevlar consist of long molecular chains produced from PPTA (poly-paraphenylene terephthalamide). There are many inter-chain bonds making the material extremely strong. Kevlar derives part of its high strength from inter-molecular hydrogen bonds formed between the carbonyl groups and protons on neighboring polymer chains and the partial pi stacking of the benzenoid aromatic stacking interactions between stacked strands. These interactions have a greater influence on Kevlar than the van der Waals interactions and chain length that typically influence the properties of other synthetic polymers and fibers such as Dyneema. The presence of salts and certain other impurities, especially calcium, could interfere with the strand interactions and caution is used to avoid inclusion in its production. Kevlar's structure consists of relatively rigid molecules which tend to form mostly planar sheet-like structures rather like silk protein.

Thermal properties

For a polymer Kevlar has very good resistance to high temperatures, and maintains its strength and resilience down to cryogenic temperatures (-196°C); indeed, it is slightly stronger at low temperatures.

At higher temperatures the tensile strength is immediately reduced by about 10-20 per cent, and after some hours the strength progressively reduces further. For example at 160°C about 10 per cent reduction in strength occurs after 500 hours. At 260°C 50 per cent reduction occurs after 70 hours.

At 450°C Kevlar sublimates.

Applications

Armor

Kevlar is well-known as a component of some bulletproof vests and bulletproof facemasks. The PASGT helmet and vest used by US military forces since the early 1980s both have Kevlar as a key component, as do their replacements. Other military uses include bulletproof facemasks used by sentries. Civilian applications include Kevlar reinforced clothing for motorcycle riders to protect against abrasion injuries and also Emergency Service's protection gear if it involves high heat (e.g., tackling a fire), and Kevlar body armor such as vests for Police officers, security, and S.W.A.T.

Rope and Cable

The fiber is used in woven rope and in cable, where the fibers are kept parallel within a polyethylene sleeve. Known as "Parafil", the cables have been used in small suspension bridges such as the bridge at Aberfeldy in Scotland. They have also been used to stabilise cracking concrete cooling towers by circumferential application followed by tensioning to close the cracks.

Sports Equipment

In many cases it is part of a composite material, used in the production of some tennis, table tennis, badminton and squash rackets, kayaks, some cricket bats and some field hockey, ice hockey and lacrosse sticks. It is used as an inner lining for some bicycle tires to prevent punctures, and due to its excellent heat resistance, is used for fire poi wicks. It is used for motorcycle safety clothing, especially in the areas featuring padding such as shoulders and elbows. It was also used as speed control patches for certain Soap Shoes models. In Kyudo or Japanese archery, it may be used as an alternative to more expensive hemp for bow strings. It is one of the main materials used for paraglider suspension lines.

Audio Equipment

It has also been found to have useful acoustic properties for loudspeaker cones, specifically for bass and midrange drive units.

Electricity Generation

It was used by scientists from Georgia Tech as a base textile for an experiment in electricity-producing clothing. This was done by weaving zinc-oxide nanowires into the fabric. If successful, the new fabric would generate about 80 milliwatts per square meter.

Drumheads

Kevlar is sometimes used as a material in high tension drum heads usually used on marching snare drums. It supposedly gives a higher pitched sound and allows for an extremely high amount of tension. There is usually some sort of resin poured onto the kevlar to prevent the air inside the drum from escaping through the head. This is one of the primary types of marching snare drum heads. Remo's falam slam patch is made with kevlar and is stuck to bass drum heads where the beater strikes to strengthen the head so it doesn't wear down as quickly.

Fiber Optic Cable

Kevlar is widely used as a protective outer sheath for fiber optic cable, as its strength protects the cable from damage and kinking.

Building Construction

A retractable roof of over 60,000 square feet (5,575 square metres) of Kevlar was a key part of the design of Montreal's Olympic stadium for the 1976 Summer Olympics. It was spectacularly unsuccessful, as it was completed ten years late and replaced just ten years later in May 1998 after a series of problems.

Sager Brakes

They were used in th 90's to create better and more sustainable brakes because Ford was having trouble those days creating efficient brakes. So Adam Sager used Kevlar in his brakes and sold the idea to Ford.

Expansion Joints and Hoses

Kevlar can be found as a reinforcing layer in rubber bellows expansion joints and rubber hoses, for use in high temperature applications, and for its high strength. It is also found as a braid layer used on the outside of hose assemblies, to add protection against sharp objects.

Composite Materials

Aramid fibers are widely used for reinforcing composite materials, often in combination with carbon fiber and glass fiber. The matrix for high performance composites is usually epoxy resin. Typical applications include monocoque bodies for F1 racing cars, and helicopter rotor blades.

5.23 Twaron

Twaron is the brandname of Teijin Aramid for a para-aramid.

History

Twaron is a heat-resistant and strong synthetic fiber developed in the early 1970s by the Dutch company AKZO, division Enka, later Akzo Industrial Fibers. The research name of the para-aramid fiber was originally Fiber X, but it was soon called Arenka. Although the Dutch para-aramid fiber was developed only a little later than DuPont's Kevlar, introduction of Twaron as a commercial product came much later than Kevlar due to financial problems at the AKZO company in the 1970s.

A short overview of the history of Twaron:

- In 1960s start of research program for Fiber X.
- In 1972 the ENKA Research laboratory develops a para-aramid called **Arenka**.
- In 1973 Akzo decides to use sulfuric acid (H_2SO_4) as a solvent for spinning.
- In 1976 a pilot plant is built. First production in 1977.
- In 1984 the product is renamed **Twaron**.
- In 1986 commercial production is started at three locations and nine plants.
- In 1987 **Twaron** is introduced as a commercial product.
- In 1989 the aramid business of Akzo becomes an independent Business Unit called Twaron BV.
- Since 2000 Twaron BV is owned by the Teijin Group, now called Teijin Twaron BV and based in Arnhem, The Netherlands. The main production facilities for Twaron are in Emmen en Delfzijl.
- In 2007 Teijin Twaron expands for the fourth time in six years and also changes its name into Teijin Aramid.

Production

Polymer Preparation

Twaron is a *p*-phenylene terephtalamide (P*p*PTA), the simplest form of the AABB para polyaramide. P*p*PTA is a product of *p*-phenylene diamine (PPD) and terephtaloyl dichloride (TDC). To dissolve the aromatic polymer Twaron used a co-solvent of N-methyl pyrrolidone (NMP) and an ionic component (Calcium Chloride ($CaCl_2$) to occupy the hydrogen bonds of the amide groups. Prior to the invention of this process by Leo Vollbracht, working at the Dutch chemical firm AKZO, no practical means of dissolving the polymer was known. The use of this system by DuPont led to a patent war between AKZO and DuPont as Dupont initially used the carcinogenic HMPT (Hexamethylphosphoramide). Despite heavy research DuPont now also uses the AKZO patent to use the less hazardous NMP in the Kevlar process.

Spinning

After the production of the Twaron polymer in Delfzijl, the polymer is brought to Emmen, where fibers are produced by spinning the dissolved polymer into a solid fiber from a liquid chemical blend. Polymer solvent for spinning PPTA is generally 100 per cent (water free) sulfuric acid (H_2SO_4). The polymer is dissolved by mixing frozen sulfuric acid in powder form with the polymer in powder form and gently heating the mixture. This process, which differs from the more difficult DuPont process, was invented by Henri Lammers and patented by AKZO.

Major Industrial Uses

Twaron is a para-aramid and is used in automotive, construction, sport, aerospace, military and industry applications, e.g., "bullet-proof" body armor, fabric, and as an asbestos substitute.

❖ Protective Gear (Heat Resistant/Ballistics):

- Flame-resistant clothing,
- Protective clothing and helmets,
- Anti-cut or heat resistant gloves,
- Sporting goods,
- Fabrics, and
- Body armor.

❖ Composites:

- composite materials
- technical paper
- asbestos replacement
- hot air filtration
- sail cloth
- speaker woofers
- boat hull material
- fiber reinforced concrete
- drumheads.

❖ Automotive:

- brake pads
- turbo hoses
- v-belts and Timing belts
- tires that incorporate Sulfron (sulfur modified Twaron)
- mechanical rubber goods reinforcement.

❖ Linear Tension:

- optical fiber cables (OFC)
- ropes, wire ropes, cables
- umbilical cables
- Electrical Mechanical Cable (EMC)
- Reinforced Thermoplastic Pipes.

5.24 Technora

Technora is the brandname of Teijin for a aromatic copolyamid.

Production Process

The manufacturing process of Technora reacts PPD and 3,4'-diaminodiphenylether (3,4'-ODA) with terephtaloyl chloride (TCl). This relatively simple process uses only one amide solvent and therefore spinning can be done directly after the polymer production.

Major Industrial Uses

❖ Automotive and other industries:
- Turbo hoses
- High pressure hoses
- Timing and V-belts
- Mechanical rubber goods reinforcement.

❖ Linear Tension:

- Optical fiber cables (OFC)
- Ropes, wire ropes and cables
- Umbilical cables
- Electrical Mechanical Cable (EMC).

❖ Drumheads

5.25 PET Film (Biaxially Oriented)

Biaxially-oriented polyethylene terephthalate (boPET) polyester film is used for its high tensile strength, chemical and dimensional stability, transparency, reflective, gas and aroma barrier properties and electrical insulation.

A variety of companies manufacture boPET and other polyester films under different trade names. In the US and Britain, the most well-known trade names are Mylar and Melinex.

History and Manufacture

Biaxially oriented PET film was developed in the mid-1950s, originally by DuPont and ICI. In 1960 and 1964 NASA launched the Echo satellites, 100 feet (30 m) diameter balloons of metallized 0.005 inches (0.13 mm) thick boPET film. The manufacturing process begins with a film of molten PET being extruded onto a chill roll, which quenches it into the amorphous state. It is then biaxially oriented by drawing. The most common way of doing this is the sequential process, in which the film is first drawn in the machine direction using heated rollers and subsequently drawn in the transverse direction, i.e. orthogonally to the direction of travel, in a heated oven. It is also possible to draw the film in both directions simultaneously, although the equipment required for this is somewhat more elaborate. Draw ratios are typically around 3 to 4 in each direction.

Once the drawing is completed, the film is "heat set" or crystallized under tension in the oven at temperatures typically above 200 °C. The heat setting step prevents the film from shrinking back to its original unstretched shape and locks in the molecular orientation in the film plane. The orientation of the polymer chains is responsible for the high strength and stiffness of biaxially oriented PET film, which has a typical Young's modulus of about 4 GPa. Another important consequence of the molecular orientation is that it induces the formation of many crystal nuclei. The crystallites that grow rapidly reach the boundary of the neighboring crystallite and remain smaller than the wavelength of visible light. As a result, biaxially oriented PET film has excellent clarity, despite its semicrystalline structure. If it were produced without any additives, the surface of the film would be so smooth that layers would adhere strongly to one another when the film wound up, similar to the sticking of clean glass plates when stacked. To make handling possible, microscopic inert inorganic particles are usually embedded in the PET to roughen the surface of the film. Biaxially oriented PET film can be aluminized by evaporating a thin film of metal onto it. The result is much less permeable to gasses (important in food packaging) and reflects up to 99 per cent of light, including much of the infrared spectrum. For some applications like food packaging, the aluminized boPET film can be laminated with a layer of polyethylene, which provides sealability and improves puncture resistance. The polyethylene side of such a laminate appears dull and the PET side shiny.

Metallized nylon (or "foil") balloons used for floral arrangements and parties are often mistakenly called "Mylar", one of the trade names for boPET film. Other coatings, such as conductive indium tin oxide (ITO), can be applied to boPET film by sputter deposition.

Uses for boPET Film

Uses for boPET polyester films include, but are not limited to:

Flexible Packaging and Food Contact Applications

- Laminates containing metallized boPET film protect food against oxidation and aroma loss, achieving long shelf life. Examples are coffee "foil" packaging and pouches for convenience foods.
- Attractive glossy or matte surfaces on the outside of packages are achieved using boPET film.
- White boPET film is used as lidding for dairy goods such as yoghurt.
- Clear boPET film is used as lidding for fresh or frozen ready meals. Due to its excellent heat resistance, it can remain on the package during microwave or oven heating.
- Roasting bags.

Covering Over Paper

- A clear overlay on a map, on which notations, additional data, or copied data, can be drawn without damaging the map.
- Metallized boPET is used as a mirror-like decorative surface on some book covers, T-shirts, and other flexible cloths.
- Protective covering over buttons/pins/badges.
- The glossy top layer of a Polaroid SX-70 photographic print.
- As a backing for very fine sandpaper.
- boPET film is used in bagging comic books, in order to best protect them during storage from environmental conditions (moisture, heat, and cold) that would otherwise cause paper to slowly deteriorate over time. This material is used for archival quality storage of documents by the Library of Congress. It should be noted that while boPET is widely (and effectively) used in this archival sense, it is not immune to the devastating effects of fire and heat and could potentially melt, depending on the intensity of the heat source, causing further damage to the encased item.
- For protecting the spine of important documents, such as medical records.

Insulating Material

- An electrical insulating material.
- As base material for audio or video magnetic recording tapes.
- Insulation for houses and tents in *cold* environments, covering the inner walls with the metallized surface facing *inward,* thus reflecting heat *back into* the space.

- Insulation for houses and tents in *hot* environments, covering the outer walls with the metallized surface facing *outward*, thus reflecting heat *away from* the space.
- Five layers of metallized boPET film in NASA's spacesuits make them radiation resistant and keep astronauts warm.
- Metallized boPET film "emergency blankets" conserve a shock victim's body heat.
- As a thin strip to form an airtight seal between the control surfaces and adjacent structure of aircraft, especially gliders.
- Light insulation for indoor gardening.
- Wildland fire shelters.
- Proximity(aluminized) suits used by AR-FF fire fighters for protection from the high amount of heat relase from fuel fires.
- Aluminized boPET films are no longer used as thermal/acoustic insulation in aircraft since they were found to have been a factor in the Swissair Flight 111 crash. The accident investigation showed that the aluminum layer prevents the film from self-extinguishing in a fire under the conditions in aircraft.

Solar and Marine Applications

- Solar sails as an alternative means of propulsion for spacecraft such as Cosmos 1
- Metallized boPET solar curtains reflect sunlight and heat away from windows.
- Aluminized, as an inexpensive solar eclipse viewer, although care must be taken, because invisible fissures can form in the metal film, reducing its effectiveness.
- High performance sails for sailboats and hanggliders

Electronic/Acoustic Applications

- Very thin boPET film is often used as the diaphragm material in electrostatic loudspeakers and electret microphones.
- boPET film has been used in the production of banjo & drumheads since 1958 due to its durability and acoustical properties when stretched over the bearing edge of the drum. They are made in single- and double-ply versions, with each ply being between 2 mils and 10 mils (0.05—0.25 mm) in thickness, with a clear or opaque surface, originally used by the company EVANS.
- boPET film is used as the substrate in practically all magnetic recording tapes.

- Metallized boPET film, along with other plastic films, is used as a dielectric in foil capacitors.
- Clear boPET bags are used as packaging for audio media such as compact discs and vinyl records.

Graphic arts

- Often engineering plans or architectural drawings are plotted onto sheets of boPET film. The boPET sheets become legal documents from which copies or blueprints are made. boPET sheets are more durable and can withstand more handling than bond paper.
- Overhead transparency film for photocopiers or laser printers (boPET film withstands the high heat).

Other

- For materials in kites.
- Covering glass to decease probability of shattering.
- In theatre effects as confetti.
- As the adhesive strip to attach the string to a teabag.
- One of the many materials used as Windsavers or valves for valved harmonicas.
- As a light diaphragm material separating gases in hypersonic shock and expansion tube facilities.
- On California farmland, highly reflective aluminized PET film ribbons are tied to the plants to create shimmers from the sun for an effect similar to a scarecrow.

External Links

- *Plastics:* Background Information for Teachers by The American Plastics Council (HTML format) or (PDF format)—1.9MB), which includes the "chasing arrow" recycling symbols (PET is #1) and a description of plastics.
- Links to most manufacturers of boPET films are available at the AMPEF website.

5.26 Nylon

Nylon is a generic designation for a family of synthetic polymers known generically as polyamides and first produced on February 28, 1935 by Wallace Carothers at DuPont. Nylon is one of the most commonly used polymers.

Nylon	
Density	1.15 g/cm^3
Electrical conductivity (σ)	10^{-12} S/m
Thermal conductivity	0.25 W/(m·K)
Melting point	463 K-624 K; 190°C-350°C; 374°F-663°F

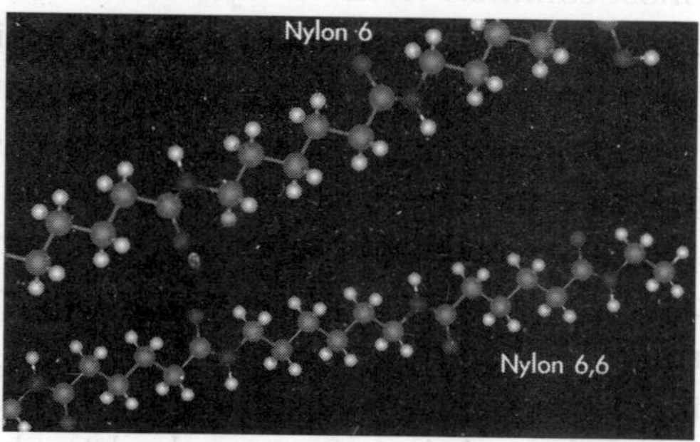

Fig. 5.14: Nylon.

Overview

Nylon is a thermoplastic silky material, first used commercially in a nylon-bristled toothbrush (1938), followed more famously by women's "nylons" stockings (1940). It is made of repeating units linked by peptide bonds (another name for amide bonds) and is frequently referred to as *polyamide* (PA). Nylon was the first commercially successful polymer and the first synthetic fiber to be made entirely from coal, water and air. These are formed into monomers of intermediate molecular weight, which are then reacted to form long polymer chains. Nylon was intended to be a synthetic replacement for silk and substituted for it in many different products after silk became scarce during World War II. It replaced silk in military applications such as parachutes, flak vests, and was used in many types of vehicle tires.

Nylon fibers are used in many applications, including fabrics, bridal veils, carpets, musical strings and rope. Solid nylon is used for mechanical parts such as gears and other low- to medium-stress components previously cast in metal. Engineering grade nylon is processed by extrusion, casting, and injection molding. Solid nylon is used in hair combs. Type 6/6 Nylon 101 is the most common commercial grade of nylon, and Nylon 6 is the most common commercial grade of cast nylon. Nylon is available in glass-filled and molybdenum sulfide-filled variants which increase structural and impact strength and rigidity or lubricity. Aramids are another type of polyamide with quite different chain structures which include groups in the main chain. Such polymers make excellent ballistic fibres.

Chemistry

Nylons are condensation copolymers formed by reacting equal parts of a diamine and a dicarboxylic acid, so that peptide bonds form at both ends of each monomer in a process analogous to polypeptide biopolymers. The numerical suffix specifies the numbers of carbons donated by the monomers; the diamine first and the diacid second. The most common variant is nylon 6-6 which refers to the fact that the diamine (hexamethylene diamine) and the diacid (adipic acid) each donate 6 carbons to the polymer chain. As with other regular copolymers like polyesters and polyurethanes, the "repeating unit" consists of one of each monomer, so that they alternate in the chain. Since each monomer in this copolymer has the same reactive group on both ends, the direction of the amide bond reverses between each monomer, unlike natural polyamide proteins which have overall directionality: C terminal → N terminal. In the laboratory, nylon 6,6 can also be made using adipoyl chloride instead of adipic It is difficult to get the proportions exactly correct, and deviations can lead to chain termination at molecular weights less than a desirable 10,000 daltons (u). To overcome this problem, a crystalline, solid "nylon salt" can be formed at room temperature, using an exact 1:1 ratio of the acid and the base to neutralize each other. Heated to 285 °C, the salt reacts to form nylon polymer. Above 20,000 daltons, it is impossible to spin the chains into yarn, so to combat this, some acetic acid is added to react with a free amine end group during polymer elongation to limit the molecular weight. In practice, and especially for 6,6, the monomers are often combined in a water solution. The water used to make the solution is evaporated under controlled conditions, and the increasing concentration of "salt" is polymerized to the final molecular weight.

DuPont patented nylon 6,6, so in order to compete, other companies (particularly the German BASF) developed the homopolymer nylon 6, or polycaprolactam—not a condensation polymer, but formed by a ring-opening polymerization (alternatively made by polymerizing aminocaproic acid). The peptide bond within the caprolactam is broken with the exposed active groups on each side being incorporated into two new bonds as the monomer becomes part of the polymer backbone. In this case, all amide bonds lie in the same direction, but the properties of nylon 6 are sometimes indistinguishable from those of nylon 6,6—except for melt temperature (N6 is lower) and some fiber properties in products like carpets and textiles. There is also nylon 9.

Nylon 5,10, made from pentamethylene diamine and sebacic acid, was studied by Carothers even before nylon 6,6 and has superior properties, but is more expensive to make. In keeping with this naming convention, "nylon 6,12" (N-6,12) or "PA-6,12" is a copolymer of a 6C diamine and a 12C diacid. Similarly for N-5,10 N-6,11; N-10,12, etc. Other nylons include copolymerized dicarboxylic acid/diamine products that are *not* based upon the monomers listed above. For example, some aromatic nylons are polymerized with the addition of diacids like terephthalic acid (→ Kevlar) or isophthalic acid (→ Nomex), more commonly associated with polyesters. There are copolymers of N-6,6/N6; copolymers of N-6,6/N-6/N-12; and others. Because of the way polyamides are formed, nylon would seem to be limited to unbranched, straight chains. But "star" branched nylon can be produced by the condensation of dicarboxylic acids with polyamines having three or more amino groups.

The general reaction is:

$$n\ \mathrm{HOOC{-}R{-}COOH} + \mathrm{H_2N{-}R'{-}NH_2} \longrightarrow \left[\overset{O}{\overset{\|}{C}}{-}R{-}\overset{O}{\overset{\|}{C}}{-}\underset{H}{N}{-}R'{-}\underset{H}{N} \right]_n$$

A molecule of water is given off and the nylon is formed. Its properties are determined by the R and R' groups in the monomers. In nylon 6,6, R' = 6C and R = 4C alkanes, but one also has to include the two carboxyl carbons in the diacid to get the number it donates to the chain. In Kevlar, both R and R' are benzene rings.

Nylon Fiber

The Federal Trade Commission's definition for Nylon Fiber: A manufactured fiber in which the fiber forming substance is a long-chain synthetic polyamide in which less than 85 per cent of the amide-linkages are attached directly (-CO-NH-) to two aliphatic groups:

- A synthetic thermoplastic fiber (Nylon melts/glazes easily at relatively low temperatures).
- Round, smooth, and shiny filament fibers.
- cross sections can be either:
  - trilobal to imitate silk; and
  - multilobal to increase staple like appearance and hand.
- Its most widely used structures are multifilament, monofilament, staple or tow and is available as partially drawn or as finished filaments.
- Regular nylon has a round cross section and is perfectly uniform. The filaments are generally completely transparent unless they have been delustered or solution dyed. Thus, they are microscopically recognized as glass rods.
- Molecular chains of nylon are long and straight variations but have no side chains or linkages.

- Cold drawing (step 18 on the model) can align the chains so they are oriented with the lengthwise direction and are highly crystalline.

❖ Nylon is related chemically to the protein fibers silk and wool.

- They both have similar dye sites but nylon has many fewer dye sites than wool.

Basic Concepts of Nylon Production

❖ The first approach: combining molecules with an acid (COOH) group on each end are reacted with two chemicals that contain amine (NH_2) groups on each end.

This process creates nylon 6,6, made of hexamethylene diamine with six carbon atoms and acidipic acid, as well as six carbon atoms.

❖ The second approach: a compound has an acid at one end and an amine at the other and is polymerized to form a chain with repeating units of (-NH-[CH2]n-CO-)x.

- In other words, nylon 6 is made from a single six-carbon substance called caprolactam.
- In this equation, if n = 5, then nylon 6 is the assigned name. (may also be referred to as polymer).

Nylon 6,6

❖ Pleats and creases can be heat-set at higher temperatures.

❖ Nylon is very easy to dye, but Nylon 6,6 is not.

Nylon 6

❖ Better dye Affinity,

❖ Softer Hand,

❖ Greater elasticity and elastic recovery, and

❖ Better weathering properties; better sunlight resistance.

Producers

The producers of nylon include: Honeywell Nylon Inc., Invista, Wellman Inc. among many others. The Dupont Company, is the most famous pioneer of the nylon we know today.

Characteristics

❖ *Variation of luster*: Nylon has the ability to be very lustrous, semilustrous or dull.

- *Durability*: Its high tenacity fibers are used for seatbelts, tire cords, ballistic cloth and other uses.
- High elongation.
- Excellent abrasion resistance.
- Highly resilient (nylon fabrics are heat-set).
- Paved the way for easy-care garments.
- High resistance to:
  - insects, fungi and animals,
  - molds, mildew, rot, and
  - many chemicals.
- Used in carpets and nylon stockings.
- Melts instead of burning.
- Used in many military applications.

Bulk Properties

Above their melting temperatures, T_m, thermoplastics like nylon are amorphous solids or viscous fluids in which the chains approximate random coils. Below T_m, amorphous regions alternate with regions which are lamellar crystals. The amorphous regions contribute elasticity and the crystalline regions contribute strength and rigidity. The planar amide (-CO-NH-) groups are very polar, so nylon forms multiple hydrogen bonds among adjacent strands. Because the nylon backbone is so regular and symmetrical, especially if all the amide bonds are in the *trans* configuration, nylons often have high crystallinity and make excellent fibers. The amount of crystallinity depends on the details of formation, as well as on the kind of nylon. Apparently it can never be quenched from a melt as a completely amorphous solid.

Nylon 6,6 can have multiple parallel strands aligned with their neighboring peptide bonds at coordinated separations of exactly 6 and 4 carbons for considerable lengths, so the carbonyl oxygens and amide hydrogens can line up to form interchain hydrogen bonds repeatedly, without interruption. Nylon 5,10 can have coordinated runs of 5 and 8 carbons. Thus parallel (but not antiparallel) strands can participate in extended, unbroken, multi-chain â-pleated sheets, a strong and tough supermolecular structure similar to that found in natural silk fibroin and the â-keratins in feathers. (Proteins have only an amino acid á-carbon separating sequential -CO-NH- groups.) Nylon 6 will form uninterrupted H-bonded sheets with mixed directionalities, but the â-sheet wrinkling is somewhat different. The three-dimensional disposition of each alkane hydrocarbon chain depends on rotations about the 109.47° tetrahedral bonds of singly-bonded carbon atoms.

When extruded into fibers through pores in an industrial spinneret, the individual polymer chains tend to align because of viscous flow. If subjected to cold drawing afterwards, the fibers align further, increasing their crystallinity, and the material acquires additional tensile strength. In practice, nylon fibers are most often drawn using heated rolls at high speeds.

Block nylon tends to be less crystalline, except near the surfaces due to shearing stresses during formation. Nylon is clear and colorless, or milky, but is easily dyed. Multistranded nylon cord and rope is slippery and tends to unravel. The ends can be melted and fused with a heat source such as a flame or electrode to prevent this.

When dry, polyamide is a good electrical insulator. However, polyamide is hygroscopic. The absorption of water will change some of the material's properties such as its electrical resistance. Nylon is less absorbent than wool or cotton.

Historical Uses

Bill Pittendreigh, DuPont, and other individuals and corporations worked diligently during the first few months of World War II to find a way to replace Asian silk with nylon in parachutes. It was also used to make tires, tents, ropes, ponchos, and other military supplies. It was even used in the production of a high-grade paper for U.S. currency. At the outset of the war, cotton accounted for more than 80 per cent of all fibers used and manufactured, and wool fibers accounted for the remaining 20 per cent. By August 1945, manufactured fibers had taken a market share of 25 per cent and cotton had dropped.

Some of the terpolymers based upon nylon are used every day in packaging. Nylon has been used for meat wrappings and sausage sheaths.

Use in Composites

Nylon can be used as the matrix material in composite materials, such as glass or carbon fiber, and yields a higher density than pure nylon.

Etymology

In 1940 John W. Eckelberry of DuPont stated that the letters "nyl" were arbitrary and the "on" was copied from the suffixes of other fibers such as cotton and rayon. A later publication by DuPont (*Context*, vol. 7, no. 2, 1978) explained that the name was originally intended to be "No-Run" ("run" meaning "unravel"), but was modified to avoid making such an unjustified claim and to make the word sound better. The story goes that Carothers changed one letter at a time until DuPont's management was satisfied. But he was not involved in the nylon project during the last year of his life, and committed suicide before the name was coined.

Two theories about the origin of the name claim that it is an acronym of "Now you've lost, Old Nippon" (N.Y.L.O.N.), or that it stands for "New York-London". In the latter case, it is claimed that these were the two cities where the product was researched and developed, or that the inspiration came from a New York to London airplane ticket. There is no evidence for the 'airline ticket' theory, though some compelling evidence of the former from contemporary researchers at Oxford University who assisted in development.

Uses

- carpet fiber
- clothing
- fishing lines
- footwear
- nylon fiber
- pantyhose
- windpants
- toothbrush bristles
- velcro
- airbag fiber
- auto parts: intake manifolds, gas (petrol) tanks
- slings and rope used in climbing gear and slacklining
- machine parts, such as gears and bearings
- parachutes
- metallized nylon balloons
- classical and flamenco guitar strings
- paintball marker bolts
- racquetball, badminton, squash, and tennis racquet strings
- Strings for String instruments
- Drumstick heads
- as filter media in sterlizing grade filters
- flexible tubing
- basketball netting
- sutures
- flags
- rings for baby slings.

5.27 Ultem

Ultem is a family of polyimide thermoplastic resins, of type amorphous polyetherimide. Ultem is a registered trademark of SABIC Innovative Plastics the principal developer and manufacturer. Ultem resins are used in medical and chemical instrumentation, due to their heat resistance, solvent resistance and flame resistance. Ultem 1000 (standard, unfilled polyetherimide) high dielectric strength, natural flame resistance, and extremely low smoke generation. Ultem has exceptionally high mechanical properties and performs in continuous use to 340°F (170°C) Rated UL 94V-0. Ultem is available in glass-reinforced grades 30 per cent (Ultem 2300), 20 per cent (Ultem 2200), 10 per cent (Ultem 2100).

5.28 Vectran

Vectran is a manufactured fibre, spun from a liquid crystal polymer created by Celanese Acetate LLC and now manufactured by Kuraray Co., Ltd. Chemically it is an aromatic polyester. These fibres are noted for thermal stability at high temperatures, high strength and modulus, low creep, and good chemical stability. They are moisture resistant and are generally stable in hostile environments. They have gold color. They are often used in combination with some polyester as a coating around Vectran core; polyurethane coating can improve abrasion resistance and act as a water barrier. Vectran has melting point of 330 °C, with progressive strength loss from 220 °C. As it has high resistance to ultraviolet radiation, it can be used outside for long term, if inspected regularly.

They are used as reinforcing (matrix) fibres for ropes, cables, and advanced composite materials, professional bike tires, and in electronics applications. Perhaps most notably, Vectran is used as one of the five layers in NASA's current space suit design, and was the fabric used for the airbags on the Mars Pathfinder and twin Mars Exploration Rovers *Spirit* and *Opportunity* missions, which allowed those craft to "soft land" on Mars in 1997 and 2004, respectively. The material is expected to be used again on NASA's 2009 Mars Science Laboratory in the bridle cables. It is also used in the inflatable spacecraft developed by Bigelow Aerospace, the first of which (*Genesis I*) has been in space since July 12, 2006.

Production

Kuraray Co., Ltd. owns 100 per cent of the worldwide Vectran production since 2005 as they acquired the Vectran business from Celanese Advanced Materials Inc. (CAMI), based in South Carolina, U.S.

In 2007 the total capacity of Vectran is about 600 tons/yr, but expanded to 1000 tons/yr in 2008.

5.29 Viton

Viton is a brand of synthetic rubber and fluoropolymer elastomer commonly used in O-rings and other moulded or extruded goods. The name is a registered trademark of DuPont Performance Elastomers L.L.C.

Fig. 5.15: Genuine Viton.

Viton fluoroelastomers are categorized under the ASTM D1418 & ISO 1629 designation of FKM. This class of elastomers is a family comprising copolymers of hexafluoropropylene (HFP) and vinylidene fluoride (VDF or VF2), terpolymers of tetrafluoroethylene (TFE), vinylidene fluoride (VDF) and hexafluoropropylene (HFP) as well as perfluoromethylvinylether (PMVE) containing specialties. The fluorine content of the most common Viton grades varies between 66 and 70 per cent.

Four Families of Polymers

- *A (Dipolymers of VF2/HFP)*: General purpose sealing. Automotive, Aerospace fuels & lubricants
- *B (Terpolymers of VF2/HFP/TFE)*: Chemical Process plant, Power Utility Seals & Gaskets
- *F (Terpolymers of VF2/HFP/TFE)*: Oxygenated Automotive fuels. Concentrated aqueous inorganic acids, water, steam
- *Viton Extreme (Copolymers of TFE/Propylene and Ethylene/TFE/PMVE)*: Automotive, Oil Exploration, Special Sealing, Ultra Harsh Environments

The performance of fluoroelastomers in aggressive chemicals depends on the nature of the base polymer and the compounding ingredients used for moulding the final products (e.g. O-rings). This performance can vary significantly when end-users purchase Viton polymer containing rubber goods from different sources. Viton is generally compatible with hydrocarbons (an advantage over Buna-N, which is the most common O-ring material), but incompatible with ketones such as acetone and organic acids such as acetic acid. O-rings made of Viton are typically color coded as black.

Viton O-rings have been used safely for some time in the SCUBA diving world for divers who dive with gas blends referred to as Nitrox. Viton is used because it has a lower probability of catching fire, even with the increased percentages of oxygen found in Nitrox. It is also less susceptible to decay under increased oxygen conditions.

5.30 Zylon

Zylon is a trademarked name for a range of thermoset polyurethane synthetic polymer materials manufactured by the Toyobo Corporation. Zylon was invented and developed by SRI International in the 1980s. Like Kevlar, Zylon is used in a number of applications that require very high strength with excellent thermal stability. Tennis racquets, Table Tennis blades, various medical applications, and some of the martian rovers are some of the more well known instances.

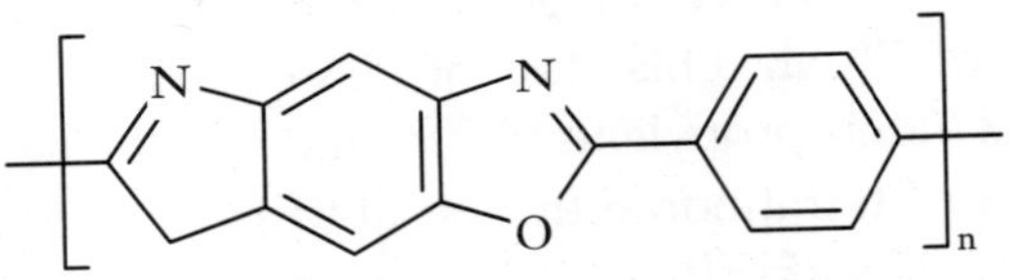

Fig. 5.16: Poly(p-phenylene-2, 6-benzobisoxazole).

IUPAC name: poly(p-phenylene-2,6-benzobisoxazole)

The particular subset that Zylon is derived from are the EBXL-TPU plastics. (From electron beam cross-linked ThermoPlastic Polyurethane).

Usage

Body Armor

Zylon also gained wide use in U.S. police officers body armor protection in 1998 with its introduction by Second Chance Body Armor, Inc. But protective vests constructed with Zylon became controversial in late 2003 when Oceanside, CA Police Officer Tony Zeppetella's and Forest Hills, PA Police Officer Ed Limbacher's vests failed, leaving Zeppetella mortally wounded and Limbacher seriously injured. Some studies subsequently reported that the Zylon vests may degrade rapidly, leaving wearers with significantly less protection than expected. Second Chance eventually recalled all of its zylon-containing vests, which led to its subsequent bankruptcy. In early 2005, Armor Holdings, Inc. first recalled its existing Zylon-based products, and decreased the rated lifespan warranty of new vests from 60 months to 30 months. In August 2005, AHI decided to discontinue manufacturing all of its Zylon-containing vests. This was largely based on the actions of the U.S. government's National Institute of Justice, which decertified Zylon for use in its approved models of ballistic vests for law enforcement.

Space Elevator Research

A competition was held in the Wirefly X Prize Cup in Las Cruces, New Mexico, US, on October 20-21, 2006. A team from the University of British Columbia entered into the Tether Challenge, using a construction made from Zylon fibers. And the house tether used by Spaceward that the other teams would have to beat in strength by 50 per cent in the 2007 strong tether challenge was made out of Zylon.

Formula One Racing

Starting in the 2007 season, the driver's cockpit must now be clad in special anti-penetration panels made of Zylon. The Indy Racing League will also use Zylon starting in 2008.

Standing Rigging

On modern racing yachts Zylon is used for parts of the standing rigging. It is used as shrouds and stays except the forestay due to torsion weakening the fiber. The PBO fiber degrade by UV and visible light, seawater and shafing and is therefore protected by a synthetic melted-on jacket. It is claimed to be 65 per cent lighter than traditional rigging and the price 110-130 per cent of rodrigging. And even the durability is claimed to be superior, but the statement is based on laboratory tests. It also has a high Young's modulus of 270 GPa, meaning that it is very stiff.

6

Observing and Measuring Polymer/Colloids Properties

6.1 Exploring Properties of Polymers

Introduction

We are surrounded by polymers in our environment, many of which are synthetic, i.e. manmade. In this synthetic category are many fabrics, including rayon, nylon, dacron, and polyester. The polyvinylchloride in some pipes and polyvinyl floor tiles are also synthetic polymers, as is the teflon in our frying pans. However, polymers have existed on earth even before life itself. In our bodies, our muscles and bones are made of the polymer we call protein. Our genetic material, DNA and RNA, is also a polymer. The cellulose we find in all plant material is a polymer, as are the complex carbohydrates we find in bread and pasta. Rubber is a good example of a naturally occurring polymer.

What is a polymer? What do all these very different materials have in common? Polymers form when many of the same molecule, or similar molecules, called monomers, are attached together like beads on a string. The beads are called monomers and the "necklace" is the polymer. The links between the beads in polymers are chemical bonds. Most polymers are comprised of many long strings of beads. To make the polymer stronger, many times these long strings are connected with "cross-linkers" to make a stronger mesh like material from the polymers.

Scientists have been able to create polymers with very different properties by using different monomers and different cross-linking reagents. In this lab we have the students form people polymers to demonstrate some of the different properties between monomers, straight chain polymers and crosslinked polymers. The students also have the opportunity to explore and create for themselves several cross-linked polymers from straight chain polymers. [While there are many great demonstrations of forming the polymers directly from monomers, I have not found any suitable for a middle school classroom. They use chemicals which cannot be handled safely in your classrooms.]

EXPERIMENT 1: PEOPLE POLYMERS

Introduction

The purpose of this particular actively is to demonstrate the different in mobility between monomers, straight-chain polymers and cross-linked polymers.

Objectives

1. To have the students learn about the cohesiveness of polymers and begin to understand how they work.
2. To have the students understand the difference of monomers and polymers.
3. To have the students get active in their learning.

Vocabulary

- Monomer
- Polymer
- Materials:
  - 10-12 students
  - 10-12 pencils

Procedure

This demonstration works best in an old-fashioned classroom with desks in rows.

1. Depending on the size of the class, invite 10-12 students to be part of the demonstration. e.g. For 12 students, form two polymers chains of 5 students each, using the pencils as the linkers between the people monomers.
2. Allow the other two students to be monomers.
3. Have all these students then walk from the front of the class room to the back. The polymers are not allowed to break any of the links. The monomers should get the back more easily than the long chains of polymers, but the polymers still slink to the back fairly easily, though a little more slowly.
4. Now use the two monomers as crosslinkers, either as one two person cross-link or two one person.

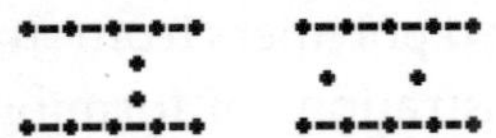

And have the whole cross-linked polymer try to move from the front of the class room to the back, again without breaking any links. This should be much more difficult that the straight-chain polymers. This demonstrates that crosslinked polymers move more slowly. They are thicker and more gel-like than straight chain polymers because of this lack of mobility.

Assessments

The students could have to explain in their own words what happened and why?

Extensions

The students could be lead in a discussion as to why this is the case. Have then talk about why it is easier to walk with only one link rather then many. Where else is this prevalent in their lives?

EXPERIMENT 2: TRASH BAGS

Introduction

The purpose of this experiment is to explore the strength of a crosslinked polymer.

Objectives

1. To have the students learn about the strength of a crosslinked polymer.
2. To let the students see the crosslinked polymer in items they recognize.
3. To let them have some fun:

- *Vocabulary*: Crosslink.
- *Materials*: Cheap trash bags, cut into 4 inch squares.

Procedure

1. Have the students hold the plastic square up to the light and report on what they see.
2. Someone should say they see lines. Bring everyone's attention to the parallel lines in the bags. These are the long chains of straight chain polymers. One cannot see the cross-linkers which connect them.
3. Have the student pull on the square in the direction of the horizontal lines.
4. Then have them pull in the direction perpendicular to the horizontal lines.

5. Ask which way is easier. It should be much easier to pull perpendicular to the horizontal lines, as you only have to break the few cross linkers to stretch the plastic. Pulling in the direction of the horizontal lines forces you to pull against the many links of the straight-chain polymer.

EXPERIMENT 3: PLAYING WITH POLYMERS

Introduction

Polymers can be made out of everyday household items and can be a lot of fun. The next three experiments are messy, creative and fun, as well as good examples of Polymers. In each of these experiments a polymer with different properties is made. In each we start with a straight chain polymer. In both the guar gum of Part 1 and the corn starch of Part 2, the monomers are simple sugar-like molecules. The monomer of corn starch is glucose, the monomers of guar gum are two other simple sugars similar to glucose. However the monomers are connected differently in the two polymers. For example one could say in the corn starch the sugars are holding hands, while in the guar gum, a hand may be holding onto a foot.

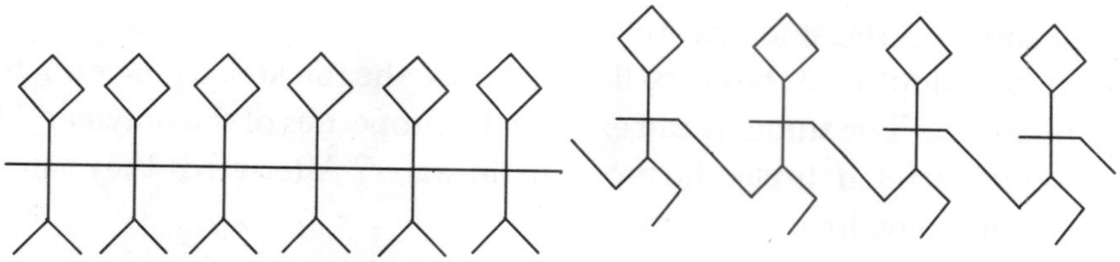

Corn starch is familiar to students as a product that is derived from the flour one gets from grinding corn kernels. Guar gum may seem foreign, but is a very similar product. It is derived from the flour (called guar flour) of the seeds of another plant, called cyamopsis. Corn starch is frequently used in cooking as a thickening agent. Guar gum is a much better thickening agent, with five to eight times the thickening power of corn starch. In fact it is a common natural food additive, especially in lower priced ice creams. You can ask your students to look for it among the ingredients in food products. Borax, is a cleaning agent, available in most super markets.

In part 3, the monomer is an alcohol, like ethanol. All of the straight chain polymers have lots of O-H groups which can hydrogen bond with water. That is why the powdery polymers gets thicker and takes on different properties in water. They are collecting many layers of structured water around them though hydrogen bonds. In Part 1 and Part 3, we add a cross-linker to the polymer which changes its properties dramatically.

Part 1: Slime

Materials: each student will need

- 9 ounce plastic cup,
- Tongue depressor or coffee stirrer,
- Water,
- Guar gum (available from us or chemical supply house),
- Borax (a saturated solution in water).

Procedure

1. Give each student a cup and tongue depressor.
2. Put about 1/2 teaspoon of guar gum into each cup.
3. Put about 1/2 cup of water into the cups (that's about 2/3 filled with water), have the students stir this to dissolve the guar gum as you add the water, and then set it aside.
5. We recommend you do Experiment 2: Corn Starch during the wait for the guar gum to thicken.
6. Add some (1 T or a good squirt.) of the saturated borax solution to each cup, and have the students stir.
7. With the addition of the cross-linking borax, the thickened guar gum turns into."slime." The students can explore the properties of the polymer. Does it bounce, stretch, break, float or sink in water? Afterwards they can take the slime home in a plastic bag.

Part 2: Corn Starch

Corn starch and water is one of the most fascinating natural polymers. If you don't tell them what it is they tend to do more experimental things with it instead of just playing. This mixture is bizarre because it has pressure dependent viscosity. This means that the more pressure you exert on it the less it flows. For example, you can pick it up by squeezing it, but if you let it sit in your hand it will drip out like a liquid. It will dry out while the students are playing with it and the corn starch powder will be all over the place. But it is well worth the mess! After letting them play for a coup of minutes, you can return to the slime activity."

Materials: Each Group Will Need

- Corn starch (a premixed solution of about 1 cup of corn starch and about 1/4 cup of water),
- Foil disposable baking pans,
- Tongue depressors.

Procedure

Give each group of students (4 -5/group) a pan of the corn starch mixture and several tongue depressors and let them explore the properties of the polymer.

Part 3: Glue/Starch Polymer

You can make just one of these polymers or try all three crosslinking reagents to compare the properties of the polymer formed.

Materials

- ❖ Plastic cups
- ❖ White school glue like "Elmers"
- ❖ Liquid laundry starch
- ❖ Saturated borax solution
- ❖ Epsom salts
- ❖ Plastic cups
- ❖ Stirrer
  - Food coloring (optional)

Procedure

1. Mix together the glue and starch(about a 1:1 ratio of glue to liquid starch, or borax solution), stir.
2. The polymer formed has neat properties. If you pull it slowly it will stretch, but it you pull it quickly it breaks (like silly putty). If you do add food coloring, be warned that it will come off on your fingers.
5. 3 For the epsom salts polymer, for one Tablespoon of glue, use about 1/2 tsp. of epsom salts dissolved in 1/2 tsp. of water.

Assessments

Have the students shows you the various polymers that they made since the main objective for this lesson was to have fun exploring the wonders of science.

Extensions

Ask the students to write a story about a polymer that they would like to invent and what it would be used for. When Laura had her students do this, she was impressed with some of the stories, Not only were the stories very creative but also they demonstrated that the students learned the key concepts of the monomers forming polymer chains which are then crosslinked by a different molecule.

There are also additional activities from *WonderScience*: pp. 177-201: Section 6, Materials Science; Unit 23, Polymers, Unit 24, Plastics, Unit 25, Rubber.

Philadelphia Science Content Standards:

SCIENCE CONTENT STANDARD #5: DESIGNED WORLD

This lesson on polymers satisfies Benchmark 1 for grades 5-8: "Discuss how and why science and technology have created synthetic (non-natural) materials".

Cross-References

This lab is a great way to get the students really excited about science. Many may be scared of the lab or the concepts that come with learning science, but this lab shows that science does not have to be frightening. It can be fun, interactive, and entertaining. The Extensions section also provides a way for the students to practice their writing skills.

6.2 Polymer Definition and Properties

The simplest definition of a polymer is a useful chemical made of many repeating units. A polymer can be a three dimensional network (think of the repeating units linked together left and right, front and back, up and down) or two-dimensional network (think of the repeating units linked together left, right, up, and down in a sheet) or a one-dimensional network (think of the repeating units linked left and right in a chain). Each repeating unit is the "-mer" or basic unit with "poly-mer" meaning many repeating units. Repeating units are often made of carbon and hydrogen and sometimes oxygen, nitrogen, sulfur, chlorine, fluorine, phosphorous, and silicon. To make the chain, many links or "-mers" are chemically hooked or polymerized together. Linking countless strips of construction paper together to make paper garlands or hooking together hundreds of paper clips to form chains, or stringing beads helps visualize polymers. Polymers occur in nature and can be made to serve specific needs. Manufactured polymers can be three-dimensional networks that do not melt once formed. Such networks are called THERMOSET polymers. Epoxy resins used in two-part adhesives are thermoset plastics. Manufactured polymers can also be one-dimensional chains that can be melted. These chains are THERMOPLASTIC polymers and are also called LINEAR polymers. Plastic bottles, films, cups, and fibers are thermoplastic plastics.

Polymers abound in nature. The ultimate natural polymers are the deoxyribonucleic acid (DNA) and ribonucleic acid (RNA) that define life. Spider silk, hair, and horn are protein polymers. Starch can be a polymer as is cellulose in wood. Rubber tree latex and cellulose have been used as raw material to make manufactured polymeric rubber and plastics. The first synthetic manufactured plastic was Bakelite, created in 1909 for telephone casing and electrical components. The first manufactured polymeric fiber was Rayon, from cellulose, in 1910. Nylon was invented in 1935 while pursuing a synthetic spider silk.

The Structure of Polymers

Many common classes of polymers are composed of hydrocarbons, compounds of carbon and hydrogen. These polymers are specifically made of carbon atoms bonded together, one to the next, into long chains that are called the backbone of the polymer. Because of the nature of carbon, one or more other atoms can be attached to each carbon atom in the backbone. There are polymers that contain only carbon and hydrogen atoms. Polyethylene, polypropylene, polybutylene, polystyrene and polymethylpentene are examples of these. Polyvinyl chloride (PVC) has chlorine attached to the all-carbon backbone. Teflon has fluorine attached to the all-carbon backbone.

Other common manufactured polymers have backbones that include elements other than carbon. Nylons contain nitrogen atoms in the repeat unit backbone. Polyesters and polycarbonates contain oxygen in the backbone. There are also some polymers that, instead of having a carbon backbone, have a silicon or phosphorous backbone. These are considered inorganic polymers. One of the more famous silicon-based polymers is Silly Putty®.

Molecular Arrangement of Polymers

Think of how spaghetti noodles look on a plate. These are similar to how linear polymers can be arranged if they lack specific order, or are amorphous. Controlling the polymerization process and quenching molten polymers can result in amorphous organization. An amorphous arrangement of molecules has no long-range order or form in which the polymer chains arrange themselves. Amorphous polymers are generally transparent. This is an important characteristic for many applications such as food wrap, plastic windows, headlight lenses and contact lenses.

Obviously not all polymers are transparent. The polymer chains in objects that are translucent and opaque may be in a crystalline arrangement. By definition, a crystalline arrangement has atoms, ions, or in this case, molecules arranged in distinct patterns. You generally think of crystalline structures in table salt and gemstones, but they can occur in plastics. Just as quenching can produce amorphous arrangements, processing can control the degree of crystallinity for those polymers that are able to crystallize. Some polymers are designed to never be able to crystallize. Others are designed to be able to be crystallized. The higher the degree of crystallinity, generally, the less light can pass through the polymer. Therefore, the degree of translucence or opaqueness of the polymer can be directly affected by its crystallinity. Crystallinity creates benefits in strength, stiffness, chemical resistance, and stability.

Scientists and engineers are always producing more useful materials by manipulating the molecular structure that affects the final polymer produced. Manufacturers and processors introduce various fillers, reinforcements and additives into the base polymers, expanding product possibilities.

Characteristics of Polymers

The majority of manufactured polymers are thermoplastic, meaning that once the polymer is formed it can be heated and reformed over and over again. This property allows for easy processing and facilitates recycling. The other group, the thermosets, cannot be remelted. Once these polymers are formed, reheating will cause the material to ultimately degrade, but not melt.

Every polymer has very distinct characteristics, but most polymers have the following general attributes:

1. Polymers can be very resistant to chemicals. Consider all the cleaning fluids in your home that are packaged in plastic. Reading the warning labels that describe what happens when the chemical comes in contact with skin or eyes or is ingested will emphasize the need for chemical resistance in the plastic packaging. While solvents easily dissolve some plastics, other plastics provide safe, non-breakable packages for aggressive solvents.
2. Polymers can be both thermal and electrical insulators. A walk through your house will reinforce this concept, as you consider all the appliances, cords, electrical outlets and wiring that are made or covered with polymeric materials. Thermal resistance is evident in the kitchen with pot and pan handles made of polymers, the coffee pot handles, the foam core of refrigerators and freezers, insulated cups, coolers, and microwave cookware. The thermal underwear that many skiers wear is made of polypropylene and the fiberfill in winter jackets is acrylic and polyester.
3. Generally, polymers are very light in weight with significant degrees of strength. Consider the range of applications, from toys to the frame structure of space stations, or from delicate nylon fiber in pantyhose to Kevlar, which is used in bulletproof vests. Some polymers float in water while others sink. But, compared to the density of stone, concrete, steel, copper, or aluminum, all plastics are lightweight materials.
4. Polymers can be processed in various ways. Extrusion produces thin fibers

or heavy pipes or films or food bottles. Injection molding can produce very intricate parts or large car body panels. Plastics can be molded into drums or be mixed with solvents to become adhesives or paints. Elastomers and some plastics stretch and are very flexible. Some plastics are stretched in processing to hold their shape, such as soft drink bottles. Other polymers can be foamed like polystyrene (StyrofoamTM), polyurethane and polyethylene.

5. Polymers are materials with a seemingly limitless range of characteristics and colors. Polymers have many inherent properties that can be further enhanced by a wide range of additives to broaden their uses and applications. Polymers can be made to mimic cotton, silk, and wool fibers; porcelain and marble; and aluminum and zinc. Polymers can also make possible products that do not readily come from the natural world, such as clear sheets and flexible films.
6. Polymers are usually made of petroleum, but not always. Many polymers are made of repeat units derived from natural gas or coal or crude oil. But building block repeat units can sometimes be made from renewable materials such as polylactic acid from corn or cellulosics from cotton linters. Some plastics have always been made from renewable materials such as cellulose acetate used for screwdriver handles and gift ribbon. When the building blocks can be made more economically from renewable materials than from fossil fuels, either old plastics find new raw materials or new plastics are introduced.
7. Polymers can be used to make items that have no alternatives from other materials. Polymers can be made into clear, waterproof films. PVC is used to make medical tubing and blood bags that extend the shelf life of blood and blood products. PVC safely delivers flammable oxygen in non-burning flexible tubing. And anti-thrombogenic material, such as heparin, can be incorporated into flexible PVC catheters for open heart surgery, dialysis, and blood collection. Many medical devices rely on polymers to permit effective functioning.

Solid Waste Management

In addressing all the superior attributes of polymers, it is equally important to discuss some of the challenges associated with the materials. Most plastics deteriorate in full sunlight, but never decompose completely when buried in landfills. However, other materials such as glass, paper, or aluminum do not readily decompose in landfills either. Some bioplastics do decompose to carbon dioxide and water, however, in specially designed food waste commercial composting facilities ONLY. They do not biodegrade under other circumstances.

For 2005[1] the EPA characterization of municipal solid waste before recycling for the United States showed plastics made up 11.8 per cent of our trash by weight compared to paper that constituted 34.2 per cent. Glass and metals made up 12.8 per cent by weight. And yard trimmings constituted 13.1 per cent of municipal solid waste by weight. Food waste made up 11.9 per cent of municipal solid waste. The characteristics that make polymers so attractive and useful, lightweight and almost limitless physical forms of many polymers designed to deliver specific appearance and functionality, make post-consumer recycling challenging. When enough used plastic items can be gathered together, companies develop technology to recycle those used plastics. The recycling rate for all plastics is not as high as any would want. But, the recycling rate for the 1,170,000,000 pounds of polyester bottles, 23.1 per cent, recycled in 2005 and the 953,000,000 pounds of high density polyethylene bottles, 28.8 per cent, recycled in 2005 show that when critical mass of defined material is available, recycling can be a commercial success[2].

Applications for recycled plastics are growing every day. Recycled plastics can be blended with virgin plastic (plastic that has not been processed before) without sacrificing properties in many applications. Recycled plastics are used to make polymeric timbers for use in picnic tabies, fences and outdoor playgrounds, thus providing low maintenance, no splinters products and saving natural lumber. Plastic from soft drink and water bottles can be spun into fiber for the production of carpet or made into new food bottles. Closed loop recycling does occur, but sometimes the most valuable use for a recycled plastic is into an application different than the original use.

An option for plastics that are not recycled, especially those that are soiled, such as used food wrap or diapers, can be a waste-to-energy system (WTE). In 2005, 13.6 per cent of US municipal solid waste was processed in WTE systems1. When localities decide to use waste-to-energy systems to manage solid waste, plastics can be a useful component.

The controlled combustion of polymers produces heat energy. The heat energy produced by the burning plastic municipal waste not only can be converted to electrical energy but also helps burn the wet trash that is present. Paper also produces heat when burned, but not as much as do plastics. On the other hand, glass, aluminum and other metals do not release any energy when burned.

To better understand the incineration process, consider the smoke coming off a burning item. If one were to ignite the smoke with a lit propane torch, one would observe that the smoke disappears. This exercise illustrates that the by-products of incomplete burning are still flammable. Proper incineration burns the material and the by-products of the initial burning and also takes care of air and solid emissions to insure public safety.

Some plastics can be composted either because of special additives or because of the construction of the polymers. Compostable plastics frequently require more intense conditions to decompose than are available in backyard compost piles. Commercial composters are suggested for compostable plastics. In 20051, composting processed 8.4 per cent of US municipal solid waste.

Plastics can also be safely land filled, although the valuable energy resource of the plastics would then be lost for recycling or energy capture. In 20051, 54.3 per cent of US municipal solid waste was land filled. Plastics are used to line landfills so that leachate is captured and groundwater is not polluted. Non-degrading plastics help stabilize the ground so that after the landfill is closed, the land can be stable enough for useful futures.

Polymers affect every day of our life. These materials have so many varied characteristics and applications that their usefulness can only be measured by our imagination. Polymers are the materials of past, present and future generations.

6.3 Introduction to Polymer Chemistry

Glass Transition Temperature (T_g)

The temperature (actually a broad range of temperatures) at which a glassy polymer softens into a viscous liquid or rubbery phase. On the molecular level, it is the temperature at which chains in amorphous (i.e., disordered) regions of the polymer gain enough thermal energy to begin sliding past one another at a noticable rate. For an amorphous polymer, the T_g reports the minimum processing temperature. The T_g is strongly dependant on polymer structure (including stereochemistry), and ranges from far below 0 °C for very flexible chains, and above 400 °C for very stiff chains.

The transition is detectable by a variety of methods, including Differential Scanning Calorimetry (DSC), and Thermal-Mechanical Analysis (TMA). Note that some authors report the T_g as the onset temperature (i.e., the beginning of the range), while others report the midpoint of the range.

Melting Temperature (T_m)

The temperature (actually a narrow range of temperatures) at which the ordered regions of a crystalline polymer melt, similar to a small molecule. Crystallization is essential for many high-performance polymers because it greatly increases the strength of the material.

Like the T_g, the T_m is detectable by DSC, TMA, and other techniques.

Decomposition Temperature

The temperature above which chemical degradation occurs. This temperature is conveniently measured by Thermogravimetric Analysis (TGA), a technique in which one simply weights the sample continuously while heating it. Once decomposition begins, small molecular fragments are released which distill away, and the sample loses weight.

Modulus

The proportionality constant between stress and strain, and therefore can be thought of as *stiffness*.

Molecular Weight

Because virtually all polymers are mixtures of many large molecules, one must resort to averages to describe molecular weight. Among many possible ways of reporting averages, three are commonly used: the *number average, weight average*, and *z-average* molecular weights. The weight average is probably the most useful of the three, because it fairly accounts for the contributions of different sized chains to the overall behavior of the polymer, and correlates best with most of the physical properties of interest.

$$\text{Number of average MW } (\overline{M}_n) = \frac{\Sigma(M_i N_i)}{\Sigma(N_i)}$$

$$\text{Weight of average MW } (\overline{M}_w) = \frac{\Sigma(M_i^2 N_i)}{\Sigma(M_i N_i)}$$

$$\text{Z average MW } (\overline{M}_w) = \frac{\Sigma(M_i^3 N_i)}{\Sigma(M_i N_i)}$$

The ratio of M_w to M_n is known as the *polydispersity index* (PDI), and provides a rough indication of the breadth of the distribution. The PDI approaches 1.0 (the lower limit) for special polymers with very narrow MW distributions, but, for typical commercial polymers, is typically greater than 2 (occasionally much greater). Here is a typical MW distribution curve, measured by Size Exclusion Chromatography (SEC):

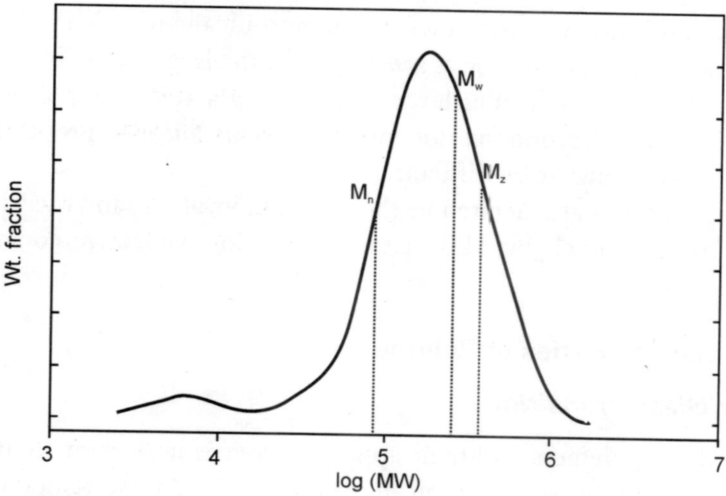

Fig. 6.1: Dependence of Polymer Properties on MW.

Many polymer properties of interest (T_g, modulus, tensile strength, etc.) follow a peculiar pattern with increasing MW. Small molecules have small values, then there is a sharp rise in properties as the chains grow to intermediate size (*oligomers*), and then the properties level off as the chains become long enough to be true polymers.

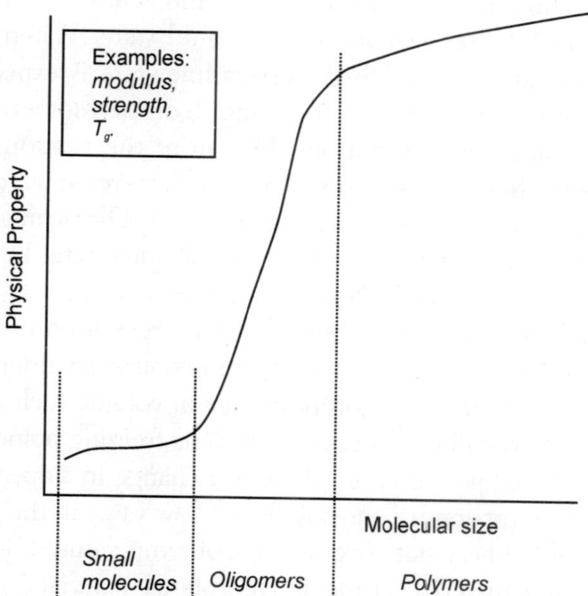

However, a few properties important for polymer processing, like melt viscosity and solution viscosity, increase monotonically with MW. This means that the goal of polymer synthesis is *not* to make the largest possible molecules, but rather, to make molecules large enough to get onto the plateau region. Increasing the MW beyond this does not improve the physical properties much, but makes processing more difficult.

A few properties are dictated by the repeat units alone, and therefore these are not changed much by MW. Examples: color, dielectric constant, and refractive index.

6.4 Thermal Properties of Polymers

Polymer Glass Transition

In the study of polymers and their applications, it is important to understand the concept of the glass transition temperature, T_g. As the temperature of a polymer drops below T_g, it behaves in an increasingly brittle manner. As the temperature rises above the T_g, the polymer becomes more rubber-like. Thus, knowledge of T_g is essential in the selection of materials for various applications. In general, values of T_g well below room temperature define the domain of elastomers and values above room temperature define rigid, structural polymers.

This behavior can be understood in terms of the structure of glassy materials which are formed typically by substances containing long chains, networks of linked atoms or those that possess a complex molecular structure. Normally such materials have a high *viscosity* in the liquid state. When rapid cooling occurs to a temperature at which the crystalline state is expected to be the more stable, molecular movement is too sluggish or the geometry too awkward to take up a *crystalline* conformation. Therefore the random arrangement characteristic of the liquid persists down to temperatures at which the viscosity is so high that the material is considered to be solid. The term glassy has come to be synonymous with a persistent non-equilibrium state. In fact, a path to the state of lowest energy might not be available.

To become more quantitative about the characterization of the liquid-glass transition phenomenon and T_g, we note that in cooling an amorphous material from the liquid state, there is no abrupt change in volume such as occurs in the case of cooling of a crystalline material through its freezing point, T_f. Instead, at the glass transition temperature, T_g, there is a change in slope of the curve of specific volume vs. temperature, moving from a low value in the glassy state to a higher value in the rubbery state over a range of temperatures. This comparison between a crystalline material (1) and an amorphous material (2) is illustrated in the figure below. Note that the intersections of the two straight line segments of curve (2) defines the quantity T_g.

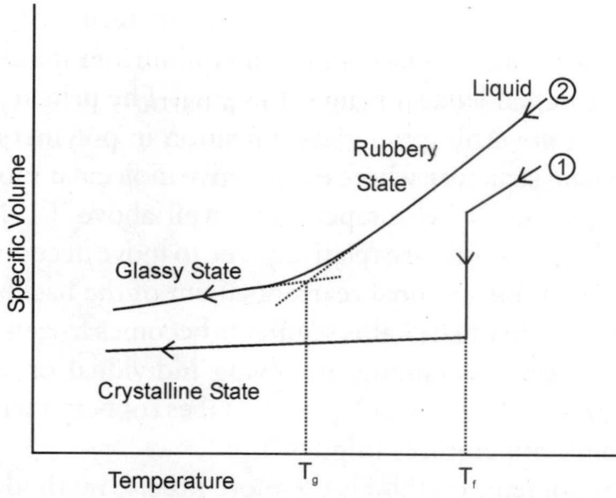

The specific volume measurements shown here, made on an amorphous polymer (2), are carried out in a dilatometer at a slow heating rate. In this apparatus, a sample is placed in a glass bulb and a confining liquid, usually mercury, is introduced into the bulb so that the liquid surrounds the sample and extends partway up a narrow bore glass capillary tube. A capillary tube is used so that relatively small changes in polymer volume caused by changing the temperature produce easily measured changes in the height of the mercury in the capillary.

The determination of T_g for amorphous materials, including polymers as mentioned above, by dilatometric methods (as well as by other methods) are found to be rate dependent. This is schematically illustrated in the figure below, again representing an amorphous polymer, where the higher value, T_{g2}, is obtained with a substantially higher cooling rate than for T_{g1}.

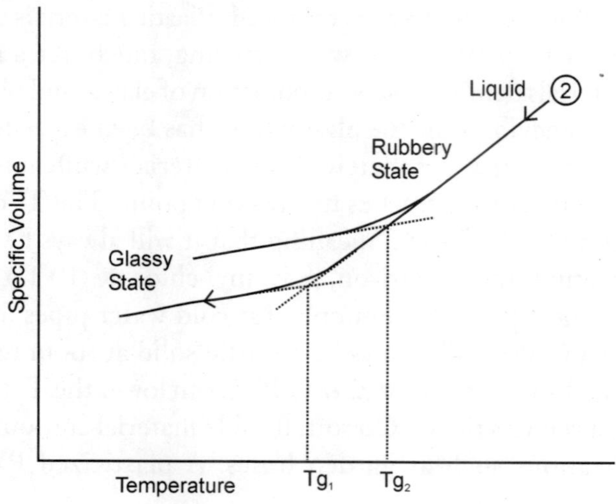

We can understand this rate dependence in terms of intermolecular relaxation processes. Since a glass is not an equilibrium phase, its properties will exhibit a time dependence, or physical aging. The primary portion of the relaxation behavior governing the glass transition in polymers can be related to their tangled chain structure where cooperative molecular motion is required for internal readjustments. At temperatures well above T_g, 10 to 50 repeat units of the polymer backbone are relatively free to move in cooperative thermal motion to provide conformational rearrangement of the backbone. Below T_g, the motion of these individual chains segments becomes frozen with only small scale molecular motion remaining, involving individual or small groups of atoms. Thus a rapid cooling rate or "quench" takes rubbery material into glassy behavior at higher temperatures (higher T_g).

While the dilatometer method is the more precise method of determining the glass transition temperature, it is a rather tedious experimental procedure and measurements of Tg are often made in a differential scanning calorimeter (DSC). In this instrument, the heat flow into or out of a small (10—20 mg) sample is measured as the sample is subjected to a programmed linear temperature change. This will be discussed in the next section. There are other methods of measurement such as density, dielectric constant and elastic modulus which are treated in texts on polymers. These methods are, of course, also rate dependent.

T_g and Mechanical Properties

Another important property of polymers, also strongly dependent on their temperatures, is their response to the application of a force, as indicated by two main types of behavior: *elastic* and *plastic*. Elastic materials will return to their original shape once the force is removed. Plastic materials will not regain their shape. In plastic materials, flow is occurring, much like a highly viscous liquid. Most materials demonstrate a combination of elastic and plastic behavior, showing plastic behavior after the elastic limit has been exceeded.

Glass is one of the few completely elastic materials while it is below its T_g. It will remain elastic until it reaches its breaking point. The T_g of glass occurs between 510 and 560 degrees C, meaning that it will always be a brittle solid at room temperature. In comparison, polyvinyl chloride (PVC) has a T_g of 83 degrees C, making it good, for example, for cold water pipes, but unsuitable for hot water. PVC also will always be a brittle solid at room temperature.

Adding a small amount of *plasticizer* to PVC can lower the T_g to—40 degrees C. This addition renders the PVC a soft, flexible material at room temperature, ideal for applications such as garden hoses. A plasticized PVC hose can, however, become stiff and brittle in winter. In this case, as in any other, the relation of the T_g to the ambient temperature is what determines the choice of a given material in a particular application.

A striking example of the rate dependence of these viscoelastic properties is furnished by Silly Putty. Slowly pulling on two parts of the Silly Putty stretches it apart until it very slowly separates. Placing the Silly Putty on a table and hitting it with a hammer will shatter it.

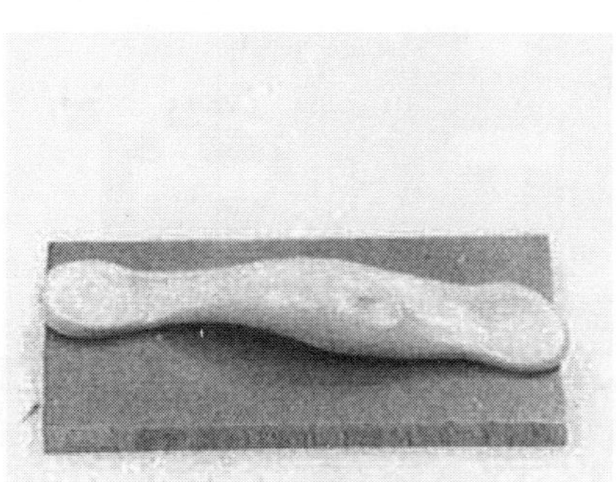

Slowly Deformed

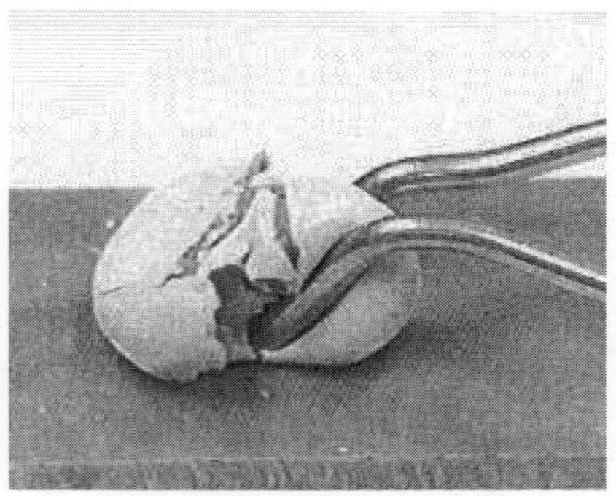

Rapidly Deformed

Photos Courtesy of Geon Corp

The above images are representative of the behavior of a material above and below its glass transition temperature. The image on the (left) is Silly Putty that has been slowly stretched. The image on the (right) is Silly Putty which has been hit with a hammer. The speed of the hammer raised the rate of the application of the force and in turn raised the T_g. This caused the Silly Putty to react as if it were below its T_g and to shatter. Even though both reactions took place at the same ambient temperature, one reaction appeared to be above the effective T_g and the other appeared to be below.

Our focus has been on amorphous polymers in the preceding discussion but we have hardly touched on their mechanical properties. A further complication arises in dealing with general polymers from their semi-crystalline morphology in which amorphous regions and crystalline regions are intermingled. This gives rise to a mixed behavior depending on the per cent crystallinity and on their temperature, relative to T_g of the amorphous regions. You are referred to texts on polymer science for basic discussion of these topic but the inhomogeneity of the material and its characteristics presents interesting analytical challenges.

Differential Scanning Calorimetry

In differential scanning calorimetry (DSC), the thermal properties of a sample are compared against a standard reference material which has no transition in the temperature range of interest, such as powdered alumina. Each is contained in a small holder within an adiabatic enclosure as illustrated below.

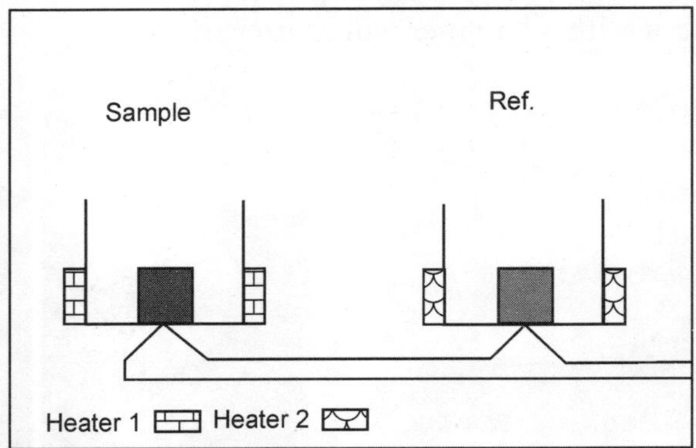

The temperature of each holder is monitored by a thermocouple and heat can be supplied electrically to each holder to keep the temperature of the two equal. A plot of the difference in energy supplied to the sample against the average temperature, as the latter is slowly increased through one or more thermal transitions of the sample yields important information about the transition, such as latent heat or a relatively abrupt change in heat capacity.

The glass transition process is illustrated in the figure below for a glassy polymer which does not crystallize and is being slowly heated from below T_g.

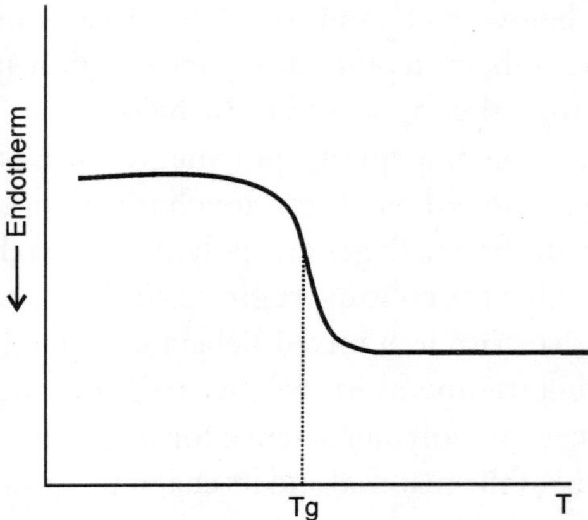

Here, the drop marked T_g at its midpoint represents the increase in energy supplied to the sample to maintain it at the same temperature as the reference material, due to the relatively rapid increase in the heat capacity of the sample as its temperature is raised through T_g. The addition of heat energy corresponds to this endothermal direction.

A melting process is also illustrated below for the case of a highly crystalline polymer which is slowly heated through its melting temperature:

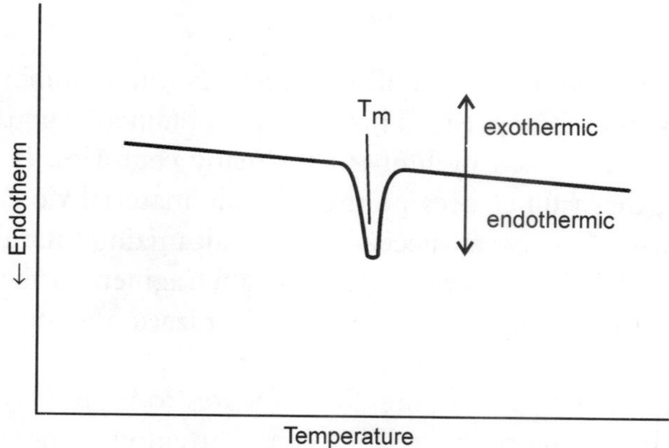

Again, as the melting temperature is reached, an endothermal peak appears because heat must be preferentially added to the sample to continue this essentially constant temperature process. The peak breadth is primarily related to the size and degree of perfection of the polymer crystals.

Note that if the process were reversed so that the sample were being cooled from the melt, the plot would be inverted. In that case, as both are being cooled by ambient conditions, even less heat would need to be supplied to the sample than to the reference material, in order that crystals can form. This corresponds to an exothermal process. Use of the DSC will be illustrated again in the section on liquid crystals in connection with the identification of their phase transitions. An interesting exercise for the reader would be to predict the general form of a DSC plot for a semicrystalline polymer which has been rapidly quenched from the melt to a temperature below T_g. In the DSC plot, assume the temperature is slowly increased from this value below T_g to a value well above, thus allowing for significant increases in the chain mobility as temperatures above T_g are reached so that some crystallization can begin, well before the melting point is reached.

Applications of Polymers

Macromolecular science has had a major impact on the way we live. It is difficult to find an aspect of our lives that is not affected by polymers. Just 50 years ago, materials we now take for granted were non-existent. With further advances in the understanding of polymers, and with new applications being researched, there is no reason to believe that the revolution will stop any time soon. This section presents some common applications of the polymer classes introduced in the section on Polymer Structure. These are by no means all of the applications, but a cross section of the ways polymers are used in industry.

Elastomers

Rubber is the most important of all *elastomers*. Natural rubber is a polymer whose repeating unit is isoprene. This material, obtained from the bark of the rubber tree, has been used by humans for many centuries. It was not until 1823, however, that rubber became the valuable material we know today. In that year, Charles Goodyear succeeded in "vulcanizing" natural rubber by heating it with sulfur. In this process, sulfur chain fragments attack the polymer chains and lead to *cross-linking*. The term vulcanization is often used now to describe the cross-linking of all elastomers.

Much of the rubber used in the United States today is a synthetic variety called styrene-butadiene rubber (SBR). Initial attempts to produce synthetic rubber revolved around isoprene because of its presence in natural rubber. Researchers eventually found success using butadiene and styrene with sodium metal as the *initiator*. This rubber was called Buna-S—"Bu" from butadiene, "na" from the symbol for sodium, and "S" from styrene. During World War II, hundreds of thousands of tons of synthetic rubber were produced in government controlled factories. After the war, private industry took over and changed the name to styrene-butadiene rubber. Today, the United States consumes on the order of a million tons of SBR each year. Natural and other synthetic rubber materials are quite important.

Plastics

Americans consume approximately 60 billion pounds of plastics each year. The two main types of plastics are thermoplastics and thermosets. Thermoplastics soften on heating and harden on cooling while thermosets, on heating, flow and cross-link to form rigid material which does not soften on future heating. Thermoplastics account for the majority of commercial usage. Among the most important and versatile of the hundreds of commercial plastics is polyethylene. Polyethylene is used in a wide variety of applications because, based on its structure, it can be produced in many different forms. The first type to be commercially exploited was called low density polyethylene (LDPE) or branched polyethylene. This polymer is characterized by a large degree of branching, forcing the molecules to be packed rather loosely forming a low density material. LDPE is soft and pliable and has applications ranging from plastic bags, containers, textiles, and electrical insulation, to coatings for packaging materials.

Another form of polyethylene differing from LDPE only in structure is high density polyethylene (HDPE) or linear polyethylene. This form demonstrates little or no branching, enabling the molecules to be tightly packed. HDPE is much more rigid than branched polyethylene and is used in applications where rigidity is important. Major uses of HDPE are plastic tubing, bottles, and bottle caps.

Other forms of this material include high and ultra-high molecular weight polyethylenes. HMW and UHMW, as they are known. These are used in applications where extremely tough and resilient materials are needed.

Fibers

Fibers represent a very important application of polymeric materials, including many examples from the categories of plastics and elastomers. Natural fibers such as cotton, wool, and silk have been used by humans for many centuries. In 1885, artificial silk was patented and launched the modern fiber industry. Man-made fibers include materials such as nylon, polyester, rayon, and acrylic. The combination of strength, weight, and durability have made these materials very important in modern industry. Generally speaking, fibers are at least 100 times longer than they are wide. Typical natural and artificial fibers can have *axial ratios* (ratio of length to diameter) of 3000 or more.

Synthetic polymers have been developed that posess desirable characteristics, such as a high softening point to allow for ironing, high *tensile strength*, adequate stiffness, and desirable fabric qualities. These polymers are then formed into fibers with various characteristics. Nylon (a generic term for polyamides) was developed in the 1930's and used for parachutes in World War II. This synthetic fiber, known for its strength, elasticity, toughness, and resistance to abrasion, has commercial applications including clothing and carpeting. Nylon has special properties which distinguish it from other materials. One such property is the elasticity. Nylon is very elastic, however after elastic limit has been exceeded the material will not return to its original shape. Like other synthetic fibers, Nylon has a large electrical resistance. This is the cause for the build-up of static charges in some articles of clothing and carpets. From textiles to bullet-proof vests, fibers have become very important in modern life. As the technology of fiber processing expands, new generations of strong and light weight materials will be produced.

Processing Polymers

Once a polymer with the right properties is produced, it must be manipulated into some useful shape or object. Various methods are used in industry to do this. Injection molding and extrusion are widely used to process plastics while spinning is the process used to produce fibers.

Injection Molding

One of the most widely used forms of plastic processing is injection molding. Basically, a plastic is heated above its glass transition temperature (enough so that it will flow) and then is forced under high pressure to fill the contents of a mold. The molten plastic in usually "squeezed" into the mold by a ram or a reciprocating screw. The plastic is allowed to cool and is then removed from the mold in its final form. The advantage of injection molding is speed; this process can be performed many times each second.

Extrusion

Extrusion is similar to injection molding except that the plastic is forced through a die rather than into a mold. However, the disadvantage of extrusion is that the objects made must have the same cross-sectional shape. Plastic tubing and hose is produced in this manner.

Spinning

The process of producing fibers is called spinning. There are three main types of spinning: melt, dry, and wet. Melt spinning is used for polymers that can be melted easily. Dry spinning involves dissolving the polymer into a solution that can be evaporated. Wet spinning is used when the solvent cannot be evaporated and must be removed by chemical means. All types of spinning use the same principle, so it is convenient to just describe just one. In melt spinning, a mass of polymer is heated until it will flow. The molten polymer is pumped to the face of a metal disk containing many small holes, called the spinneret. Tiny streams of polymer that emerge from these holes (called filaments) are wound together as they solidify, forming a long fiber. Speeds of up to 2500 feet/minute can be employed in spinning.

Following the spinning process, as noted in the section on Polymer Morphology, fibers are stretched substantially—from 3 to 8 or more times their original length to produce increased chain alignment and enhanced crystallinity in order to yield improved strength.

Polymer Morphology

Molecular shape and the way molecules are arranged in a solid are important factors in determining the properties of polymers. From polymers that crumble to the touch to those used in bullet proof vests, the molecular structure, conformation and orientation of the polymers can have a major effect on the macroscopic properties of the material. The general concept of self-assembly enters into the organization of molecules on the micro and macroscopic scale as they aggregate into more ordered structures. Crystallization, discussed below, is an example of the self-assembly process as is the orientational organization of liquid crystals to be discussed later.

Crystallinity

We need to distinguish here, between *crystalline* and *amorphous* materials and then show how these forms coexist in polymers. Consider a comparison between glass, an amorphous material, and ice which is crystalline. Despite their common appearance as hard, clear material, capable of being melted, a difference is apparent when viewed between crossed polarizers, as illustrated below:

> The highly ordered crystalline structure of ice changes the apparent properties of the polarized light, and the ice appears bright. Glass and water, lacking that highly ordered structure, both appear dark.

The amorphous morphology of glass leads to very different properties from crystalline solids. This is illustrated in the heating process where the application of heat to glass turns it from a brittle solid-like material at room temperature to a viscous liquid, as discussed later in more detail under *Thermal Properties of Polymers*. In contrast, the application of heat to ice turns it from solid to liquid. Crystalline melting leads to striking changes in optical properties during the melting process when observed through crossed polarizers. This is illustrated in the following movie of the melting of an organic crystalline material. Note that while the temperatures are not recorded, the entire process occurs over a very narrow temperature range.

The reasons for the differing behaviors lie mainly in the structure of the solids. Crystalline materials have their molecules arranged in repeating patterns. Table salt has one of the simplest atomic structures with its component atoms, Na^+ and Cl^-, arranged in alternating rows and the structure of a small cube. Salt, sugar, ice and most metals are crystalline materials. As such, they all tend to have highly ordered and regular structures. Amorphous materials, by contrast, have their molecules arranged randomly and in long chains which twist and curve around one-another, making large regions of highly structured morphology unlikely.

The morphology of most polymers is semi-crystalline. That is, they form mixtures of small crystals and amorphous material and melt over a range of temperature instead of at a single melting point. The crystalline material shows a high degree of order formed by folding and stacking of the polymer chains. The amorphous or glass-like structure shows no long range order, and the chains are tangled as illustrated below:

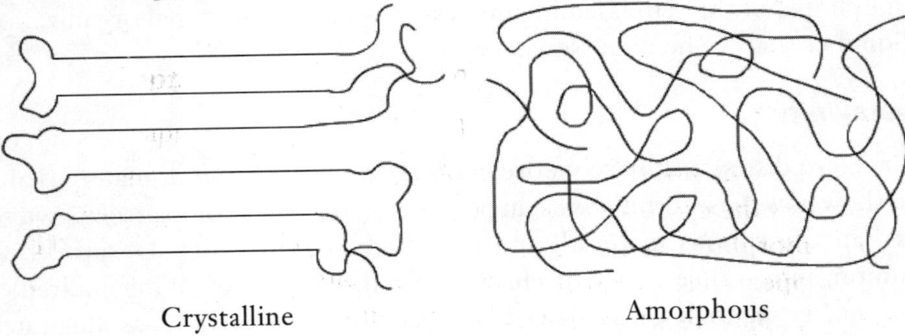

Crystalline Amorphous

There are some polymers that are completely amorphous, but most are a combination with the tangled and disordered regions surrounding the crystalline areas. Such a combination is shown in the following diagram.

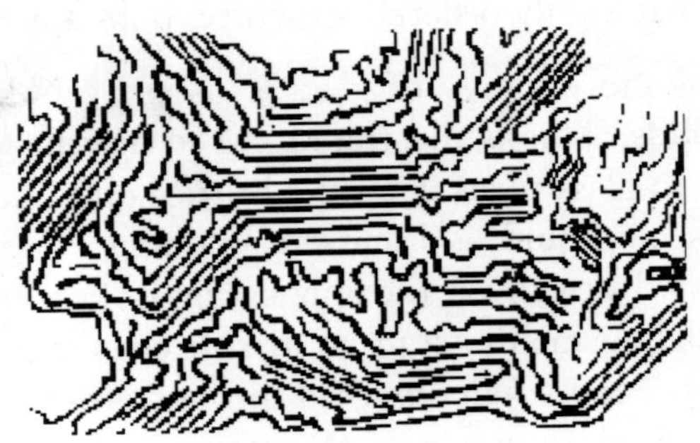

An *amorphous* solid is formed when the chains have little orientation throughout the bulk polymer. The *glass transition temperature* is the point at which the polymer hardens into an amorphous solid. This term is used because the amorphous solid has properties similar to glass.

In the crystallization process, it has been observed that relatively short chains organize themselves into crystalline structures more readily than longer molecules. Therefore, the *degree of polymerization* (DP) is an important factor in determining the crystallinity of a polymer. Polymers with a high DP have difficulty organizing into layers because they tend to become tangled.

The cooling rate also influences the amount of crystallinity. Slow cooling provides time for greater amounts of crystallization to occur. Fast rates, on the other hand, such as rapid quenches, yield highly amorphous materials. For a more complete discussion, see the section on thermal properties. Subsequent annealing (heating and holding at an appropriate temperature below the crystalline melting point, followed by slow cooling) will produce a significant increase in crystallinity in most polymers, as well as relieving stresses.

Low molecular weight polymers (short chains) are generally weaker in strength. Although they are crystalline, only weak Van der Waals forces hold the lattice together. This allows the crystalline layers to slip past one another causing a break in the material. High DP (amorphous) polymers, however, have greater strength because the molecules become tangled between layers. For uses and examples of high and low DP polymers, see the section on Polymer Applications. In the case of fibers, stretching to 3 or more times their original length when in a semi-crystalline state produces increased chain alignment, crystallinity and strength.

In most polymers, the combination of crystalline and amorphous structures forms a material with advantageous properties of strength and stiffness.

Also influencing the polymer morphology is the size and shape of the monomers' substituent groups. If the monomers are large and irregular, it is difficult for the polymer chains to arrange themselves in an ordered manner, resulting in a more amorphous solid. Likewise, smaller monomers, and monomers that have a very regular structure (e.g. rod-like) will form more crystalline polymes.

7

Polymer Chemistry and Chemical Properties of Polymers

7.1 Polymer Chemistry

Polymer chemistry or macromolecular chemistry is a multidisciplinary science that deals with the chemical synthesis and chemical properties of polymers or macromolecules. According to IUPAC recommendations, macromolecules refer to the individual molecular chains and are the domain of chemistry. Polymers describe the bulk properties of polymer materials and belong to the field of polymer physics as a subfield of physics.

- Biopolymers produced by living organisms:
  - *Structural proteins*: collagen, keratin, elastin …
  - *Chemically functional proteins*: enzymes, hormones, transport proteins…
  - *Structural polysaccharides*: cellulose, chitin…
  - *Storage polysaccharides*: starch, glycogen…
  - *Nucleic acids*: DNA, RNA
- Synthetic polymers used for plastics—fibers, paints, building materials, furniture, mechanical parts, adhesives:
  - *Thermoplastics*: polyethylene, Teflon polystyrene, polypropylene, polyester, polyurethane, polymethyl methacrylate, polyvinyl chloride, nylon, rayon, celluloid, silicone, fiberglass…
  - *Thermosetting plastics*: vulcanized rubber, Bakelite, Kevlar, epoxy…

Polymers form by polymerization of monomers. A polymer is chemically described by its degree of polymerisation, molar mass distribution, tacticity, copolymer distribution, the degree of branching, by its end-groups, crosslinks, crystallinity and thermal properties such as its glass transition temperature and melting temperature. Polymers in solution have special characteristics with respect to solubility, viscosity and gelation.

History

The work of Henri Braconnot in 1777 and Christian Schönbein in 1846 led to the discovery of nitrocellulose, which, when treated with camphor produced celluloid. Dissolved in ether or acetone, it is collodion, used as a wound dressing since the U.S. Civil War. Cellulose acetate was first prepared in 1865. In 1834, Friedrich Ludersdorf and Nathaniel Hayward independently discovered that adding sulfur to raw natural rubber (polyisoprene) helped prevent the material from becoming sticky. In 1844 Charles Goodyear received a U.S. patent for vulcanizing rubber with sulfur and heat. Thomas Hancock had received a patent for the same process in the UK the year before.

In 1884 Hilaire de Chardonnet started the first artificial fiber plant based on regenerated cellulose, or viscose rayon, as a substitute for silk, but it was very flammable. In 1907 Leo Baekeland invented the first synthetic polymer, a thermosetting phenol-formaldehyde resin called Bakelite.. Cellophane was invented in 1908 by Jocques Brandenberger who squirted sheets of viscose rayon into an acid bath. In 1922 Hermann Staudinger was the first to propose that polymers consisted of long chains of atoms held together by covalent bonds. He also proposed to name these compounds macromolecules. Before that, scientists believed that polymers were clusters of small molecules (called colloids), without definite molecular weights, held together by an unknown force. Staudinger received the Nobel Prize in Chemistry in 1953. Wallace Carothers invented the first synthetic rubber called neoprene in 1931, the first polyester, and went on to invent nylon, a true silk replacement, in 1935. Paul Flory was awarded the Nobel Prize in Chemistry in 1974 for his work on polymer random coil configurations in solution in the 1950s. Stephanie Kwolek developed an aramid, or aromatic nylon named Kevlar, patented in 1966.

There are now a large number of commercial polymers, including composite materials such as carbon fiber-epoxy, polystyrene-polybutadiene (HIPS), acrylonitrile-butadiene-styrene (ABS), and other such materials that combine the best properties of their various components, including polymers designed to work at high temperatures in automobile engines.

7.2 RAFT

RAFT or Reversible addition–fragmentation chain transfer is a form of living radical polymerization. Reversible addition–fragmentation chain transfer polymerization was discovered by the CSIRO in 1998. This is a new method for the synthesis of living radical polymers that may be more versatile than atom transfer radical polymerization (ATRP) or nitroxide-mediated polymerization (NMP). RAFT polymerization uses thiocarbonylthio compounds, such as dithioesters, dithiocarbamates, trithiocarbonates, and xanthates in order to mediate the polymerization via a reversible chain-transfer process.

This allows access to polymers with low polydispersity and high functionality. RAFT also allows for the production of complex architectures such as block, star, graft, comb, and brush (co)polymers. RAFT is also known for its compatibility with a great variety of monomers.

7.3 Concepts of Polymer Chemistry

1. Annealing
2. Boltzmann's Entropy Change of Mixing
3. Bulk Modulus
4. Chain Expansion
5. Coordination Number
6. Crystallinity
7. Differential Scanning Calorimetry
8. Differential Thermal Analysis
9. Dynamic Mechanical Analysis
10. End Group Titration
11. End to End Distance
12. Enthalpy Change of Mixing
13. Entropy Change of Mixing
14. Fox Flory Glass Transition Temperature
15. Flory Solvent
16. Flory Temperature
17. Free Volume
18. Gibbs Free Energy Change of Mixing
19. Glass Transition Temperature
20. Good Solvent
21. Light Scattering
22. Mark Houwink
23. Molecular Weight
24. Osmotic Pressure
25. Poiseuille's Law
26. Radius of Gyration
27. Random Coil

28. Relative Viscosity
29. Rigid Rods
30. Specific Viscosity
31. Sphere
32. Stirling's Approximation
33. TGA-FTIR
34. Thermogravimetric analysis
35. Thermomechanical analysis
36. Theta Solvent
37. Viscosity
38. Volume Fractions

Polymer chemistry or macromolecular chemistry is a branch of chemistry that deals with the preparation and properties of polymers and macromolecules, and one of several fields that play an important role in polymer science and technology. Due to their commercial importance, most research in polymer chemistry is concerned with synthetic organic polymers, such as plastics or fibers. Many polymer chemists, however, work on problems related to medicine, biology and biochemistry, or materials science.

Basic Concepts in Polymer Chemistry

Polymers and Macromolecules

In chemistry, the terms "polymer molecule" and "macromolecule" are used interchangeably. A polymer molecule has a high molecular mass and is comprised of many smaller, repeating subunits or *monomers*. Polymers may be found in nature, such as the DNA and proteins found in living cells, or created in laboratories or factories.

Polymer molecules come in many shapes and sizes. A polymer molecule may be a long chain of a single monomer repeated over and over again or a complex network containing dozens of different types of monomers. The identity, variety, and arrangement of monomers in a polymer molecule affect the chemical and physical properties of the polymer molecule.

Polymer Synthesis

An important area of research in polymer chemistry is finding new or better ways to prepare a polymer molecule from a stock of smaller monomers. In most cases, polymers are prepared using principles of organic chemistry. Polymer chemists are especially interested in techniques that allow them to precisely control the size and structure of the end product.

Polymer chemists are also investigating polymerization methods outside the scope of organic chemistry. One area of interest involves preparing polymers by imitating the biological processes used to create biopolymers such as proteins or cellulose. Other areas of study involve using plasma or electricity to initiate polymerization reactions.

Physical Polymer Chemistry

Physical polymer chemistry is the study of how a polymer molecule's structure relates to the behavior of the bulk substance. Physical polymer chemistry is closely related to the field of polymer physics and also overlaps with polymer research in materials science. Physical polymer chemists use analytical techniques such as light scattering and spectroscopy to characterize the size and structure of polymers.

Other areas of interest in physical polymer chemistry include the study of polymers in solution, the mechanical properties of polymers, and understanding phase transitions in polymer substances. There are also many researchers using principles of theoretical chemistry to better understand the structure and properties of polymer molecules.

History of Polymer Chemistry

Naturally occurring polymers such as amber and rubber have been used by humans for millennia. Early Mesoamericans are perhaps the true pioneers in polymer chemistry, having discovered methods for treating natural rubber that were not reproduced until thousands of years later.

The earliest work in modern polymer chemistry involved the chemical modification of naturally occurring polymers. The reaction between nitric acid and cellulose, studied by Henri Braconnot in 1832 and later by Christian Schönbein, led to the discovery of nitrocellulose and celluloid. The ensuing years saw the preparation of other cellulose derivatives, such as collodion, used as a wound dressing since the U.S. Civil War, and cellulose acetate, first prepared in 1865.

Other early work in polymer chemistry involved the modification of natural rubber to improve durability. In 1834, Friedrich Ludersdorf and Nathaniel Hayward independently discovered that adding sulfur to raw natural rubber (polyisoprene) helped prevent the material from becoming sticky. In 1844 Charles Goodyear received a U.S. patent for vulcanizing rubber with sulfur and heat. Thomas Hancock had received a patent for the same process in the U.K. the year before.

In 1884 Hilaire de Chardonnet started the first artificial fiber plant based on regenerated cellulose, or viscose rayon, as a substitute for silk, but it was very flammable. In 1907 Leo Baekeland invented the first wholly synthetic polymer, a thermosetting phenol-formaldehyde resin called Bakelite. Cellophane was invented in 1908 by Jocques Brandenberger who squirted sheets of viscose rayon into an acid bath.

The work of Wallace Carothers in the 1930s demonstrated that polymers of desired chain length and composition could be synthesized rationally from constituent monomers, laying the foundations of modern polymer chemistry and laying the framework for the now burgeoning polymer industry. Carothers is credited with the development of neoprene (1931), a synthetic rubber, the first polyester, and nylon (1935), a true silk replacement. The work of Ziegler and Natta in the 1950s laid the basis for stereospecific polymer synthesis. Stephanie Kwolek developed an aramid, or aromatic nylon named Kevlar, patented in 1966.

There are now a large number of commercial polymers, including composite materials such as carbon fiber-epoxy, polystyrene-polybutadiene (HIPS), acrylonitrile-butadiene-styrene (ABS), and other such materials that combine the best properties of their various components, including polymers designed to work at high temperatures in automobile engines.

Working in Polymer Chemistry

The American Chemical Society estimates that 50 per cent of chemistry professionals will work in a polymer-related field for some portion of their career. Though polymer chemists typically earn an advanced degree in synthetic organic chemistry, some institutions offer specialized degree programs in materials science and polymer science to meet the evolving needs of the polymer industry. Given the current commercial importance of synthetic polymers, most jobs in polymer chemistry are industrial jobs.

Current areas of active interest in polymer chemistry include the following:

1. Fundamental research into controlled syntheses and novel polymerization reactions.
2. Development of new molecular architectures, such as supramolecular polymer complexes.
3. Development of molecular architectures suited for molecular sensor technology.
4. Syntheses of polymers with energy and charge transport properties.
5. Biomedical applications, such as novel protein design and synthesis and targeted drug delivery.

There is also emerging interest in green polymer chemistry. Most artificial hydrocarbon-based polymers are formed from petroleum products. Substantial research efforts are devoted to improved recycling methods, renewable sources of raw materials, and biodegradable polymer materials.

Nobel Prizes in Polymer Chemistry

2000 Alan G. MacDiarmid, Alan J. Heeger, and Hideki Shirakawa for work on electroactive polymers contributing to the advent of molecular electronics

1974 Paul J. Flory for contributions to theoretical polymer chemistry.

1963 Giulio Natta and Karl Ziegler for contributions in polymer synthesis. (Ziegler-Natta catalysis).

1953 Hermann Staudinger for contributions to the understanding of macromolecular chemistry.

The 1991 Nobel Prize in physics was awarded to Pierre-Gilles de Gennes for developing a generalized theory of phase transitions which has been particularly important for polymer chemistry.

7.4 Polymerization

In polymer chemistry, polymerization is a process of reacting monomer molecules together in a chemical reaction to form three-dimensional networks or polymer chains. There are many forms of polymerization and different systems exist to categorize them.

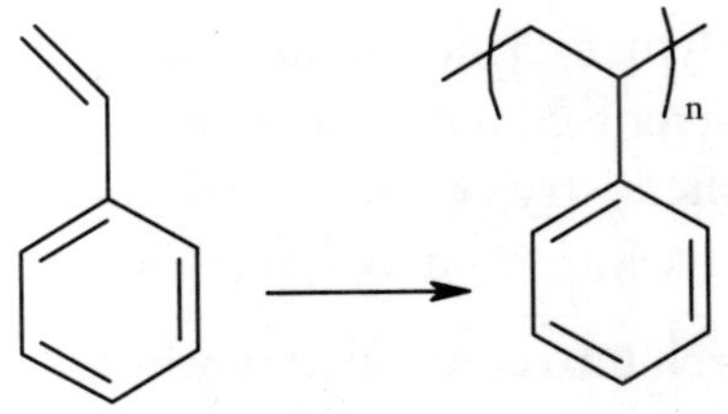

An example of alkene polymerization, in which each monomer unit's double bond reforms as a single bond with another styrene monomer and forms polystyrene.

Introduction

Single-monomer formed polymers:

$$A + A + A\ldots \rightarrow AAA\ldots$$

Co-polymers:

$$A + B + A\ldots \rightarrow ABA\ldots$$

In chemical compounds, polymerization occurs via a variety of reaction mechanisms which vary in complexity due to functional groups present in reacting compounds and their inherent steric effects explained by VSEPR Theory. In more straightforward polymerization, alkenes, which are a relatively stable due to ó bonding between carbon atoms form polymers through relatively simple radical reactions; conversely, more complex reactions such as those that involve substitution at the carbonyl atom require more complex synthesis due to the way in which reacting molecules polymerize.

As alkenes can be formed in somewhat straightforward reaction mechanisms, they form useful compounds such as polyethylene and polyvinyl chloride (PVC) when undergoing radical reactions, which are produced in high tonnages each year due to their usefulness in manufacturing processes of commercial products, such as piping, insulation and packaging. Polymers such as PVC are generally referred to as "singular" polymers as they consist of repeated long chains or structures of the same monomer unit, whereas polymers that consist of more than one molecule are referred to as "co-polymers".

Other monomer units, such as formaldehyde hydrates or simple aldehydes, are able to polymerize themselves at quite low temperatures (>-80°C) to form trimers; molecules consisting of 3 monomer units which can cyclize to form ring cyclic structures, or undergo further reactions to form tetramers, or 4 monomer-unit compounds. Further compounds either being referred to as oligomers in smaller molecules. Generally, because formaldehyde is an exceptionally reactive electrophile it allows nucleophillic addition of hemiacetal intermediates, which are generally short lived and relatively unstable "mid stage" compounds which react with other molecules present to form more stable polymeric compounds.

Polymerization that is not sufficiently moderated and proceeds at an undesirably fast rate can be very hazardous. This phenomenon is known as Hazardous polymerization and can cause fires and explosions.

Chain-Growth

Chain-growth polymerization or addition polymerization involves the linking together of molecules incorporating double or triple chemical bonds. These unsaturated *monomers* (the identical molecules which make up the polymers) have extra internal bonds which are able to break and link up with other monomers to form the repeating chain. Addition polymerization is involved in the manufacture of polymers such as polyethylene, polypropylene and polyvinyl chloride (PVC). A special case of addition polymerization leads to living polymerization.

In the polymerization of ethylene, its pi bond is broken and these two electrons rearrange to create a new propagating center like the one that attacked it. The form this propagating center takes depends on the specific type of addition mechanism. There are several mechanisms through which this can be initiated. The free radical mechanism was one of the first methods to be used. Free radicals are very reactive atoms or molecules which have unpaired electrons. Taking the polymerization of ethylene as an example, the free radical mechanism can be divided in to three stages: chain initiation, chain propagation and chain termination.

Fig. 7.1: Polymerization of Ethylene.

Free radical addition polymerization of ethylene must take place at high temperatures and pressures, approximately 300°C and 2000 At. While most other free radical polymerizations do not require such extreme temperatures and pressures, they do tend to lack control. One effect of this lack of control is a high degree of branching. Also, as termination occurs randomly, when two chains collide, it is impossible to control the length of individual chains. A newer method of polymerization similar to free radical, but allowing more control involves the Ziegler-Natta catalyst especially with respect to polymer branching.

Other forms of addition polymerization include cationic addition polymerization and anionic addition polymerization. While not used to a large extent in industry yet due to stringent reaction conditions such as lack of water and oxygen, these methods provide ways to polymerize some monomers that cannot be polymerized by free radical methods such as polypropylene. Cationic and anionic mechanisms are also more ideally suited for living polymerizations, although free radical living polymerizations have also been developed.

Step-Growth

Step growth polymers are defined as polymers formed by the stepwise reaction between functional groups of monomers. Most step growth polymers are also classified as condensation polymers, but not all step growth polymers (like polyurethanes formed from isocyanate and alcohol bifunctional monomers) release condensates. Step growth polymers increase in molecular weight at a very slow rate at lower conversions and only reach moderately high molecular weights at very high conversion (*i.e.* >95%).

To alleviate inconsistencies in these naming methods, adjusted definitions for condensation and addition polymers have been developed. A condensation polymer is defined as a polymer that involves elimination of small molecules during its synthesis, or contains functional groups as part of its backbone chain, or its repeat unit does not contain all the atoms present in the hypothetical monomer to which it can be degraded.

7.5 Polymer Chemistry

Is about Optimizing Technology?

A polymer is a chain of small molecules joined together in a repeating fashion to form a single layer molecule. Chemists develop polymers so they can be used to make ingredients for products with unique physical and chemical properties. They manipulate large, complex molecules and capitalize on the connections between their molecular structure and the properties that make them useful. Polymer products can be lightweight, hard, strong, and flexible and have special thermal, electrical, and optical characteristics; they include products from the fiber, communication, packaging, and transportation industries.

The big boom in polymer chemistry occurred largely in the first part of the twentieth century with the advent of polymer materials such as nylon and Kevlar. Today, most work with polymers focuses on improving and fine-tuning existing technologies. Still, there are opportunities ahead for polymer chemists. They work in many industries, creating a variety of synthetic polymers such as Teflon and special application plastics and developing new polymers that are less expensive or that outperform traditional materials and replace those that are scarce.

"The world is changing," says James Shepherd, a research associate in polymer chemistry at Hoechst Celanese. "New demands for polymer materials will be coming down the line. What we have learned over the past ten years will enable us to fulfill new needs. We may not discover a new polyethylene," he says, "but we may find smaller-volume and potentially more cost-effective materials."

Is about Research and Business?

There has been a shift in the economic emphasis and focus of polymer chemistry. Shepherd says when he began working in the field, many projects were purely exploratory. "Only later would we worry about the product." Now, projects are evaluated at the outset on the basis of what they will do for the company and what end-use improvements they will deliver. Therefore, industrial polymer chemists are increasingly in contact with the sales and marketing divisions of their companies and its customers.

This shift has placed a premium on good communication and interpersonal skills. It means chemists must adopt a business outlook in their work. Other skills and disciplines also come into play. "It helps if you are engineering-minded," says Kate Faron, a senior research chemist at DuPont. John Droske, professor of chemistry at the University of Wisconsin–Stevens Point, agrees. "This is a field for people who are comfortable looking at the end use as well as the preparation."

Polymer chemistry is product-oriented. However, this does not eliminate the availability of positions outside of industry. Some polymer chemists pursue their research interests in addition to their teaching and administrative responsibilities through employment at colleges and universities.

Touches Many Scientific Disciplines

Polymer chemistry touches many scientific disciplines and is vital in fields that develop products such as plastics and synthetic fibers; agricultural chemicals; paints and adhesives; and biomedical applications such as artificial skin, prosthetics, and the nicotine patch that helps smokers overcome their smoking habit. It is estimated that as many as 50 per cent of all chemists will work in polymer science in some capacity during their careers. Because they work in a field that is so broad, polymer chemists must be flexible and be able to interact and communicate with others in a variety of disciplines.

Because polymer chemistry today is product oriented, it has some overlap with materials science. However, polymer chemists emphasize that the most important aspect of their work is in the organic synthesis of materials. Most Ph.D. chemists now in the field were trained in organic chemistry. They acknowledge the strengths of degree programs in polymer science, but many say they would still choose to obtain a solid background in organic chemistry before entering the polymer science area. "You should take polymer classes, but not without a strong foundation in organic chemistry," says Jim Mason, senior chemist in the polymers division at the Bayer Corporation. "You learn a lot on the job," he adds. He says that an employer can teach you about polymers, but the fundamentals should be learned while in school. Mason adds, "Traditional training may also provide you with more long-term job security." Faron says, "It helps to be a generalist. If you go into certain polymer programs, you could be specializing too soon."

Is a Field Open to Change?

"Things in the area of polymer chemistry will be changing dramatically over the next five to ten years because of the emphasis on 'green' products," explains

Shulman. "Ingredients will have to be environmentally friendly, and there will be an emphasis on making polymers biodegradable. Concern about the effects of detergent products on the environment has brought new activity to a relatively mature area of polymer science." Polymer is an exciting field with new frontiers to be discovered.

Jim Mason

Blending and Compounding

Today, there is almost no development of entirely new polymer materials. However, blending and compounding existing polymers has become an important part of a polymer chemist's work. "Compounding is both an art and a science," says Jim Mason, a senior chemist in the polymers division of Bayer Corporation. "Quite a bit of chemistry is involved. You need to understand the reactions that go on in an extruder–the machine used to compound." Compounding is often done to achieve certain properties needed for a particular application.

One of Mason's recent projects was to develop a compounded polymeric material to be used in the manufacture of lawn tractors. "Body panels for these tractors are typically made of sheet metal," Mason explains. "But many tractor manufacturers would like plastics instead of sheet metal in the body panels. Existing plastics were not suited for this application, so we tried blending different materials to come up with the right combination of properties."

Mason worked directly with the tractor manufacturer to find out what was needed. "One of the big requirements was high impact strength," he says. "Then if the tractor hit a tree, the material would not shatter. We also needed to develop a material that had good weatherability and could be mixed with pigment to get the right color." The chemist and the customer collaborated to develop a material that is suited for this application.

James Shepherd

Liquid–Crystal Polymers

In the 1970s and 1980s, some of the most lucrative polymer breakthroughs occurred in the area of high-performance polymers. These materials were expensive to produce but had new and desired qualities, such as high strength and temperature resistance.

James Shepherd, research associate for polymer chemistry at Hoechst Celanese, focuses on developing new high-performance polymers. One example of successful high-performance polymer technology is liquid-crystal polymers (LCPs). LCPs are polymers that are highly oriented; that means that when they are heated above a certain temperature, the molecules align and flow easily. The LCP can then be pushed through a spinnaret on a spinning machine and made into a fiber, or can be injected into a small mold for an intricate part used in the electronics business. One of the benefits of LCPs is that they produce a material that is generally much stronger than those produced from traditional plastics. One aspect that Shepherd stresses is the importance of teamwork. "How to make the polymer is a team decision. There are so many different aspects to bring together and different talents required," he says. "It is not just the people in the lab making the decision. We monitor the changing needs of our customers and then determine how to produce the material with the properties they want."

An illustration of this, says Shepherd, was when economics and environmental issues required the molding industry to reuse the scrap material that is a normal byproduct of molding plastic parts. "One problem with doing this," he explains, "is that repeated processing cycles can adversely affect the properties of the material." However, by adjusting the chemistry and processing characteristics of a new LCP formulation, chemists in Shepherd's lab developed a material that maintains its properties after many processing cycles. In this way, basic researchers brought together applications, chemists, and customers to find a mutually beneficial solution.

Terry St. Clair

Polymers for Aerospace/NASA

Polymer development at NASA occupies a unique position in the field of polymer chemistry. Terry St. Clair, head of the polymeric materials branch at NASA Langley, explains that his work both serves NASA's polymer materials needs and functions as an incentive for the rest of industry to make the best polymer materials possible. "Our mission," he explains, "is to make sure that the materials that aircraft companies need are available. This, in some cases, forces industry to offer a more optimized product than the one they might want to push.

"We do a lot of the same type of polymer work as is done in industry, but in a broader and freer structure that is not confined by cost," he says. In some ways, this makes NASA a competitor with other polymer makers, the difference being that NASA does not actually manufacture large quantities of polymer materials. In some cases, St. Clair will work directly with an aircraft company to develop the products it needs. "When they have endorsed the material, we both go out into the market to try to find someone to make it," he says.

Another aspect of his job is to develop polymers for highly focused applications, such as the scientific instruments used in the space program. Staff in St. Clair's lab were asked to make a polymer used in the window of an X-ray telescope. "Only about five pounds of this material was needed annually," he says. "We were in a position to develop an exotic polymer where cost was not a factor. They would have been happy to use platinum or gold if it would work."

Jan Shulman

Detergent Polymers

Jan Shulman, a formulation chemist at Rohm and Haas, works with polymers used in automatic dishwashing detergent products. "In this field, polymers are key to making products work better," he says. One problem Shulman deals with is the effects of sodium carbonate--or soda ash–in a dishwashing detergent. "Soda ash is a main ingredient in many formulations," he says. "But when it reacts with water, it can form a chalky film on glasses. When you add polymers to a formula, you can prevent this from happening." Shulman says his lab facility includes dishwashers, plates, and glasses. "We basically replicate what the consumer does at home and conduct research to determine the most effective detergent formulation."

Shulman also works on formulation changes that are needed to address concerns about the effects of traditional detergent ingredients on the environment. "Most of this work involves reformulations around chlorine and phosphates," he explains. Beginning in Europe in 1990, he says, there was a move toward phosphate-free systems and an effort to replace chlorine bleach with oxygen bleaches, such as those found in Clorox 2. "Because this trend was making its way to the United States, it became my job to work on reformulating products," he says. "When you remove phosphates, you need to add a polymer to enhance performance. We try to determine which polymers will fit the bill."

Kate Faron

Fibers

The general public is familiar with Lycra, or spandex, the stretch fabric that goes into leggings, fitness wear, and bathing suits. Kate Faron, a senior research chemist at DuPont, knows Lycra on the polymer level. Part of Faron's job is to improve Lycra spandex for continued use in successful fashion items. "When Lycra first replaced rubber thread, the market grew in every segment," she explains. "We need to continue making some changes to the fiber to keep growing. Most of the changes that we make are incremental changes to existing products; but we still seek step-change improvements, and there is always a lot of chemistry involved."

By changing the structure of the polymer or by using chemical additives, Faron changes the properties of the spandex polymer. For example, how it responds to light and heat can be modified chemically.

"Certain fabrics that do not already include Lycra might benefit from elastic properties," she says. "We change the polymer to meet the specifications of these new materials. It is macromolecular engineering," she says. "You need to know how to synthesize small molecules and incorporate them into larger ones. You also have to understand the structure you build and the properties you expect it to have."

John Droske

Polymer Education

"Approximately 50 per cent of all chemists will work with polymers at some time in their careers," says John Droske, professor of chemistry at the University of Wisconsin–Stevens Point and director of the POLYED National Information Center for Polymer Education. "Because polymer science touches on many areas, it is important for chemists to be trained in polymer science." The POLYED has been working with a National Science Foundation grant to develop materials for polymer chemistry courses at the undergraduate level.

Droske teaches courses on the synthesis and characterization of polymers and the physical chemistry of polymers. He also conducts a polymer lab course. The aspect of polymer science that Droske enjoys most is the challenge of working with large molecules. "There is something unique about studying polymers," he says. "Macromolecules have a greater complexity than do small molecules. Over the years, our understanding of these large molecules has increased so much that, although they remain complex, we have tools that provide us with a better understanding of their properties, enabling us to make connections between their structure at the molecular level and their properties at the use level."

Work Description

Polymer chemists are concerned with the study and synthesis of large, complex molecules. They manipulate the molecular structure of a material to develop functional characteristics in an end product by chemical processing or through other processing conditions.

Places of Employment

Polymer chemists are employed in industry, government, and academia. However, most jobs are in industry where products are made. Opportunities for polymer chemists in industry exist in areas where adhesives, coatings, synthetic rubber, synthetic fibers, agricultural chemicals, packaging, automotive, aircraft, aerospace, and biomedical industries are made.

Personal Characteristics

A polymer chemist's work is interdisciplinary in nature. Individuals should be able to communicate with others in a number of fields. Those who are interested in materials and the end use of polymers as well as their synthesis will be particularly well suited to the field. This is also true for individuals who like hands-on work as opposed to purely theoretical thinking.

Education and Training

Most people employed in polymer chemistry have a Ph.D. and were trained as organic chemists. They stress the importance of a solid education in the fundamentals of chemistry. However, they acknowledge the value of the interdisciplinary degree available through programs in polymer science.

Job Outlook

Because polymer science is product-oriented, hiring can be expected to follow the economy. Polymer chemists stress the need to remain as broad-based and as flexible as possible for long-term employment security, but creative and well-trained individuals should be able to find positions in this field. Most major chemical companies have made deep cuts in their central research divisions, and industry is still in a downsizing mode. The field remains highly competitive, but some say they think these dynamics are cyclical and that the job market will improve.

Salary Information

To find out what a person in this type of position earns in your area of the country, please refer to the ACS Salary Comparator. Use of the ACS Salary Comparator is a member-only benefit. General information about salaries in chemical professions can be obtained through published survey results.

8

Chemistry of High Polymers

8.1 High Polymer

Definition

A polymer with molecules of high molecular weight, sometimes arbitrarily designated as greater than 10,000. or A polymer of a given series is considered a high polymer if its physical properties (especially its viscoelastic properties) do not vary markedly with molecular weight.

8.2 High Performance Polymers

Classification of high performance polymers (compounds):

- ❖ Hardwearing, and
- ❖ High temperature resistant and high mechanical strength.

High Performance polymers have a thermal resistance >150°C. Examples of hard wearing high performance polymers are:

- ❖ PEEK—Polyetheretherketon.
- ❖ PES—Polyethersulfon.
- ❖ PI—Polyimide.

Engineering polymers are classified by a temperature resistance within 100°C and 150°C. Examples of engineering polymers are:

- ❖ PA6.6—Polyamide
- ❖ PA6 G—Cast Polyamide (LFX with AF lubricant)
- ❖ POM—Poly Oxy Methylene (acetal)
- ❖ PETP—Poly Ethylene Terephtalate (PET)
- ❖ UHMWPE—Ultra High Molecular Weight Polyethylene
- ❖ PBTP—

Standard polymers have a thermal resistance below <100°C and less suitable for sliding/rolling surfaces. Examples of these polymers are:

- HDPE—High Density Poly Ethilene
- ABS
- PMMA
- PVC

Most high performance polymers (compounds) are reinforced by fibres or/and filled with internal anti friction lubricants.

Traditional reinforcements and internal AF lubricants are:

- PTFE
- Silicone oil
- Graphite
- MoS2
- Aramide
- Carbon fibre
- Glass fibre
- Alloy-/blend technology e.g. HWPE

Effects of reinforcements and internal AF lubricants are:

- PTFE reduces the coefficient of friction. Is effective at high pressure. Creates a PTFE film between the compound and the counterface.
- Silicone oil reduces coefficients of friction. Migrates to the wear surface. Offers lubricity at start up and at high speeds. Not effective at high surface pressures.
- PTFE with silicone oil improves tribological properties over a broad velocity range. Excellent for oscillatory motion (less "slip stick"). Considerable improvement in LPV at high speeds.
- Molybdenum Sulphide (MoS2) enhance the crystallization of PA (surface hardening). Reduction of "slip stick" effect. Moderate improvement of wear factor.
- Graphite Powder as boundary lubricant often used in aqueous moisture environment.
- Aramide for improvement of the wear factor. Low counter face wear against soft metal, e.g. Cu, Al. Low plastic-on-plastic wear (identical partner). Low noise. Low abrasive wear. Dimensional stability.
- Carbon fibres improve mechanical performance. Results in higher LPV-value and in reduction of wear factor. Reduce wear of both surface and mating surface as compared to glass fibres. Statically dissipate/conductive.
- Glass fibres improve mechanical properties. Higher LPV-value. Reduction of wear factor. Increase wear of mating surface.

The listing below is restricted to polymers that are "well" described by industrial suppliers. Both the physical properties as well as the composition is described. Unfortunately, tribology data is still limited.

Listing of "high performance" commercial polymers:

- PEEK
- PEEK-BG (zwart), Carbon Fibre, PTFE and Graphite
- PEEK-GF30, 30 per cent Glass Fibre
- PEEK-CA30, 30 per cent Carbon Fibre (not listed)
- PPS + , fibre reinforced and solid AF lubricants
- PES
- PI + 15 per cent Graphite

Listing of "engineering" commercial polymers:

- PA6.6, Polyamide (Nylon)
- PA6.6 GF30, 30 per cent Glass Fibre reinforced
- POM C, Copolymer, Acetal
- POM H, Homopolymer, Acetal, Delrin
- POM H-TF, Homopolymer + Teflon
- PETP
- PETP TX

Criteria for material selection:

- wear factor
- dynamic and static coefficients of friction
- limiting PV value
- counter face wear
- material costs

Material combinations and consequences:

- polymer—metal
- polymer—polymer

Internal lubricated compounds versus metal:

- Up to 90 per cent of frictional heating is conducted by the metal part.
- fibre reinforced polymers wear the metal surface.

Plastics against plastics:

- Dissimilar polymers results in a reduction of the static coefficient of friction.
- Use of PTFE lubrication dramatically reduces wear rates in both similar and dissimilar resins.

- Against a fibre reinforced compound, the mating material should contain PTFE lubrication
- Aramide compounds have excellent wear characteristics.

Table 8.1: Impression of friction and wear for different combinations for comparison purposes only.

mat.1	*mat.2*	*k1*	*k2*	*μstat.*	*μdyn.*
PA6	Steel	4	-	0.xx	0.28
POM	Steel	1.3	-	0.14	0.21
POM + 20%PTFE	Steel	0.3	-	0.07	0.15
PA6	PA6	50	22	0.06	0.07
POM	POM	280	200	0.19	0.15
POM + 20%PTFE	POM + 20%PTFE	11	12	0.19	0.17
POM	PA6	1.2	1.0	0.04	0.06
POM + 20%PTFE	PA6	0.4	0.7	0.05	0.06
POM + 20%PTFE	PA6 + 20%PTFE	0.5	0.24	0.03	0.04

* Data from LNP, with permission. Thrust Washer measurements, mat.1: moving sample, mat.2 stationary counter face, $k \cdot 10^{-15}$ m^2/N.

* Data presented for impression, values strongly depends on pressure, velocity, temperature, roughness, macro geometry etc.

Temperature Effects

Else than for metals where the mechanical strength and the tribological properties are constant (<250°C) the properties of polymers change much with the temperature.

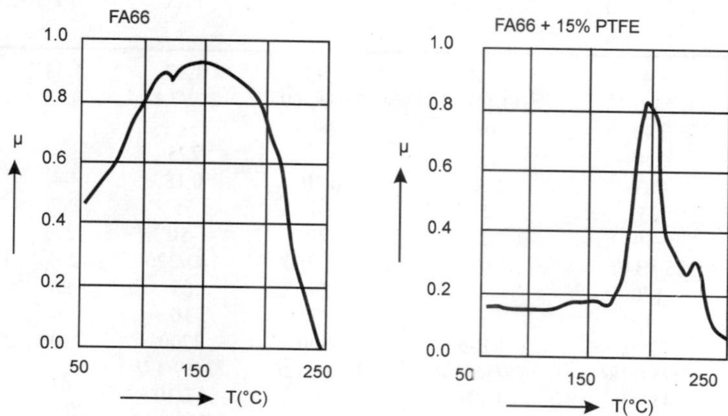

Coefficient of friction; PA6.6-steel, Ra = 0.1μm, v = 0.1 m/s, p = 1.5 MPa. Honselaar e.a., TNO, Constructeur 1989/8. The curve of PA6 + 15 per centPTFE is very smooth compared to the neat PA6.6.

Closure

High performance polymers can have excellent tribological properties however, disappointment may occur if the fundamentals are not fully understood. For example, glass fibre reinforcements wear the steel counter surface. A combination of dissimilar polymers is necessary to have acceptable wear resistance and low static friction. A bad heat conduction results in higher contact temperatures with degradation of the tribological aspects. For some combinations a very smooth counter surface is required. Running in conditions must be taken into account etc.

Fig. 8.1: Fiber reinforced polymer versus steel.

High Performance" Polymers

PEEK	*PEEK-BG*	*PEEK-GF30*	*PPS +*	*PES*	*PI*	*PI + 15%Gr + 10%PTFE*
1.32	1.47	1.48	1.43	1.37	1.43	1.65
0.2/0.45	0.14/0.30	0.14/0.30	0.03/0.09	0.9/2.1	0.72	0.62
340	340	340	280	-	—	—
150	150	150	—	225	—	—
0.25	0.24	0.43	0.30	0.18	—	—
50	25	25	50	55	55	55
-60	-30	-20	-20	-50	-273	-273
250/-/310	250/310	250/310	220/260	180/220	245	245
110	120	130	75	85	86	45
20	3	3	5	10	5	—
4200	7700	8100	4400	2700	3100	3100
32/30/25/ 20/8	55/53/48/ 45/24	58/56/50/ 47/26	36/33/12/ 9/7	23/19/17/ 13/10	25	—
M105/230	M99/240	M108/275	M84/180	M104/160	—	—
0.3-0.5	0.15-0.25	0.3-0.45	0.25-0.4		0.3-0.4	0.25-0.35
0.4-0.55	0.15-0.3	0.35-0.5	0.25-0.4		0.3-0.5	0.25-0.45
9.3	0.7	2.3	1.7		2.3	0.33
ERTA	ERTA	ERTA	ERTA	ERTA	ERTA	ERTA

PA6.6	*PA6G-LFX*	*POM C*	*POM H*	*POM H-TF*	*PETP*	*PETP-TX*	*PE-UHMW*	*PTFE*
1.14	1.135	1.41	1.43	1.5	1.39	1.44	0.945	2.18
2.4/8	2/6.3	0.2/0.85	0.2/0.85	0.17/0.72	0.25/0.5	0.23/0.47	<0.01	0.01
255	220	165	175	175	255	255	135	327
50	50	-60	-60	-60	75	75	-110	
0.28	0.28	0.31	0.31	0.31	0.29	0.29	0.40	0.23
80	80	110	95	105	60	65	200	120
-30	-20	-50	-50	-20	-20	-20	-40	-200
80/95/180	90/105/165	100/15/140	90/105/150	90/105/150	100/115/160	100/115/160	80/90	200/260/280
55	40	70	80	55	90	78	22	10
>100	>50	30	30	8	15	8	400	300
1800	1700	3000	3300	2900	3400	3200	800	490
8	8	14	16	13	26	23		8
M89/170	M82/145	M86/140	M90/160	M85/140	M96/170	M94/160	D62/45	D60/25
0.25-0.5	0.2-0.3	0.35-0.45	0.35-0.45	0.15-0.25	0.2-0.3	0.15-0.25	0.15-0.25	0.05-0.15
0.3-0.6	0.25-0.45	0.2-0.3	0.2-0.3	0.1-0.15	0.25-0.35	0.2-0.3	0.15-0.25	0.05
5.7	1	>30	>30	4.7	1.3	0.7	2.7	>100
ERTA	ERTA	ERTA	ERTA	ERTA	ERTA	ERTA	ERTA	Eriks

Physical Properties of Polymers (Indicative Values)

- ❖ Density
- ❖ Water adsorption; at saturation in air of 23°C and 50 per cent RH/at saturation in water of 23°C

Thermal Properties

- ❖ Melting point
- ❖ Glass transition temperature
- ❖ Thermal conductivity at 23°C
- ❖ Coeff. of lin. thermal expansion; average value between 23 and 100°C
- ❖ Tmin; indicative value based on impact strength.
- ❖ Tmax; high performance polymers 20.000 h/1 h, thermal oxidative degradation causes a reduction of properties, there is a decrease in tensile strength of about 50 per cent/1h, for short periods (few hours) with very low load applied.
  - Tmax, engineering polymers 20.000 h/5000 h/1 h

Mechanical properties at 23°C:

- Tensile strength, at test speed: 5 mm/min.
- Elongation at break, at test speed: 5 mm/min.
- Modulus of elasticity, at test speed: 1 mm/min. Fibre reinforced polymers are anisotropic (properties differs to extrusion direction).
- Tensile creep test; S(1/1000), 1 per cent elongation in 1000 h at 23/60/100/125/150°C.
- Hardness Rockwell, 10 mm thick specimens/Ball indentation H 358/30 or H961/30.
- Dynamic Coefficient of Friction at 23/75°C, RV 50 per cent, 0.33 m/s, 3 MPa, dry friction, steel counter surface, H = 1600 MPa, Ra = 0.8-1μm.
- Static Coefficient of Friction, same conditions as for dynamic friction.
- Wear, k [10^{-15} m^2/N], same conditions as for friction.

8.3 Process of Making Colored High Temperature Polymers

High melting polymers such as polyester, polyamide or polysulfone are uniformly internally colored by blending therewith a minor amount of easily colored lower-softening thermoplastic additive polymer such as polyethylene, containing an effective amount of coloring material uniformly blended therethrough. High temperature polymers colored in this manner provide shaped articles such as films, fibers, and machine parts having a visually uniform color.

US Patent References

2512459	Dispersion of pigments in ethylene polymers	June 1950	Hamilton
2884663	Process for producing improved polymeric terephthalate film	May 1959	Alles
3352952	Method of coloring thermoplastics	November 1967	Marr
3361848	Injection molded polyester/polyolefin blends	January 1968	Siggel *et al.*
3375219	Pigmenting polycarbonamide by means of an ethylene copolymer carrier resin	March 1968	Robb
3384693	Method for mixing plastic compositions	May 1968	Roe
3397169	Abrasion-resistant mineral-filled thermo-setting molding composition	August 1968	Wilkinson
3405198	Process for making impact resistant injection molded polyethylene terephthalate products	August 1968	Rein *et al.*
3409585	Pigment concentrate	November 1968	Hagemeyer, Jr. *et al.*
3413249	Coloring of polystyrene	November 1968	Luftglass *et al.*
3433853	POLYOLEFIN COMPOSITIONS CONTAINING A BASIC POLYAMIDE DYE SITE ADDITIVE	March 1969	Earle *et al.*
3435093	POLYMER BLENDS OF POLYETHYLENE TEREPHTHALATE AND ALPHA-OLEFIN, ALPHA, BETA-UNSATURATED CARBOXYLIC ACID COPOLYMERS	March 1969	Cope

3503922	PROCESS FOR PRODUCING DISPERSIONS OF FINELY—DIVIDED SOLIDS IN ISOTACTIC POLYPROPYLENE	March 1970	Carton
3534120	COPOLYESTER FILAMENTS CONTAINING MINOR AMOUNT OF POLYOLEFINS	October 1970	Ando *et al.*
3546319	BLENDS OF POLYAMIDES, POLYESTER AND POLYOLEFINS CONTAINING MINOR AMOUNTS OF ELASTOMERS	December 1970	Prevorsek *et al.*
3637906		May 1972	Parathoen

Other References

Renfrew *et al.*, Polythene, (Interscience), (N.Y.) (1960) (2nd ed.), pp. 419-421.
High Polymers Vol. XX, part 2, p. 411 (Interscience, 1964).

- ❖ *Primary Examiner:* Welsh M. J.
- ❖ *Assistant Examiner*: Fletcher H. H.
- ❖ *Attorney, Agent or Firm*: Alexander, Sell, Steldt & DeLahunt
- ❖ *Claims*: What is claimed is?

A process of forming an oriented film having a visually uniform coloring by coextruding a film consisting essentially of a normally uncolored, relatively high melting, difficulty colorable, thermoplastic polymer selected from the group consisting of polyesters, polysulfones and polyamides with a minor amount of a uniformly colored, lower softening easily colorable thermoplastic additive polymer selected from the group consisting of polyethylene, polypropylene, polystyrene, and acrylonitrile-butadiene-styrene copolymers, whereby the colored additive polymer is uniformly dispersed throughout the difficultly colorable polymer, and thereafter orienting said film.

Description

Background of the Invention

This invention relates to a process of uniformly internally coloring high temperature polymers and the resultant uniformly colored products. In modern society, there has been a commercial demand for uniformly internally colored shaped articles for such applications as colored dishware, toys, household goods, etc. Uniformly internally colored films, particularly oriented films, have been sought for such applications as overhead projector transparencies, plastic laminates, flexible packaging, electrical insulation, plastic tapes, etc., while uniformly internally colored fibers are greatly desired for clothing and carpeting. External coloration of these items by means of coatings is not satisfactory because the coatings tend to wear away or flake off the article.

While it has been possible to uniformly internally color such thermoplastic polymers as polyethylene and polypropylene, it has been very difficult to internally color thermoplastic polymers which melt at high temperatures or which have a high melt viscosity. This is particularly true if the coloring is done at the time a shaped article is extruded. The several methods known to the art for internally coloring polymers have not been found satisfactory for coloring high melting polymers.

For example, it is generally necessary to use a heated extruder hopper to keep high melting polymers dry. Many coloring agents soften, melt, or decompose at a temperature below the melting point of the high temperature polymer. When dry blended with polymer is a heated extruder hopper, these coloring materials melt before the polymer does, running down the hopper sidewalls to form a "slug" which produces unpredictable and erratic coloring of the polymer. Additionally, when the coloring material melts before the polymer it often wets and lubricates the extruder screw, preventing feeding of the polymer into the extruder. The temperatures necessary to extrude high melting polymers also decompose and ruin many coloring materials. Further, the different particle sizes of polymer and coloring agent tend to cause them to classify and stratify as the mixture flows into the extruder, resulting in uneven coloration.

The technique of dissolving coloring materials in solvent, introducing hot polymer into the solvent to flash it off, and thereafter extruding the color coated polymer has not been found satisfactory for coloring high melting polymers. As with dry blending, the coloring material melts before the polymer does, flows down the hopper walls and forms undesirable color "slugs," or wets the extruder screw and prevents flow of the polymer. The use of heated rubber mills and high shear mixers to blend coloring material into polymer has not been found feasible with high melting polymers, the necessary elevated temperatures decomposing many coloring materials. Further, extensive time-consuming clean-up is necessary between runs of different colors and it is difficult to control the coloring agent concentration with any degree of precision.

Since the requisite amount of coloring material to yield a given color is inversely related to the thickness of the final article, it is impractical to incorporate coloring agent into a high melting polymer during polymerization. Further, this technique necessitates extensive clean-up between each lot, and exposes the coloring material to high temperatures for excessively long periods of time.

In short, a satisfactory method of uniformly coloring high temperature polymers has not been available in the art. Presently available methods are expensive, messy, provide non-uniform coloration, and decompose many coloring materials.

Summary

This invention provides a method of providing high temperature polymers with visually uniform internal coloration. The colored polymers are suitable for extrusion into uniformly colored shaped articles such as dishware, toys, self-supporting films, and fibers. The coloration is accomplished in a manner which prevents colorant "slugs," unpredictable and erratic polymer coloring, extruder screw lubrication, and prolonged exposure of coloring material to high temperatures, thereby preventing decomposition. With the method of this invention it is possible to change from one color to another at will without extensive clean-up between lots. It is also easy to vary the color concentration so as to maintain constant color intensity in articles of various thickness.

In accordance with the invention, high temperature polymers are uniformly internally colored by blending therewith a minor amount of an easily colored lower-softening thermoplastic additive polymer which contains an effective amount of coloring material uniformly blended therethrough.

Polyesters, polyamides, and polysulfones soften and melt at temperatures above about 400°F. and have been found difficult to color uniformly. It has been found that a visibly uniform color may be imparted to these high temperature polymers by first blending the coloring material into an easily colored polymer to form a color concentrate and subsequently blending the colored polymer into the high melting polymer. The coloring materials are first blended with the easily colored additive polymer by any one of the commonly accepted techniques used in the art. For example, the blending can readily be accomplished on a heated rubber mill or high shear blender. Alternatively, if the additive polymer is soluble, the blend can be made by dissolving the polymer and coloring material in solvent and thereafter recovering the colored additive polymer. Also, it is possible to dry blend some polymers and coloring materials, thereafter coextruding them to obtain the colored additive concentrate. Polymers which have been found to be readily colorable and to perform effectively in practicing the invention include high and low density polyethylene, polypropylene, polystyrene, acrylonitrile-butadiene-styrene copolymers, cellulose acetate polymers, and polycarbonate. Suitable additive polymers are those which are capable of being physically mixed with the high-melting polymer without chemically reacting therewith and thereafter extruded from a die to form smooth continuous self-supporting films. These additive polymers may contain any desired effective concentration of coloring material, ranging from a trace amount in relatively thick articles to about 60 per cent by weight in thin articles such as films. A 50 per cent coloring material concentration is economical, easily prepared, and preferred.

Following coloration, the additive polymer is blended with the high temperature polymer to provide a visually uniform color. In some instances it is possible to dry blend the additive polymer with the high temperature polymer and thereafter extrude them together; however, it is preferred to combine them by coextrusion. A preferred method of coextrusion is to "tap" the barrel or feed zone of an extruder for high temperature polymer, attaching a second extruder at this location to permit feeding the colored additive polymer. The two polymers are then rapidly and intimately mixed by the screw of the high temperature extruder to provide a visually uniform color. This method combines the advantages of rapid and intimate mixing in the extruder screw, simplicity, greatly reduced possibility of detrimental interaction between the polymers, and short exposure of the coloring material to high temperatures.

When a colored additive polymer is blended with high temperature polymer by means of "tapping" the barrel of the high temperature extruder, the tap is preferably made at a point which is far enough from the hopper feed zone so that additive polymer is not forced out the extruder hopper. Conversely, the tap must be sufficiently far from the point of discharge from the extruder so that mixing is complete. It has been found that the tap should be made at a distance of about three to about nine, preferably about five, screw diameters in front of the leading edge of the hopper throat. If desired, several taps can be made on one extruder barrel and several additive polymers simultaneously blended into the high temperature polymer.

It has been found that addition of colored additive polymer to a high temperature polymer does not significantly affect its physical properties such as tensile strength, elongation, and break strength, as would normally be expected when a high temperature polymer is adulterated. As much as 20 per cent by weight of additive polymer having a relatively low melting point can be added without significant detrimental effect, a smooth homogeneous blend being obtained. While physical properties begin to decrease when more than about 20 per cent additive polymer is included, it is possible to make polymer blends containing up to about 70 per cent by weight additive polymer. The additive polymer is preferably the minor component and is less than 50 per cent by weight of the final blend. Conversely, the high temperature polymer is preferably the major component and is more than 50 per cent by weight of the blend. If desired, the technique of providing uniform coloration in high temperature polymers, by blending therein a colored additive polymer, can also be utilized with virtually any polymer, regardless of softening or melting point.

The coloring materials which have been found to perform effectively in practicing the invention are virtually unlimited and typically include such organic pigments as phthalocyanine blues and greens; azo reds; quinacridone reds, oranges, scarlets, and violets; alizarin yellows and crimsons; anthraquinone blues, violets, oranges, and yellows. Typical inorganic pigments are such as ultramarine blues, purples, and violets; cadmium reds, yellows and oranges; chromium greens and oxides; iron oxide browns and reds. Typical dyes are solvent, plastic, and oil soluble dyes such as anthraquinone, azo, triarylmethane, azine, xanthene, nitro, nitroso, thiazine, polymethine, oxazine, and acridene. Typical fillers are such as carbon black, titanium dioxide, calcium carbonate, talc, clay, molybdenum disulfide, graphite, etc. The main criterion for selecting a coloring material to be blended into the additive polymer is the stability of the coloring material at the softening or melting temperature of the additive polymer. The particle size of the coloring material should be relatively small so as to provide uniform coloration, less than 10 microns being preferred.

When coloring polymers and shaped articles in accordance with the invention, it is possible to obtain a given color intensity in articles of varying thickness by merely varying the quantity of additive polymer blended into the high temperature polymer. For example, when coextruding, a simple change in screw speed of the additive polymer extruder accomplishes the desired color change. It is also possible to vary color and color intensity by dry blending uncolored additive polymer with colored additive polymer and thereafter combining the blend with high temperature polymer. It is also possible to dry blend any desired number of colored or uncolored additive polymers and subsequently combine the blend with high temperature polymer. Various colors can be blended with each other to provide any desired color and color intensity. It is possible to change from one color to another at will, without timeconsuming clean-up, by feeding a second additive polymer into the extruder, the second polymer purging the first.

The shaped articles colored in accordance with the invention are virtually transparent when the refractive indices of the polymers in the blend are about equal. Blending together polymers of different refractive indices will provide a finished article having a translucent haze. The degree of translucency is directly proportional to the amount of additive polymer in the blend. Shaped articles containing titanium dioxide, carbon black, clay, and other similar colorants are substantially opaque. Further, shaped articles containing a significant amount of inorganic fillers have a matte surface which is readily receptive to marking by pen or pencil.

The following examples, in which all parts are by weight unless otherwise noted, illustrate preparation of the uniformly colored shaped articles of the invention, without limiting the scope thereof.

Description of Preferred Embodiments

Example 1

This example illustrates the coloration of a biaxially oriented polyethylene terephthalate film. Fifty parts of "Pigment Yellow 35" commercially available from Hercules Powder Company under the trade name Golden Cadmium Yellow, X-2283, was combined with 50 parts of high density polyethylene, passed through a 1-inch extruder, and pelletized to provide a uniformly colored additive polymer. A 21/2-inch extruder was tapped by drilling a 3/8 inch diameter hole in the barrel five screw diameters in front of the hopper throat and a 1-inch diameter extruder was piped to feed into the tap. The screw of the 21/2-inch extruder was relieved to a depth of one-sixteenth inch for a distance of one-half inch on each side of the tap to reduce the back pressure and permit continuous introduction of additive polymer. Polyethylene terephthalate was processed through the 21/2-inch primary extruder, the colored additive polymer introduced through the 1-inch extruder, the blended polymers cast into a 10-mil film, and the film biaxially oriented 3 X in both the machine and cross directions. The resultant 0.9 mil film which contained 2 per cent coloring material and 2 per cent polyethylene was uniformly colored and had a 42 per cent light transmission as measured by a haze meter (Gardner Model UX-10A, Gardner Instrument Company), a 27,000 psi tensile strength at an ultimate elongation of 114 per cent.

8.4 VESTAKEEP—Polyether Ether Ketone

With the high-temperature polymer polyether ether ketone (PEEK), registered as VESTAKEEP®, High Performance Polymers has expanded its product range of high-performance polymers at the end of 2005. VESTAKEEP® is produced at JIDA Evonik High Performance Polymers (Changchun) Co., Ltd. in China, and is marketed globally by High Performance Polymers. VESTAKEEP® compounds are particularly characterized by the following material properties:

- very high heat resistance,
- high rigidity,
- low water absorption and therefore high dimensional stability,
- high hardness,
- good strength,
- excellent sliding friction behavior, minimal abrasion,
- good electrical characteristics,
- excellent chemical resistance,
- excellent hydrolytic stability,

- good processability, and
- low tendency to form stress cracks.

This properties profile makes VESTAKEEP® compounds suitable for a wide range of applications. It is used, for example, to produce chip carriers and connectors for the electronics industry and bearings, seals and cables for the automotive and aerospace industries and for rail cars. The machine and apparatus construction industry and the food processing industry have discovered that polyether ether ketone is good for valves and parts that are subjected to severe stress. Medical technology uses it for hoses and instrument handles.

9

Polymer Physics and Physical Properties of Polymers

9.1 Polymer Physics

Polymer physics is the field of physics associated to the study of polymers, their fluctuations, mechanical properties, as well as the kinetics of reactions involving degradation and polymerisation of polymers and monomers respectively. While it focuses on an aspect of the study of condensed matter physics, the field of polymer physics has developed as a branch of statistical physics. Polymer physics and polymer chemistry are part of the wider field of polymer science. Disordered polymers are too complex to be described using a deterministic method. However statistical approaches can yield results and are often pertinent since large polymers (that is to say, polymers which contain a large number of monomers) can be described efficiently as systems at the thermodynamic limit. Thermal fluctuations continuously affect the shape of polymers in liquid solutions, and modelling their effect requires a recourse to the principles of statistical mechanics. As a corollary temperature strongly affects the physical behavior of polymers in solution. The statistical approach to polymer physics is based on an analogy between a polymer and either a brownian motion, or some other type of random walk. The simplest possible polymer model is presented by the ideal chain, corresponds to homogeneous random walk. The Russian and Soviet schools of physics have been particularly active in the development of polymer physics.

Models

Models of polymer chains are split into two types: "ideal" models, and "real" models. Ideal chain models assume that there are no interactions between chain monomers. This assumption is valid for certain polymeric systems, where the positive and negative interactions between the monomer effectively canceled out. Ideal chain models provide a good starting point for investigation of more complex systems and is better suited for equations with more parameters.

Ideal Chains

The freely-joined chain is the simplest model of a polymer. In this model, fixed length polymer segments are linearly connected, and all bond and torsion angles are equiprobable. The polymer can therefore be described by a simple random walk and ideal chain. The freely-rotating chain improves the freely-jointed chain model by taking into account that polymer segments make a fixed angle to neighbouring units because of specific chemical bonding. Under this fixed angle the segments are still free to rotate and all torsion angles are equally likely. The hindered rotation model assumes that the torsion angle is hindered by a potential energy. This makes the probability of each torsion angle proportional to a Boltzmann factor:

$$P(\theta) \propto \exp(-U(\theta) / kT)$$

In the rotational isomeric state model the allowed torsion angles are determined by the positions of the minima in the rotational potential energy. Bond lengths and bond angles are constant.

The Worm-like chain is a more complex model. It takes the persistence length into account. Polymers are not completely flexible, bending it causes bending energy. At the length scale below persistence length, the polymer behaves more or less like a rigid rod.

Real Chains

Interactions between chain monomers can be modelled as excluded volume. This causes a reduction in the conformational possibilities of the chain, and leads to a self-avoiding random walk. Self-avoiding random walks have different statistics to simple random walks.

Solvent and Temperature Effect

The statistics of a single polymer chain depends on the solvent. For good solvent the chain is more expanded while for bad solvent the chain segments stay close to each other. In the limit of a very bad solvent the polymer chain merely collapses to form a hard sphere, while in good solvent the chain swells in order to maximize the number of polymer-fluid contacts. For this case the radius of gyration is approximated using Flory's mean field approach which yields a scaling for the radius of gyration of:

$$R_g \sim N^{\nu}$$

where R_g is the radius of gyration of the polymer, N is the number of bond segments (N, which is the degree of polymerization) of the chain.

For good solvent, $\nu = 3/5$; for bad solvent, $\nu = 1/3$. Therefore polymer in good solvent has larger size and behaves like a fractal object. In bad solvent it behaves like a solid sphere. In the so called θ solvent, $\nu = 1/2$, which is the result of simple random walk. The chain behaves as if an ideal chain.

The quality of solvent depends also on temperature. For a flexible polymer, low temperature may correspond to poor quality and high temperature makes the same solvent good. At a particular temperature called theta (θ) temperature, the solvent behaves as if an ideal chain.

Excluded Volume Interaction

Ideal chain model assumes that polymer segments can be overlapped with each other as if it is a phantom chain. In reality, two segments cannot occupy the same space at the same time. This interaction between segments is called excluded volume interaction. The simplest formulation of excluded volume is the self-avoiding random walk, a random walk that cannot repeat its previous path. A path of this walk of N steps in three dimensions represents a conformation of a polymer with excluded volume interaction. Because of the self-avoiding nature, the number of possible conformation is significantly reduced. The radius of gyration is generally larger than that of ideal chain.

Flexibility

Whether a polymer is flexible or not depends on the scale of interest. For example, the persistence length of double-stranded DNA is about 50nm. Looking at length scale smaller than 50nm, it behaves more or less like a rigid rod. At length scale much larger than 50nm, it behaves like a flexible chain.

9.2 Experiment: Plastics the Second Time Around

Physical Properties of Polymers

Objective: The objective of this experiment is to test and compare the physical properties of thermoplastic polymers.

Review of Scientific Principles

Plastics are long chain molecules. Depending upon the monomers, the plastic will have different physical and chemical properties. Chemical properties are difficult to test for and usually call for the destruction of the plastic through incineration. Burning plastics can give off toxic fumes. This is one of the reasons firemen wear a self contained breathing apparatus when entering a burning building. It is easier and safer to check the physical properties. Different plastics look, feel, and behave differently. Some are clear and colorless, while others are opaque. Some feel soft, while others feel slimy, slippery, or tacky. Some are more rigid than others. Each plastic has a unique density. Each plastic has a temperature at which it softens and/or melts as we saw in the last laboratory experiment.

The densities (in g/ml) of the plastics you will be checking are:

- HDPE 0.952 to 0.965,
- LDPE 0.917 to 0.940,
- PET 1.29 to 1.4,
- PP 0.900 to 0.910,
- PS (in solid form) 1.04 to 1.05,
- PS (in foam form) variable but always less than 1,
- PVC (rigid) 1.30 to 1.58,
- PVC (flexible) 1.16 to 1.35.

The melting or softening point we discussed in the last investigation is important when recycling, because when a plastic is softened or melted, it will adhere (stick) to itself. Generally when two or more plastics are softened or melted, the plastics will not adhere to one another. This is one of the reasons why recycled plastics must be sorted. This is an expensive process that adds to the costs of recycling.

Time: To perform this experiment and answer the questions will require 35-40 minutes.

General Safety Guidelines

- Care should be taken in minimizing contact with the solutions used in the buoyancy portion of this experiment.
- Aprons and goggles should be worn during the experiment.
- The ethanol solutions and calcium chloride solutions are to be returned to the container provided by your instructor.
- Hands should be washed after the experiment has been completed.

Materials and Supplies

Samples of the Following Plastics

- HDPE (high density polyethylene),

- LDPE (low density polyethylene),
- PET (polyethylene terephthalate),
- PP (polypropylene),
- PS [in solid form] (polystyrene),
- PS [in foam form], and
- PVC (polyvinyl chloride).

Ethanol/Water Solutions of Various Concentrations

- 52 per cent ethanol (density = 0.911),
- 38 per cent ethanol (density = 0.9408), and
- 24 per cent ethanol (density = 0.9549).

Calcium Chloride/water Solutions of Various Concentrations

- 6 per cent $CaCl_2$ (density = 1.0505),
- 32 per cent $CaCl_2$ (density = 1.3059),
- 40 per cent $CaCl_2$ (density = 1.3982), and
- 250 ml beaker.

Procedure

1. Obtain a sample of each type of plastic, noting the letter on each piece. The letters are used to reference each sample.
2. Examine each sample and write a visual description in the proper location in the data table.
   (a) Is the sample clear? Is the sample opaque? Does it have color?
   (b) In the data table describe how the sample feels. Is the sample smooth or rough? Does it have a pattern?
   (c) Flex each sample through an angle of 10° to 30°. Note in the data table how easy it was to flex the sample. Is it flexible or rigid? You might want to compare the various samples.
3. Pour 50 ml of the 40 per cent $CaCl_2$ solution into a 150 ml beaker. Place each of the plastic samples in the solution. Note which samples sink (S) or float (F) in the DATA TABLE.
4. Return the solution to the appropriate container and dry out the beaker.
5. Dry off your samples.
6. Repeat Step #3-5 with:
   - 32 per cent calcium chloride,
   - 6 per cent calcium chloride,
   - 24 per cent ethanol,
   - 38 per cent ethanol,

- 52 per cent ethanol, and
- Return all plastics to the recycling box after usage.

Data Table

Test	A	B	C	D	E	F	G
Visual Description							
Surface Appearance							
Rigidity							
Float/Sink Ethanol mix 52%							
38%							
24%							
Float/Sink $CaCl_2$ mix 6%							
32%							
40%							

Questions

1. From the data given in the REVIEW OF SCIENTIFIC PRINCIPLES section identify each of the plastics by proper recycling number and proper name.
2. Sample A
3. Sample B
4. Sample C
5. Sample D
6. Sample E
7. Sample F
8. Sample G
9. When testing for the density of the plastic samples, why did some of the samples stick out of the solution more than other samples?
10. If you were given two plastic samples, how would you identify them?
11. Which of the polymer (plastics) would be used as a material in making each of the following?
12. Use letters and names to identify each polymer.

    (i) A covering to go around a sandwich?
    (ii) A replacement for a picture window ?
    (iii) As a covering for a plastic bowl?
    (iv) As a replacement for the lead sinkers used in fishing?
    (v) As a clip board to write on?

13. Which polymer was most flexible?
14. Which polymer was most transparent?

Teacher Notes

Time: This lab is expected to take 40 minutes.

Materials and Supplies

Samples of the following plastics:

- HDPE (high density polyethylene)
- LDPE (low density polyethylene)
- PET (polyethylene terephthalate)
- PP (polypropylene)
- PS [in solid form] (polystyrene)
- PS [in foam form]
- PVC (polyvinyl chloride)
- Three different ethanol solutions of various concentrations are used. 50 ml of each solution is needed for every group.
- 52 per cent (density = 0.911)—to 619.1 ml of 95 per cent ethanol add distilled water to make 1000.0 ml
- 38 per cent (density = 0.941)—to 467.2 ml of 95 per cent ethanol add distilled water to make 1000.0 ml
- 24 per cent (density = 0.965)—to 302.7 ml of 95 per cent ethanol add distilled water to make 1000.0 ml
- Three different calcium chloride solutions of various concentrations are used. 50 ml of each solution is needed for every group.
- *Caution*: The hydration of $CaCl_2$ is an exothermic reaction. The solutions should be made ahead of time to allow the solution to cool. One day in advance is recommended.
- 6 per cent (density = 1.0505)—6 g of $CaCl_2$ (anhyd.) plus 94 ml distilled water or 7.95 g of $CaCl_2$*2 H_2O plus 92.05 ml of distilled water.
- 32 per cent (density = 1.306)—32 g of $CaCl_2$ (anhyd.) plus 68 ml of distilled water or 42.39 g of $CaCl_2$*2 H_2O plus 58.61 ml of distilled water.
- 40 per cent (density = 1.398)—40 g of $CaCl_2$ (anhyd.) plus 60 ml of distilled water or 52.99 g of $CaCl_2$*2 H_2O plus 47.06 ml of distilled water.
- 250 ml beaker.

General Safety Guidelines

- Students should wear lab aprons and goggles at all times during this experiment.

- Students should wash their hands after the lab has been completed.

Procedure

1. Prepare the solutions given above and ask the students to return them to the original containers after use. An alternative procedure might be to provide a number of beakers each containing one solution. Identify the beakers so the students can simply take these beakers to their desk for usage and exchange beakers among lab groups after their use.
2. Using the polymers brought by the students in the previous lab, cut the polymers into 2 cm by 4 cm pieces. With an indelible marker, write a letter A-G on each type of polymer. These letters provide a technique for identifying and talking about the polymers used. A polymer can be identified using the recycling code on the bottom of the plastic. The recycling codes are: 1 = PET, 2 = HDPE, 3 = PVC, 4 = LDPE, 5 = PP, and 6 = PS. Students should be familiar with the code numbers found in the 3 bent arrow recycling symbol. At this time you should not identify any of the polymers used in this experiment nor their recycling numbers.
3. Students should get one sample of each type of polymer to use in this experiment. Should the indelible ink be leached from a polymer, a fresh identification marking should be reapplied.
4. When the experiment is finished, students are to place their polymers in the recycling box.
   - Students will obtain good results with the densities except for the PET and PVC samples. These two polymers are hard to distinguish using their densities.
   - Students will be able to identify Styrofoam from its appearance. After the lab, ask the class about the appearance, color, feel, and rigidity of the samples. These are extrinsic, physical properties. Density, melting, and softening point are intrinsic physical properties. Depending upon the placement of this lab in the school year, these terms might be new or a review for the students.
   - Review the manufacturer's recycling code for the plastics with the students.

9.3 Hands-on Activity: Physical Properties

Since the physical properties of polymers are a big part of why they are useful, it is important for students to learn about those properties firsthand. Since polymers are often used to replace metals, this activity compares the physical properties of polymers with those of some common metals.

This is really five activities investigating five different physical properties. The students will be divided into five groups. Five tables will be set up, each a station for carrying out one of the activities. The five groups will rotate from station to station so that all students can perform all five activities.
This activity requires five tables because five different physical properties will be investigated. Label each table with the name of the property to be investigated there. The five physical properties are:

1. Mass
2. Strength
3. Ductility
4. Conductivity
5. Temperature stability

Mass

At this table, provide the students with containers containing about equal volumes of steel nuts or washers, a plastic, and a polystyrene or polyurethane foam. Students will simply pick them up and feel the relative masses. Students should conclude that plastics and foams are much lighter than steel.

Strength

At this table, provide pieces of steel, and a brittle plastic such as polystyrene and a high-strength plastic such as polycarbonate. Label them polymer A and polymer B. The students will be asked to break each piece. The students should conclude that steel is strong, while plastics can be strong or not so strong.

Ductility

At this table, provide the students with an aluminum can or steel wire, or some other ductile piece of metal. Also provide a piece of a soft plastic and a piece of a stiffer plastic. Label the plastics polymer A and polymer B. The students will be asked to try to bend and reshape the samples. They should conclude that metals are ductile, while polymers can be ductile or not ductile.

Conductivity

At this table, provide an apparatus consisting of a 12-volt battery wired to a light bulb and a sample of steel and one each of any two common plastics, as shown in the photographs and the schematic below. Label the plastics polymer A and polymer B. The students will be asked to try to complete the circuit using the steel and the plastic. The students should conclude that steel conducts electricity while polymers generally do not.

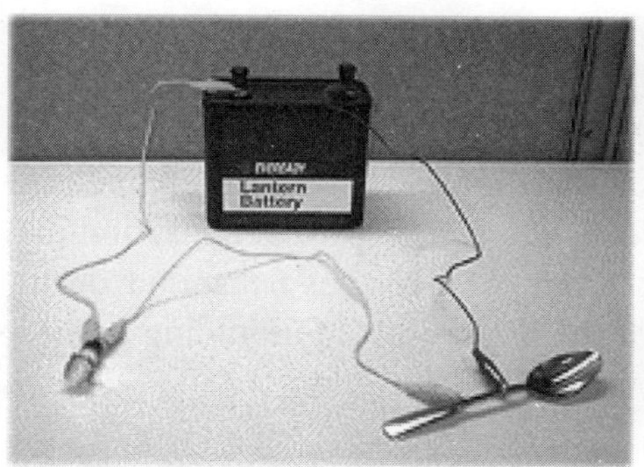

Metals conduct electricity.

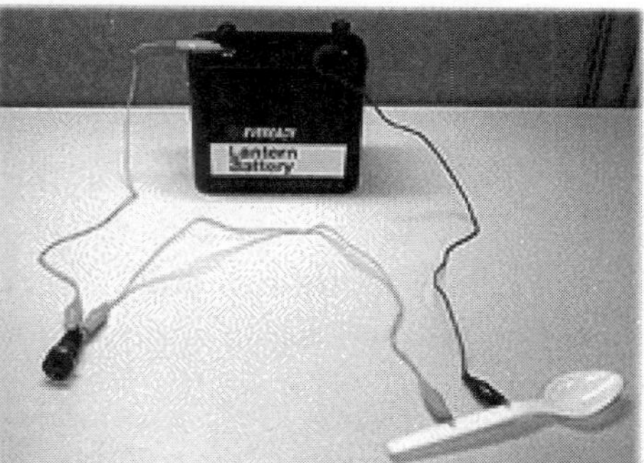

Most polymers do not.

This apparatus can be built with materials found at any electronics store. For wires it is best to use jumper leads (insulated wires with alligator clips at each end) so you won't have to solder any of the connections. Also, be sure to use a 12V light bulb, and that it has terminals to which alligator clips can be attached. Such bulbs are available form Radio Shack, cat. no. 272-336.

1. 12-volt battery,
2. Insulated wire,
3. Alligator clip,
4. Light bulb,
5. Sample (steel or polymer).

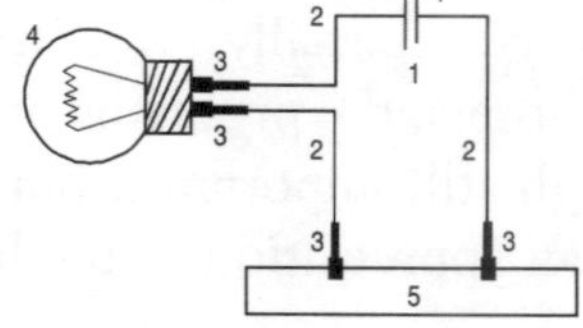

Schematic for the conductivity apparatus.

Temperature Stability

At this table, provide the students with samples of steel, polystyrene, and polycarbonate or nylon, a hot pot in which to boil water, and tongs. Label the polymer samples polymer A and polymer B. Using the tongs, the students should place each sample in the boiling water for several minutes. The steel will be unaffected, as will the polycarbonate. The polystyrene will undergo its glass transition, and will become flexible rather than rigid. The students should conclude that while steel is very resistant to high temperatures, polymers may or may not be temperature resistant. Note: Because of the risk involved with using boiling water, you may want to carry out the hands-on work at this table yourself for the students to watch.

Physical Properties of Polymers

The manufacture of polymer products is controlled by two often conflicting demands: the quality of the finished article in terms of its response to its environment and the ease or difficulty of processing it to shape. Both factors are controlled by what is termed viscoelasticity, namely, the behaviour of the polymer in response to applied stress or strain, and temperature. It is important to appreciate the duality in terms of the elastic and viscous responses of polymer solids and polymer melts, especially for thermoplastics used in engineering applications. For thermosets, the problems of creep and stress relaxation may be less critical (although still important), but their viscous behaviour during processing is vital for an appreciation of the limitations of moulding them to shape.

9.4 Behaviour of Polymers

The manufacture of polymer products is controlled by two often conflicting demands: the quality of the finished article in terms of its response to its environment and the ease or difficulty of processing it to shape. Both factors are controlled by what is termed viscoelasticity, namely, the behaviour of the polymer in response to applied stress or strain, and temperature. It is important to appreciate the duality in terms of the elastic and viscous responses of polymer solids and polymer melts, especially for thermoplastics used in engineering applications. For thermosets, the problems of creep and stress relaxation may be less critical (although still important), but their viscous behaviour during processing is vital for an appreciation of the limitations of moulding them to shape.

9.5 Viscoelasticity and Master Curves

An immediate consequence of the viscoelasticity of polymers is that their deformations under stress are time dependent. If the imposed mechanical stress is held constant then the resultant strain will increase with time, i.e. the polymer creeps. If a constant deformation is imposed then the induced stress will relax with time (stress relaxation). Figure 9.1 shows the creep strain response to a constant stress followed by unloading. Note that in the recovery stage the strain has still not returned to zero even a considerable time after the stress has been removed.

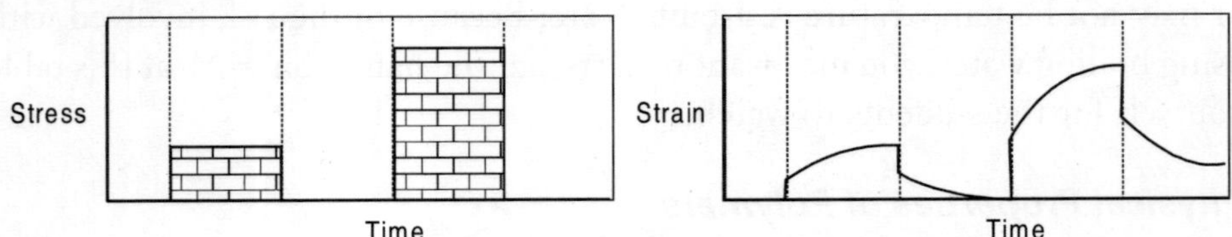

Fig. 9.1: The creep strain response of a viscoelastic material subjected to a constant stress of finite duration.

There are many examples of products which creep in service. For example, the plastic tub of an automatic washing machine will successively creep and recover as it is loaded and unloaded several times during the washing cycle. Viscoelastic stress analysis during the design of the tub ensures that the maximum strain due to this loading pattern is well within the strain limits for the material and the application.

The creep and stress relaxation properties of a polymer can be described by the time-dependent moduli $E_c(t)$ and $E_R(t)$. The creep modulus is the ratio of an imposed constant stress σ_0 to the time-dependent strain $\varepsilon(t)$, while the stress relaxation modulus is the ratio of the time-dependent stress $\sigma(t)$ to an imposed constant strain σ_0, i.e.

$$E_c(t) = \frac{\sigma_0}{\varepsilon(t)} \text{ and } E_R(t) = \frac{\sigma(t)}{\varepsilon_0}$$

Practical examples of the need to design for stress relaxation are in seals where the sealing force must remain adequate under conditions of constant deformation, or when a metal peg needs to be held in a plastic block by push fitting into an undersized hole. The oversize peg results in a constant hoop strain in the plastic. The corresponding hoop stress will decay with time but must always be sufficient to hold the peg in place. Both creep and stress relaxation are factors that have to be considered in design, although they are not necessarily always deleterious.

$E_R(t)$ is characteristic of the polymer concerned at a particular strain and temperature. It is the tensile stress relaxation modulus if the corresponding strains and stresses are tensile. Alternatively if the strains and stresses relate to shear or hydrostatic changes then the corresponding material parameters are the shear and bulk relaxation moduli respectively. As noted earlier, for polymers, the tensile modulus should not be referred to as Young's modulus. Young's modulus is the limiting case of the tensile modulus when the induced strains can be considered infinitesimal and independent of time.

Time-Temperature Superposition

For amorphous polymers above their T_gs, there is a convenient approximation which makes experiments easier. It is known as time-temperature superposition, and it relates time to temperature for viscoelastic materials. A sequence of measurements of $E_R(t)$ is performed at different temperatures at a fixed initial strain. The time scale might be limited between several seconds and say 100 hours. The curves obtained on uncrosslinked polyisobutylene (PIB) are shown in the lefthand portion of Figure 9.2, with temperatures of measurement varying from -80.8 °C up to + 50 °C. The curves span many decades of modulus, reflecting the change in behaviour of the material. At the lowest temperatures PIB is becoming glass-like, so $E_R(t)$ is very high. As the material passes through the transition region, the modulus drops rapidly—the material is becoming rubbery in its response to the applied stress. The onset of true elastomeric behaviour is marked by the so-called rubbery plateau. This is followed by another steep fall in modulus where viscous flow occurs as the temperature is raised further.

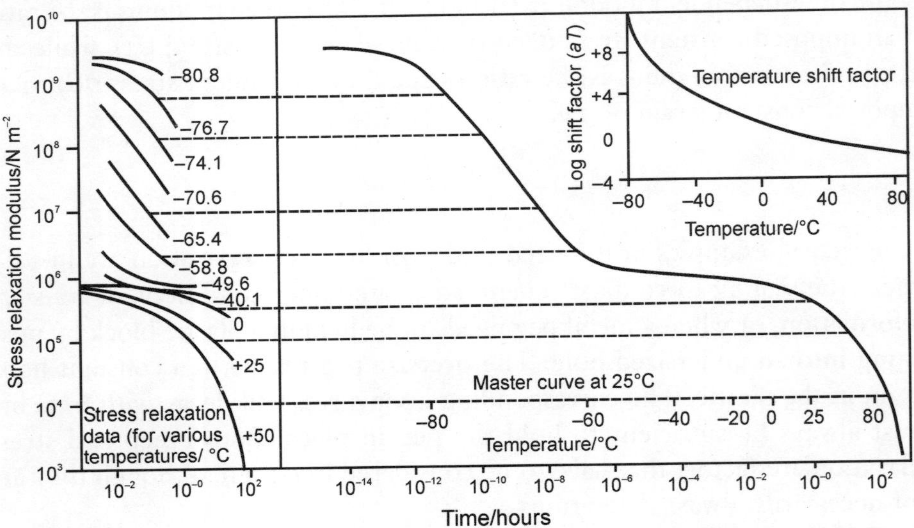

Fig. 9.2: Construction of the viscoelastic master curve for PIB at 25 °C reference temperature by shifting stress relaxation curves obtained at different temperatures horizontally along the time axis. The shift factor, a_T varies with temperature as shown in the inset at upper right.

Why creep and stress relaxation are often needed: It is usually said of engineering polymers that they are 'difficult' and 'problematic' materials for designers to use in stressed applications. Such comments often come from engineers who have been brought up with mild steel as the material of choice for *any* application, since design calculations are normally much more straightforward when the extra variable of time can be neglected (not that you can neglect time when corrosion is possible!). But a brief consideration of many consumer products shows that creep and stress relaxation are sometimes necessary for the correct functioning of many products and devices.

Consider leather shoes and boots where the stress relaxes under the constant strain of the foot, or clothes which relax from the strain applied by the human body. Rigid and inflexible materials would be quite inappropriate for protecting the human body comfortably. There are other examples too, from civil engineering, where viscoelastic timbers in a bridge or building creep and relax to accommodate imposed stresses and strains. In old buildings, such shape changes are regarded as of positive value in assessing its age and integrity.

Mastic polymer is used to seal joints in buildings where some movement is likely: its function depends on being able to flow and 'give' when strained, so maintaining a good seal.

Nevertheless, in many applications, creep and relaxation must be faced and appreciated; too much distortion may cause the part to drop away and so the product ceases to function. Loss of sealing stress in rubber joints can cause leakage of oil from engines, and hence failure. It is only by careful design in matching product function and material that such problems can be solved.

The curves can be fitted together like a jigsaw puzzle, as shown in the right-hand portion of figure 9.2 to form a 'master' curve, a single curve which represents the stress relaxation behaviour of PIB at a reference temperature of 25°C. The two scales, time, t, and temperature, T, are shown for the master curve. The derived relation between time and temperature a_T is plotted in the upper right-hand inset of the figure and can be modelled by the Williams-Landel-Ferry (WLF) equation

$$\log a_T = \frac{-17.44(T - T_g)}{51.6 + T - T_g}$$

when the reference temperature for the shift is chosen to be T_g and the numerical factors are the fitting parameters for this grade of PIB.

The master curve of Figure 9.2 indicates how PIB responds to stress at much shorter times than are accessible directly by this type of experiment and also re-emphasises the common polymeric behaviour of both rubbers and thermoplastics. They differ only in their glass transition temperatures: an amorphous thermoplastic is rigid at ambient temperature because its T_g lies above ambient temperature, and conversely a rubber is flexible because its T_g lies below ambient temperature. Any rubber will become rigid or glasslike if the temperature of the environment is low enough, and any thermoplastic will become rubbery at high temperatures.

Effects of Structure on Viscoelasticity

If a single measurement of $E_R(t)$ is taken at an arbitrary but fixed interval of time, say 10 seconds, then it will vary with temperature in a way rather similar to the viscoelastic master curve. Such a curve for atactic polystyrene is shown in Figure 9.3, where the various zones of behaviour are identified. The effect of lightly crosslinking the material is to eliminate flow of any kind, extending the region of elastomeric behaviour to higher temperatures. Heavier crosslinking will move the rubbery plateau upwards and the same net effect is achieved by crystallisation. Although atactic PS is inherently non-crystallisable, the isotactic form can crystallise and its melting point (T_m) is about 230 °C. The stress relaxation modulus of this polymer drops much less at 7V, from about 5000 MN irr^2 to about 100 MN m^{-2} at 140 °C. At the same temperature, atactic polystyrene is completely rubbery with a modulus of about 0.5 MN m^{-2}. Thereafter, the modulus of the crystalline polymer drops.

Molecular mass also affects the stress relaxation spectrum, particularly at intermediate and higher temperatures (>100°C). For high molecular masses, only the rubbery plateau and viscous flow regions differ markedly; the glassy and transition regions are unaffected. Increasing molecular mass clearly improves some mechanical properties, but nowhere near as effectively as crystallisation.

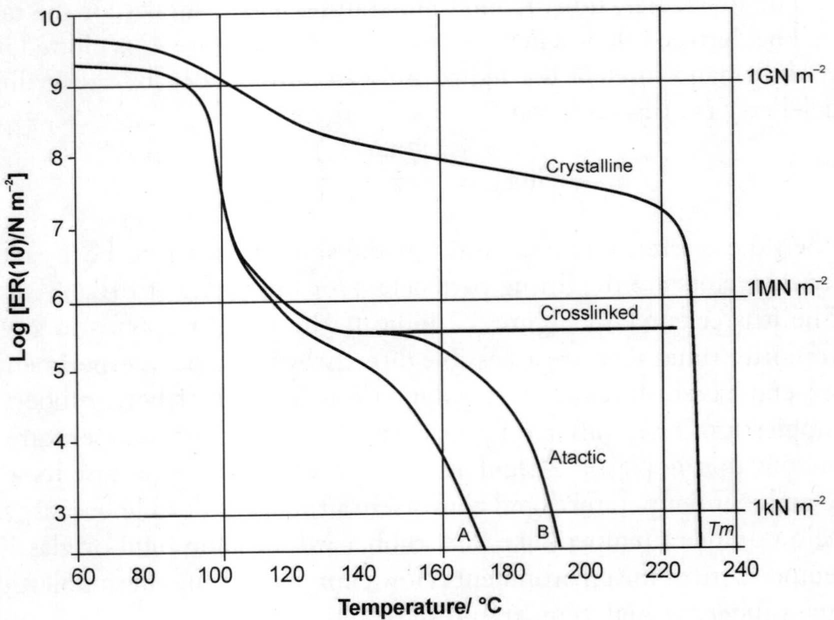

Fig. 9.3: Variation of stress relaxation modulus of polystyrene with temperature. The top curve represents the behaviour of partially crystalline isotactic PS with melting point T_m.

In Fig. 9.3 the main lower curve (B) is that for atactic PS of $\overline{M}_n$ = 217 000 and the lower subsidiary curve (A) for a $\overline{M}_n$ = 140 000. Crosslinking eliminates the rubber and viscous flow regions.

Self Assessment Question

1. Uncrosslinked silicone rubber treated with boron oxide exhibits curious mechanical properties. If hit hard with a hammer, it shatters like glass, but if dropped from a height of 1 metre it bounces like a rubber ball. When kneaded in the hand it behaves like putty but when left on a flat surface it flows like a liquid.
2. Sketch the viscoelastic master curve for the material. Assume that the hammer is travelling at a maximum velocity of 20 m s^{-1} and that the acceleration due to gravity g is 9.8 m s^{-2}. At what velocity would a well-vulcanized NR ball cease to bounce? Assume that both NR and silicone balls are 2 cm in diameter, and that the terminal velocity v of an object dropped from rest over a distance s is $\sqrt{2gs}$.

Viscoelasticity of Polymers

The simplest models for the deformation behaviour of an ideal material are those of Hookean linear elasticity in the solid state, and Newtonian linear viscosity in the liquid state. The end point of elastic deformation is either fracture or plastic flow, with the latter taking place at a constant yield stress in the ideal case. Whilst the behaviour of many real materials does approximate to these idealised models, that of polymers deviates markedly from them. In particular, their solid state deformation is time-dependent and nonlinear and so resembles some combination of elastic and viscous responses, whilst their melt rheology is also significantly nonlinear. To start with, let's consider polymer behaviour in the context of the idealised models of elastic deformation and of viscous flow.

Elastic and Viscoelastic Behaviour

When an elastic (*not* elastomeric, or long range elastic) material is stressed, there is an immediate and corresponding strain response. Figure 9.4 illustrates this by showing schematically the strain response to a particular stress history. Note that when the stress is removed the strain also returns to zero. So in a perfectly elastic material all the deformation is returned to the forcing agency. If this energy had not been stored elastically then it would have been dissipated as either heat or sound. Tyre squeal and the heat build-up in the sidewalls of car tyres are good examples of such dissipation.

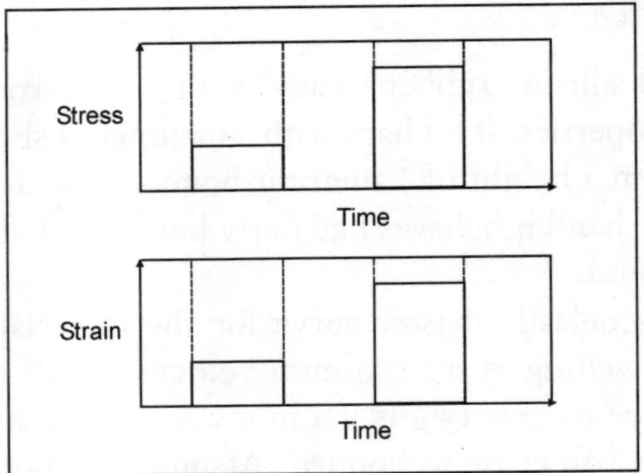

Fig. 9.4: When stressed, a perfectly elastic material deforms in proportion to the applied stress and returns to its original state when the stress is released.

If the material is linear and elastic then the applied stress σ directly proportional to the strain ε. Then, for simple tension,

$$\sigma = E\varepsilon \tag{1}$$

where E is a constant known as Young's modulus, and is considered to be a property of the material. For polymers, due to time-dependence and nonlinearity, E is not a constant and the term tensile modulus is used to reflect this. Various values of tensile modulus are tabulated in the Data Book for a range of polymers together with those of Young's modulus for other materials. The stress-strain curves for PS, HIPS and two types of rubber are shown in Figure 9.5. While polystyrene apparently obeys Hooke's law (Equation 1.), HIPS yields and necks before failing. By contrast, rubbers exhibit long-range elasticity and fail only at many hundred per cent strain.

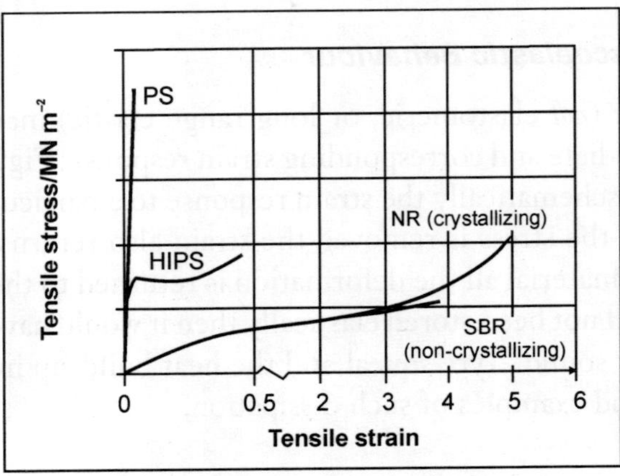

Fig. 9.5: Tensile stress-strain curves for some rubbers and plastics.

So some plastics like HIPS yield, but others fracture in a brittle manner like polystyrene. Rubbers do not yield, but at high strains some of them crystallise and hence stiffen. When the stresses are removed from a polymeric material before fracture, the strain recovery path is not necessarily identical to that of the loading part of the deformation cycle. So energy must have been dissipated during the deformation of such materials—another indication of deviation from perfect elasticity. Both the deformation and the subsequent recovery are time-dependent, suggesting that some part of their behaviour is viscous. In fact solid polymers show a combination of elastic and viscous behaviour known as viscoelasticity. The degree of viscoelasticity is strongly dependent upon the temperature of test and the rate at which the polymer is deformed, as well as such structural variables as degree of crystallinity, crosslinking, and molecular mass.

Viscous Behaviour

Viscous flow is not recoverable. When the stress is removed from a viscous fluid the strain remains. Hence the work energy is not returned to the forcing agency and has to be otherwise dissipated. Fig. 9.6 illustrates this schematically by showing the strain response in such a viscous material when a simple stress history has been imposed upon it.

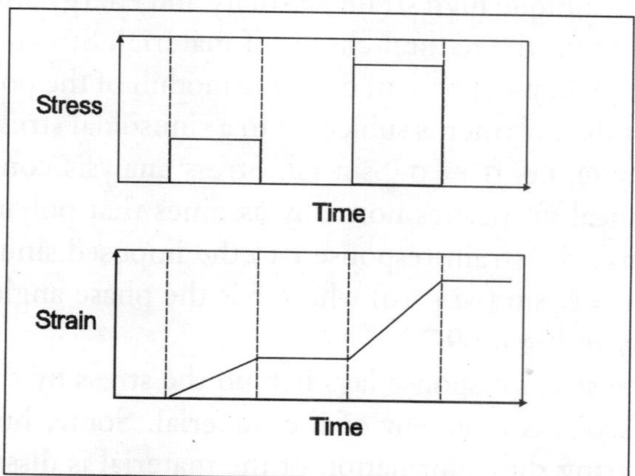

Fig. 9.6: A viscous fluid is deformed permanently by an applied stress and continues to deform if stressed again.

For a linear viscous material, the rate of change of shear strain with time, $\bar{\gamma}$ is directly proportional to the imposed tensile stress σ

$$\sigma = 3\eta\bar{\gamma} \qquad (2)$$

where η is the Newtonian viscosity of the fluid. In polymer melts, η is not a constant, but depends on the shear strain rate $\bar{\gamma}$, with, in general, η decreasing as $\bar{\gamma}$ increases. The viscosity is also directly dependent on molecular mass and so inversely related to melt flow index. These factors are thus of direct importance to a processor or moulder.

Viscous properties dominate during the earlier stages of processing but elastic effects are also important. Molten thermoplastics are obviously highly viscous as are partially polymerized and uncured thermosets and masticated and unvulcanized rubbers. However, observable features such as the swelling of extrudates as they leave dies are manifestations of melt elasticity.

In essence, elasticity in molten polymers arises from the entropy factor or the tendency of the macromolecules to coil into their configurations of maximum disorder. This is in opposition to the stretching which occurs as the material flows under the influence of the stress field.

9.6 Dynamic Mechanical Properties

Viscoelasticity is not experienced just under quasi-static conditions, i.e. when the imposed stresses and strains are constant or change only slowly. Polymers, and particularly rubbers, are often deliberately selected for products which are to be subjected to dynamic mechanical loading. Tyres are an obvious example where the unique high strain elasticity and energy absorbing qualities of rubbers make them the natural choice of material. Stress analysis involves the use of the frequency-dependent dynamic moduli of the polymers. Assume, for example, that the polymer is subjected to a sinusoidal stress σ of amplitude σ_o and frequency ω, i.e. $\sigma = \sigma_0 \sin \omega t$. Stress analysis concerned with the dynamic mechanical properties normally assumes that polymers are linearly viscoelastic. Hence the strain response *e* to the imposed sinusoidal stress can be described as $\varepsilon = \varepsilon_0 \sin (wt — \delta)$ where δ is the phase angle. This is shown diagrammatically in Figure 9.7.

Note that the strain response lags behind the stress by the phase angle—owing to the viscous component of the material. Some, but not all, of the energy stored during the deformation of the material is dissipated. Since the material is assumed to be linear, the stress is proportional to the strain at all times, i.e. $\sigma = E\varepsilon$, but *E* is a function of the frequency ω. Because the stress and strain are not in phase, *E* must be treated as a complex function:

$$E^* = E' + iE'' \qquad (3)$$

where $i = \sqrt{-1}$, and *E*" and *E*" are the in-phase and out-of-phase components of the modulus.

From the above definitions of the dynamic moduli and by manipulation of the linear relationship between the sinusoidal stress and the corresponding strain response, the phase angle δ can be expressed as follows:

$$\tan \delta = E'' / E' \tag{4}$$

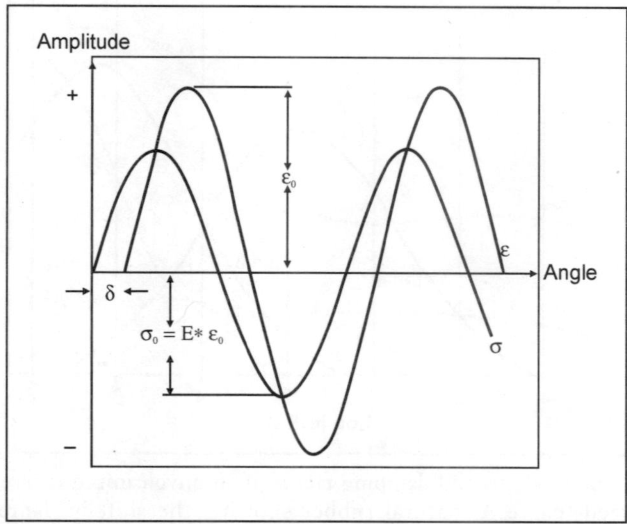

Fig. 9.7: The sinusoidal stress cr and corresponding strain ε response for a linear viscoelastic material. The imposed stress and the material response do not coincide, and the phase angle δ is the difference between the two curves.

Tan δ is commonly called the loss tangent or damping factor. *E*" and tan δ are the most commonly measured dynamic properties of rubbers, representing the elastic stiffness and damping or hysteresis properties respectively. Fig. 9.8 is a schematic representation of *E*' and tan δ as a function of frequency for natural rubber, and shows the effect of crosslinking on damping and hence heat dissipation. Different rubbers will have different curves, and some care is needed in rubber product design to match the material with expected imposed frequencies. Fillers such as carbon black will also affect the shape and position of the damping maximum. Sometimes the 'argument' of the complex modulus $|E|$ is used instead of E^*, and is given by the equation

$$|E| = \sqrt{(E)^2 + (E'')^2} \tag{5}$$

At very high frequencies (ù = 10^4-10^8 cycles s^{-1} or Hz) rubber is very stiff with a glass-like modulus. At these frequencies the polymer molecules do not have time to react in response to the forcing oscillations. The damping factor is then small but it increases to a maximum value in the 'leathery' transition region between the glassy modulus and the usual (low) modulus which is characteristic of rubbers that are deformed slowly ($\omega < 1$ cycle s^{-1} or Hz).

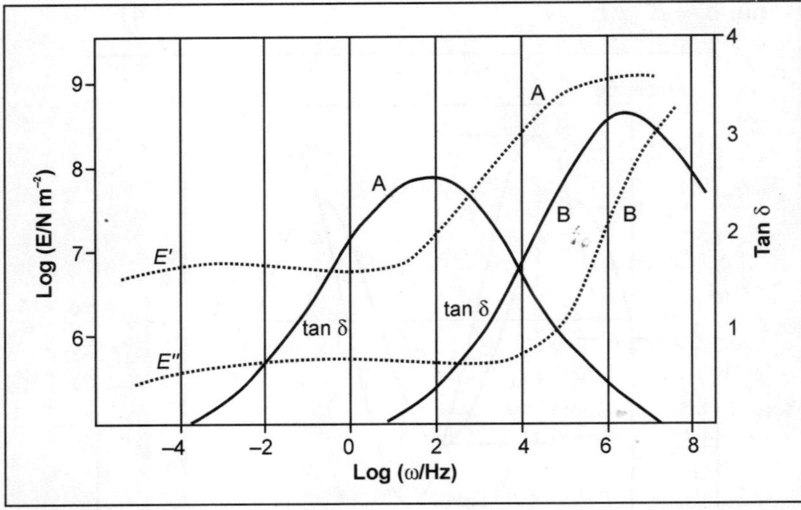

Fig. 9.8: The elastic modulus and damping factor of an unvulcanized (curve B) and a well-vulcanized (curve A) natural rubber showing the shift in damping caused by crosslinking. Increasing the frequency of oscillation is equivalent to decreasing the temperature.

It will be noticed from Fig. 9.8 that decreasing the frequency of oscillation imposed on the material is equivalent to increasing the temperature (glass-like behaviour is experienced at high frequencies). This is another manifestation of viscoelasticity, in which the effect of the three basic variables, time (t), temperature (T) and frequency (ω), on a polymer are all closely interrelated. Since decreasing the time scale of an experiment is equivalent to decreasing the temperature, so decreasing the frequency is equivalent to *increasing* the time scale (and hence *increasing* temperature).

Self Assessment Question

1. What is the effect of increasing the speed of a car on the rubber material of the tyre treads? Use the WLF equation and time-temperature superposition principle to justify your argument.
2. Assuming that a car tyre is 0.3 m in diameter, evaluate the effect of increasing car speed on the dynamic properties of the tread material. What other factors might you need to take into account to evaluate the dynamic properties?

9.7 Orientation in Polymers

Viscoelasticity, like thermodynamics, is concerned with the correlation of controllable variables and bulk, macroscopic phenomena. But one unique feature of polymeric materials is that the molecular unit, the polymer chain, can be highly anisotropic, i.e. the chain can be fully extended, or curled up in an amorphous equilibrium state without any net orientation. In fact, unoriented polymer is rarely encountered in manufactured products because of the different ways it is processed to shape. By its very nature, forcing viscous polymer fluid into cool moulds or through dies by extrusion gives some molecular orientation depending on the stresses to which it has been subjected during manufacture. The control of orientation (and the related effect of crystallisation) during shaping is the key to product quality and the properties that product will exhibit in service. A related problem concerns non-uniform distribution of filler particles in a polymer matrix.

Non-Uniform Mixtures

Moulded rubbers and plastics are compounds of a polymer matrix and a variety of additives. The mixing history of the material before and during the moulding process can have a critical influence upon the final product properties. If mixing is done badly then the microstructure of the moulding can be non-uniform. Lack of uniformity can cause variations of strength and other physical properties within the moulding. The degree of dispersion or distribution of relatively minor quantities of additives can have a significant effect upon the properties of the product. This is illustrated in Fig. 9.9 which shows a thin slice of polyethylene tubing of diameter 5 cm. PE masterbatch, heavily pigmented with carbon black, has been added to the unpigmented granules in the hopper of the extruder, in order to improve UV resistance. The poor state of mixing gives a laminated section to the tube, very like a rolled-up newspaper. The outside surface appears black and so gives some measure of protection against sunlight, but the maldistribution of carbon black weakens the material by concentrating stress locally at clusters of particles. Although the apparent degree of orientation is high, the polymer molecules in reality have a relatively low degree of orientation.

Many plastics composites use fibres as the reinforcing agents within a polymer matrix. The distribution of orientations of the fibres then determines the overall anisotropy of the components. Suppose for example that a laminate is made from sheets of resin-impregnated glass fibre cloth, plied together so that the warp and weft directions of successive layers coincide.

Fig. 9.9: Thin slice of part of a PE tube 5 cm outer diameter (o.d.) showing the poor distribution of carbon black masterbatch during extrusion. The carbon black is added to give protection from sunlight but because of poor mixing can weaken the product substantially. The quadrants are created by spiders in the extrusion head which divide the mixing polymer melt. *Source:* RAPRA.

The properties in the orthogonal warp and weft directions are different from each other and are different again from the properties through the thickness of the laminates. Similar symmetry applies to the biaxially oriented stretched films which are used as outer wrappings for many consumer goods and supermarket foodstuffs, as well as PET bottles, although in these examples the orientation is molecular in origin.

Molecular Orientation

As polymers are processed and shaped by flowing into moulds the shear stress fields induce preferred orientations in the molecules. The hydrostatic components of the stress field cause packing. These orientation and packing effects will relax with time if the temperatures are high enough, but the moulding cycle is frequently such that they are 'frozen-in' by cooling or perhaps fixed into the structure because the material has been crosslinked. The consequent moulded-in or residual stresses and strains may:

- Subsequently warp the moulding and
- Can increase the likelihood of fracture or cracking, particularly in the presence of some hostile chemicals.

Molecular orientation due to moulding sometimes results in physical properties which vary significantly with direction, i.e. the properties are anisotropic. This can be beneficial and is therefore sometimes induced deliberately. For example, when synthetic fibres are spun they are oriented uniaxially to increase the strength in the fibre direction. In this way, the high potential strength and stiffness of carbon-carbon bonds in the backbone chain of linear high polymers can be achieved if the chains can be fully aligned along the fibre axis. This has been achieved in aramid fibres and other aliphatic fibres also show significant improvements although nowhere near as great as with aramid fibres like *Kevlar*. Its tensile modulus of 60-124 GN m^{-2} may be compared with that for carbon fibre of 200-500 GN m^{-2}, glass fibre (about 70-80 GN m^{-2}) and PET fibre (5500 MN m^{-2}). However, when sheet polymer is oriented by cold drawing, enhancement of stiffness in and orthogonal to the drawing direction is compensated by a reduction of stiffness from an angle of about 20° to 70° to the draw direction. The dotted line in the figure denotes the unoriented value of the tensile modulus. Another kind of orientation effect results from high strains in crystallising rubbers. In this case, orientation gives rise to crystallisation with a consequent rise in tensile strength over and above that for non-crystallising rubbers like SBR. Unlike cold drawing, the effect is reversible except for low temperatures (below 0 °C) when crystallisation occurs at zero strain and so called stark rubber is formed.

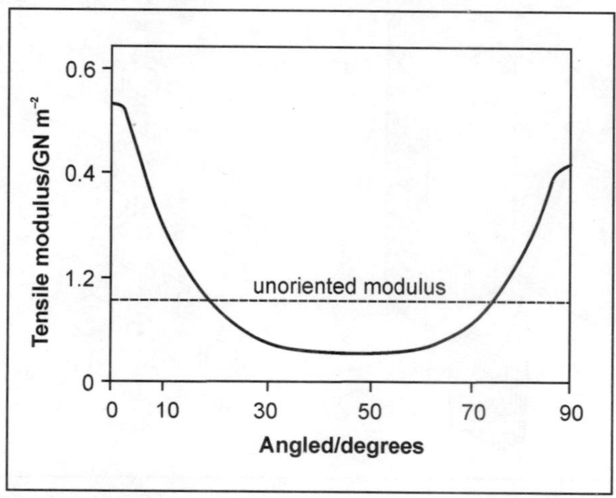

Fig. 9.10: Variation of the tensile modulus with angle to the stretch direction of a sheet of LDPE which has been cold drawn.

9.8 Crystallization of Polymers

The major benefits of crystallisation of chain molecules to end users are: (a) Since $T_m > T_g$, the maximum service temperatures are higher than with amorphous polymers; (b) Above T_g, the modulus of a crystalline polymer is higher than that of an amorphous polymer owing to reinforcement and physical crosslinking of the amorphous matrix by the crystallites.

Both effects are evident by comparing the stress relaxation curves of atactic and isotactic PS of Figure 9.10. However, crystallisation morphology and kinetics differ from polymer to polymer as a result of structural and energetic factors.

Morphology of Polymer Crystallites

The fundamental unit of structure formed by crystalline polymers which is accessible using the optical microscope is the spherulite. Isolated spherulites are formed easily at relatively slow spherulite growth rates such as those exhibited by polypropylene and isotactic polystyrene. Unlike aramid fibres where the degree of crystallisation is close to 100 per cent (Figure 9.11), most crystalline polymers contain significant amounts of amorphous polymer either between spherulites or present at crystallite boundaries within individual spherulites.

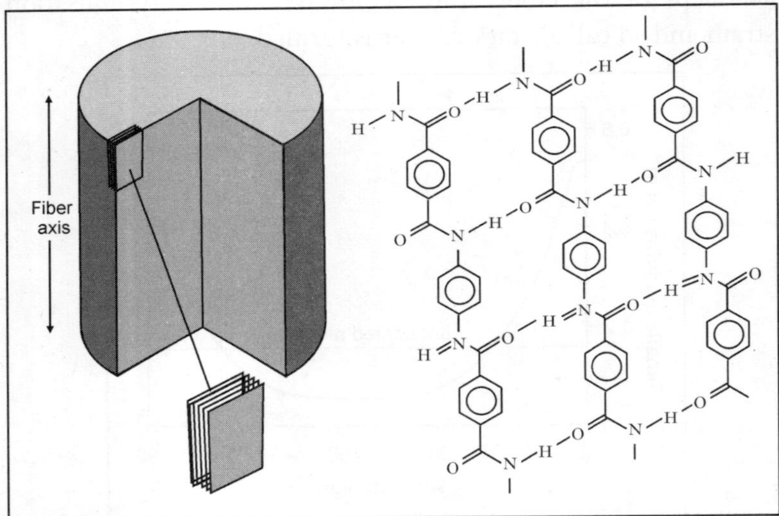

Fig. 9.11: Crystalline structure of aramid fiber.

By cooling solutions of polymers in organic solvents, it is possible to obtain minute platelets (lamellae) which represent the smallest crystal elements within much more complex structures like spherulites (Figure 9.11). Such lamellae vary in thickness between 10 and 40 nm and may be several micrometres (ìm) wide. The chains in each crystal, which in polyethylene are in a zig-zag conformation [Figure 9.12(a)], are folded one upon the other as shown schematically in Figure 9.12(b). It is thought that similar folds occur in bulk polymers, although there is probably more disorder at the fold surface where chains interconnect with other lamellae [Figure 9.12(c)]. Clearly, such chain segments represent non-crystalline polymer; although their modulus is lower they provide the essential links between crystals and so bind the material together like a composite.

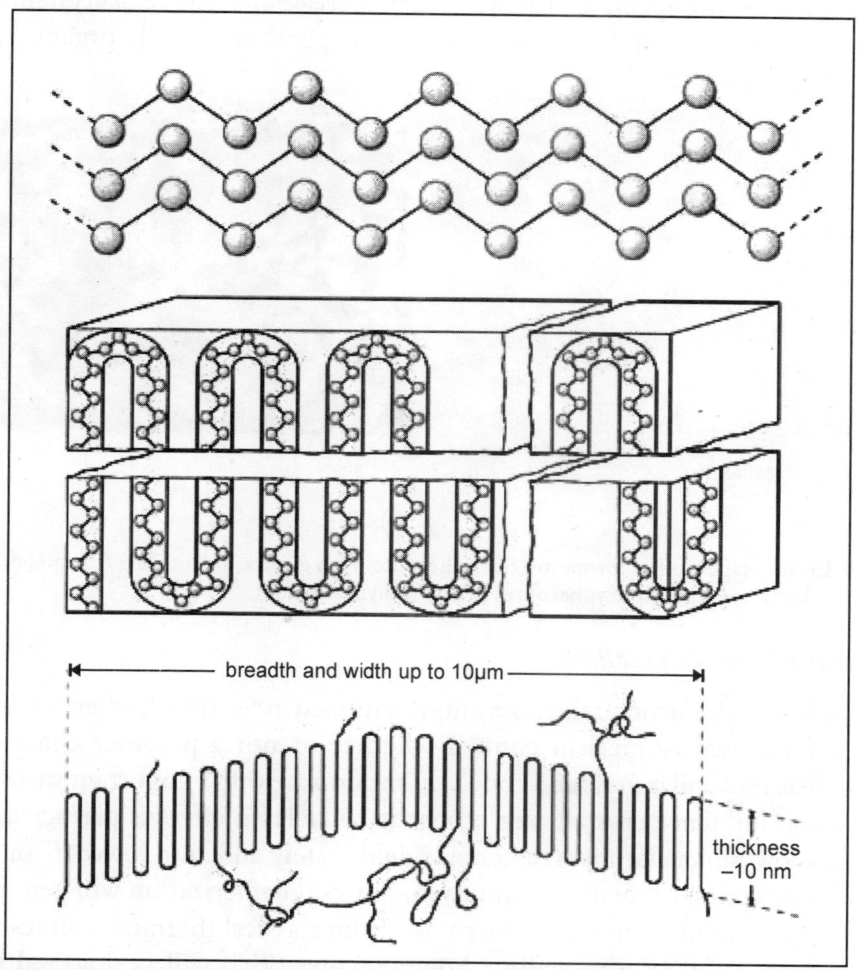

Fig. 9.12: (a) Zig-zag conformation of HDPE, with stacking in crystal; (b) chain folded model for HDPE in lamella; (c) dimensions of HDPE single crystal showing non-crystalline tie molecules.

A spherulite can grow from a single crystal nucleus very much in the way suggested by the sequence shown in Fig. 9.13(a). Growth occurs when chains continue the folding action and crystal defects lead to lamellar twisting and branching. Fibrillar structures are thus formed which successively twist round to form a 'wheatsheaf'. Continued growth ultimately yields a spherulite which then grows uniformly as a sphere. For HDPE and nylon 6,6, nucleation and growth are very rapid, radial growth rate for HDPE is about 5000 μm min^{-1}), so that spherulite impingement is the norm [Fig. 9.13(b)]. In polymers with lower growth rates, such as polypropylene (growth rate about 20 μm min^{-1}, specific nucleating agents are often added to increase the degree of crystallinity developed in objects during processing. Such agents are usually organic salts like calcium stearate.

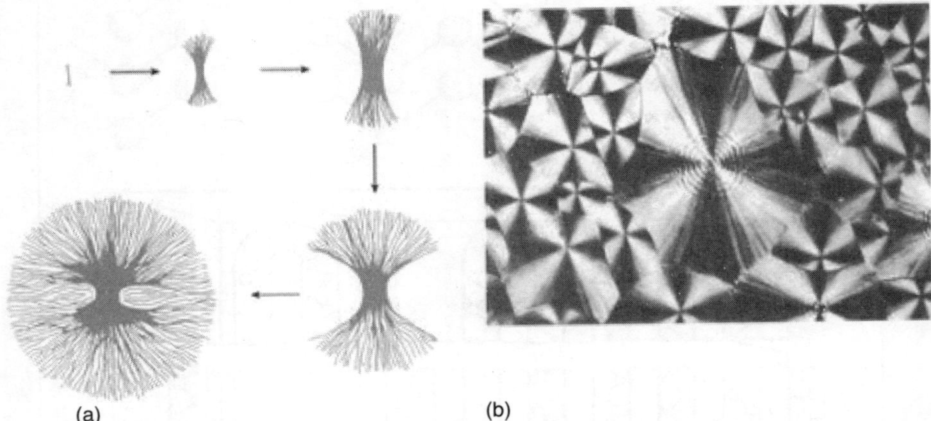

(a) (b)

Fig. 9.13: (a) Stages in the formation of a spherulite from a stack of lamellae; (b) a polarised-light micrograph of spherulites in poly(ethylene oxide).

Structure and Crystallinity

In addition to the structural constraints mentioned in Section 2, where tacticity and geometrical isomerism control whether or not a polymer chain can crystallise, molecular mass and copolymerization are other important variables which can influence crystallising properties. A related effect is plasticization where a low molecular mass material is deliberately added to lower T_g or T_m.

The lowering of melting point caused by copolymerization will generally affect T_g in a similar way. It can be shown from classical thermodynamics that the melting point of a crystalline homopolymer (T_m°) will be lowered to a value T_m (in K) by the presence of x mole fraction of randomly copolymerized repeat units (B) according to the equation:

$$\frac{1}{T_m} - \frac{1}{T_m^0} = \frac{R}{\Delta H_f} x_B \tag{5}$$

where R is the gas constant and ΔH_f the heat of fusion per mole of homopolymer repeat unit. The equation predicts that the melting point depression is directly proportional to the mole fraction of the second component (B) and inversely proportional to the heat of fusion.

Since chain ends can also be regarded as 'impurities' in a chain which will lower the melting point, a similar equation can be derived for the effect of molecular mass:

$$\frac{1}{T_m} - \frac{1}{T_m^0} = \frac{2R}{\Delta H_f}\frac{M_R}{M_n} \quad (6)$$

Since most polymers show less than 100 per cent crystallinity, *observable* heats of fusion will be proportionally lower depending on the exact degree of crystallisation. However, the melting points are unaffected by the presence of amorphous material. A further quantity of interest for crystalline polymers is the entropy of fusion, ΔS_f. It is simply related to $T_m^{\,0}$ by the equation

$$T_m^{\,0} = \frac{\Delta H_f}{\Delta S_f} \quad (7)$$

Self assessment Question

(a) At what number-average molecular mass will the melting point of polypropylene be 99.9 per cent of that for very high molecular mass polymer?

(b) In material of molecular mass such that chain ends are not important, what is the effect on the crystalline melting temperature T_m of randomly copolymerizing 10 mole per cent ethylene into the homopolymer? What would be the practical consequences in terms of processing the copolymer to shape and the service properties of the copolymer?

Conformation and Crystallinity

If there are key connections between the chain configuration and crystallisation, you might also expect some more subtle effects from rotation about chain bonds. After all, polymer chains must be able to twist into the regular conformation demanded for crystal structures [Fig. 9.13(a)]. And what influence will rotation have on the precise conformation adopted by the chain?

Polyethylene crystallises into the most stable conformation represented by the linear zig-zag, but when substituents are present, as in propylene, then there is substantial steric hindrance from the pendant methyl group at each alternate carbon atom. In polypropylene and related polyolefins, the chains adopt a helical conformation (Fig. 9.14) where the extra side groups are accommodated on the outside of the helix by regular twisting of the whole chain. And as you might expect, the larger and longer the side group, the larger the diameter [Fig. 9.14(b) and (c)]. With large aromatic side groups, the pitch of the helical conformation grows [Fig. 9.14(d)], but the diameter actually shows a slight decrease. A similar effect occurs in PTFE,—[CF_2–CF_2,]—where the fluorine atoms present on *all* the carbon atoms, pack neatly together into the helix (Fig. 9.15). The Fig. is based on a so-called space filling rather than outline model, so that the true atom sizes are taken into account. This is useful, because it demonstrates that no part of the backbone chain itself is visible. It is thought that this is one of the reasons for the exceptionally low coefficient of friction of PTFE ì?0.04), because fluorine has extremely weak van der Waals bonds with others around it, whether other fluorine atoms or foreign atoms at an interface.

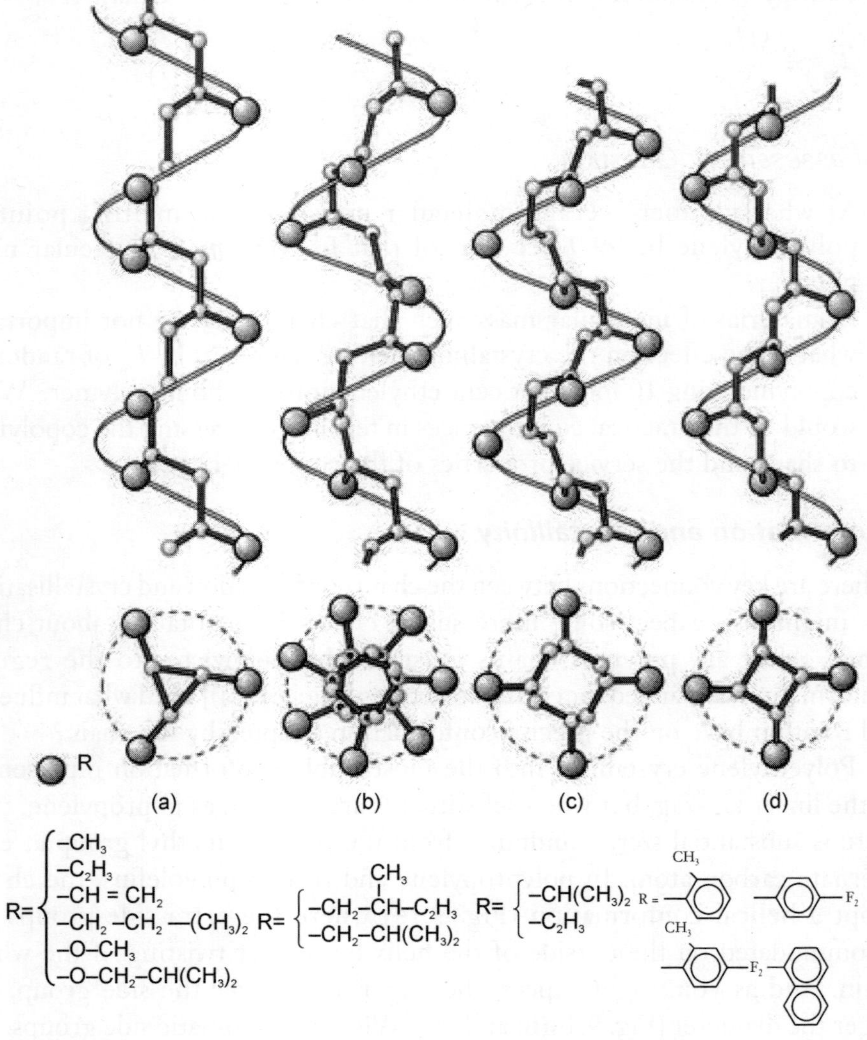

Fig. 9.14: Helical conformations of isotactic vinyl polymers (Gaylord and Mark, 1959).

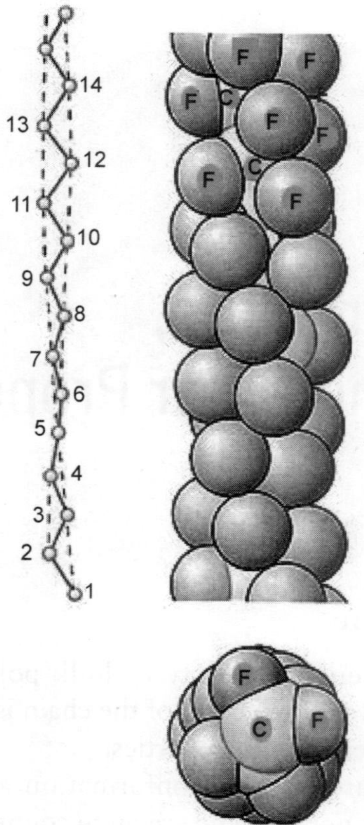

Fig. 9.15: Structure of the molecule of PTFE showing a helical conformation.

There is also the problem of rate of crystallisation. One might expect hindered repeat units to crystallise only slowly. This explains why polypropylene crystallises so slowly when compared with polyethylene.

10

Polymer Structure, Tacticity and Other Properties

10.1 Polymer Structure

Although the fundamental property of bulk polymers is the degree of polymerization, the physical structure of the chain is also an important factor that determines the macroscopic properties.

The terms configuration and conformation are used to describe the geometric structure of a polymer and are often confused. Configuration refers to the order that is determined by chemical bonds. The configuration of a polymer cannot be altered unless chemical bonds are broken and reformed. Conformation refers to order that arises from the rotation of molecules about the single bonds. These two structures are studied below.

Configuration

The two types of polymer configurations are *cis* and *trans*. These structures can not be changed by physical means (e.g. rotation). The *cis* configuration arises when substituent groups are on the same side of a carbon-carbon double bond. *Trans* refers to the substituents on opposite sides of the double bond.

```
H\        /H         -CH2\        /H
   C==C                  C==C
  /      \              H/      \
-CH2      CH2-                  CH2-
      cis                   trans
```

Stereoregularity is the term used to describe the configuration of polymer chains. Three distinct structures can be obtained. is an arrangement where all substituents are on the same side of the polymer chain. A polymer chain is composed of alternating groups and is a random combination of the groups. The following diagram shows two of the three of polymer chain.

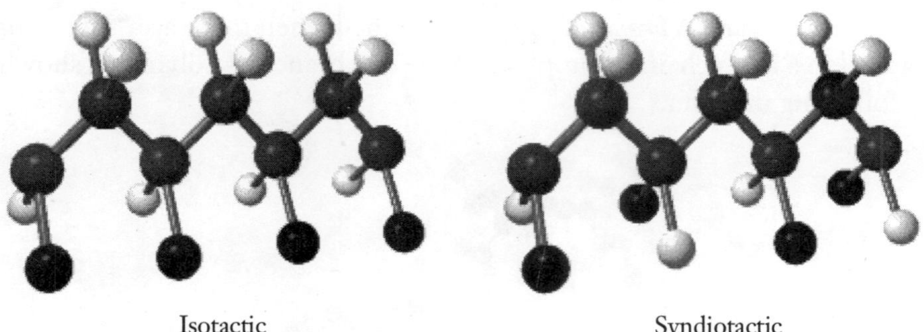

Isotactic Syndiotactic

Conformation

If two atoms are joined by a single bond then rotation about that bond is possible since, unlike a double bond, it does not require breaking the bond.

The ability of an atom to rotate this way relative to the atoms which it joins is known as an adjustment of the torsional angle. If the two atoms have other atoms or groups attached to them then configurations which vary in torsional angle are known as conformations. Since different conformations represent varying distances between the atoms or groups rotating about the bond, and these distances determine the amount and type of interaction between adjacent atoms or groups, different conformation may represent different potential energies of the molecule. There several possible generalized conformations: Anti (Trans), Eclipsed (Cis), and Gauche (+ or -).

Conformation Lattice Simulation

Like the polymer growth simulation, the conformation lattice simulation takes a statistical approach to the study of polymers. Probabilities of the different conformations are assigned which produces a polymer chain with many possible shapes.

Other Chain Structures

The geometric arrangement of the bonds is not the only way the structure of a polymer can vary. A *branched polymer* is formed when there are "side chains" attached to a main chain. A simple example of a branched polymer is shown in the following diagram.

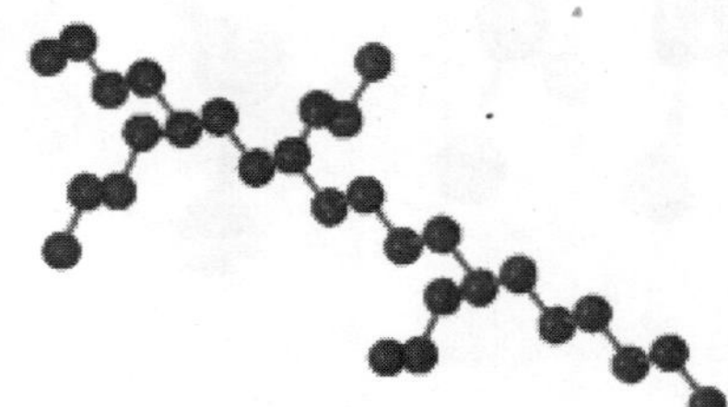

There are, however, many ways a branched polymer can be arranged. One of these types is called "*star-branching*". Star branching results when a polymerization starts with a single monomer and has branches radially outward from this point. Polymers with a high degree of branching are called *dendrimers* Often in these molecules, branches themselves have branches. This tends to give the molecule an overall spherical shape in three dimensions.

A separate kind of chain structure arises when more that one type of monomer is involved in the synthesis reaction. These polymers that incorporate more than one kind of monomer into their chain are called *copolymers*. There are three important types of copolymers. A *random copolymer* contains a random arrangement of the multiple monomers. A *block copolymer* contains blocks of monomers of the same type. Finally, a *graft copolymer* contains a main chain polymer consisting of one type of monomer with branches made up of other monomers. The following diagram displays the different types of copolymers.

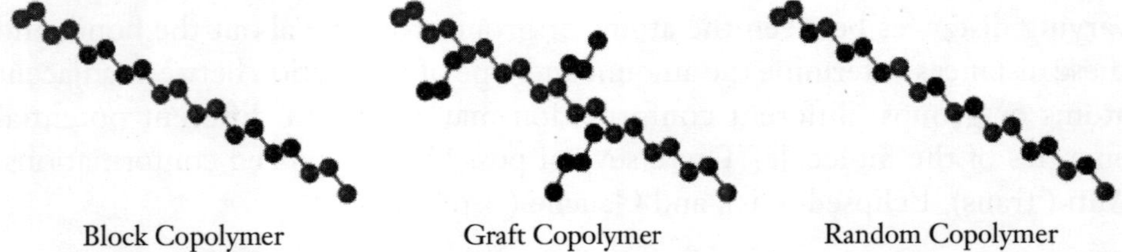

Block Copolymer Graft Copolymer Random Copolymer

An example of a common copolymer is Nylon. Nylon is an alternating copolymer with 2 monomers, a 6 carbon diacid and a 6 carbon diamine. The following picture shows one monomer of the diacid combined with one monomer of the diamine:

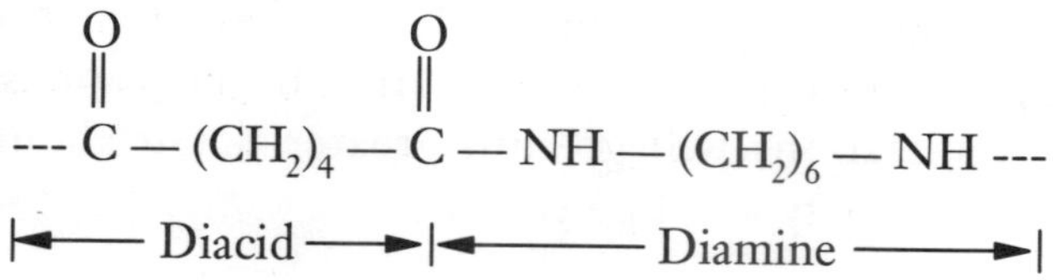

$$
\begin{array}{c}
\overset{\displaystyle O}{\overset{\|}{\text{---} C}} - (CH_2)_4 - \overset{\displaystyle O}{\overset{\|}{C}} - NH - (CH_2)_6 - NH \text{---}
\end{array}
$$

|← Diacid →|← Diamine →|

Cross-Linking

In addition to the bonds which hold monomers together in a polymer chain, many polymers form bonds between neighboring chains. These bonds can be formed directly between the neighboring chains, or two chains may bond to a third common molecule. Though not as strong or rigid as the bonds within the chain, these *cross-links* have an important effect on the polymer. Polymers with a high enough degree of cross-linking have "memory." When the polymer is stretched, the cross-links prevent the individual chains from sliding past each other. The chains may straighten out, but once the stress is removed they return to their original position and the object returns to its original shape.

One example of cross-linking is *vulcanization*. In vulcanization, a series of cross-links are introduced into an *elastomer* to give it strength. This technique is commonly used to strengthen rubber.

Classes of Polymers

Polymer science is a broad field that includes many types of materials which incorporate long chain structure of many repeat units as discussed above. The two major polymer classes are described here. *Elastomers*, or rubbery materials, have a loose cross-linked structure. This type of chain structure causes elastomers to possess memory. Typically, about 1 in 100 molecules are cross-linked on average. When the average number of cross-links rises to about 1 in 30 the material becomes more rigid and brittle. Natural and synthetic rubbers are both common examples of elastomers. *Plastics* are polymers which, under appropriate conditions of temperature and pressure, can be molded or shaped (such as blowing to form a film). In contrast to elastomers, plastics have a greater stiffness and lack reversible elasticity. All plastics are polymers but not all polymers are plastics. Cellulose is an example of a polymeric material which must be substantially modified before processing with the usual methods used for plastics. Some plastics, such as nylon and cellulose acetate, are formed into fibers (which are regarded by some as a separate class of polymers in spite of a considerable overlap with plastics). As we shall see in the section on liquid crystals, some of the main chain polymer liquid crystals also are the constituents of important fibers. Every day plastics such as polyethylene and poly(vinyl chloride) have replaced traditional materials like paper and copper for a wide variety of applications. The section on Polymer Applications will go into greater detail about the special properties of the many types of polymers.

10.2 The Growth of Polymers

Polymers, or materials composed of long molecular chains, are now well-accepted for a wide variety of applications, both structural and non-structural, and for mass-manufactured as well as one-off speciality products. The growth in their use has continued in the last two decades or more, despite the effects of several recessions in industrial activity (Figure 10.1). In the same period the demand for traditional materials like metals, ceramics and glasses has remained static or even fallen. Steel usage in the UK, for example, has fallen from about 14 million tonnes in the 1970s to about 12 million tonnes in the 1990s, while that of aluminium has stayed at about 600 000 tonnes. The growth in use of polymers is forecast to continue into the next millennium, with consumption approaching 4 million tonnes in the UK. In one of the most active areas, that of thermoplastic polymers, consumption is divided between packaging, building, and a wide range of other applications (Figure 10.2).

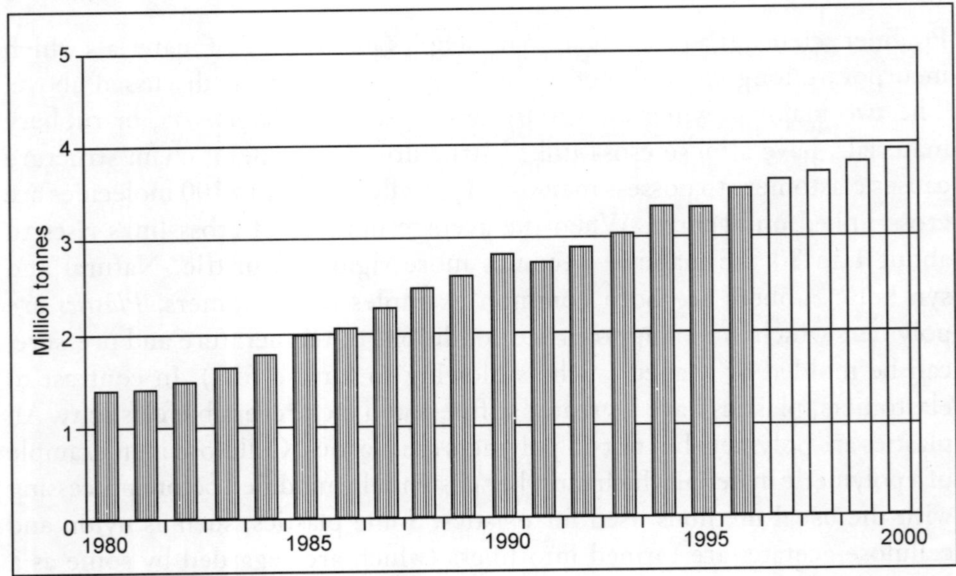

Source: data from British Plastics Federation.

Fig. 10.1: Growth in demand for polymers in the UK, 1980-2000.

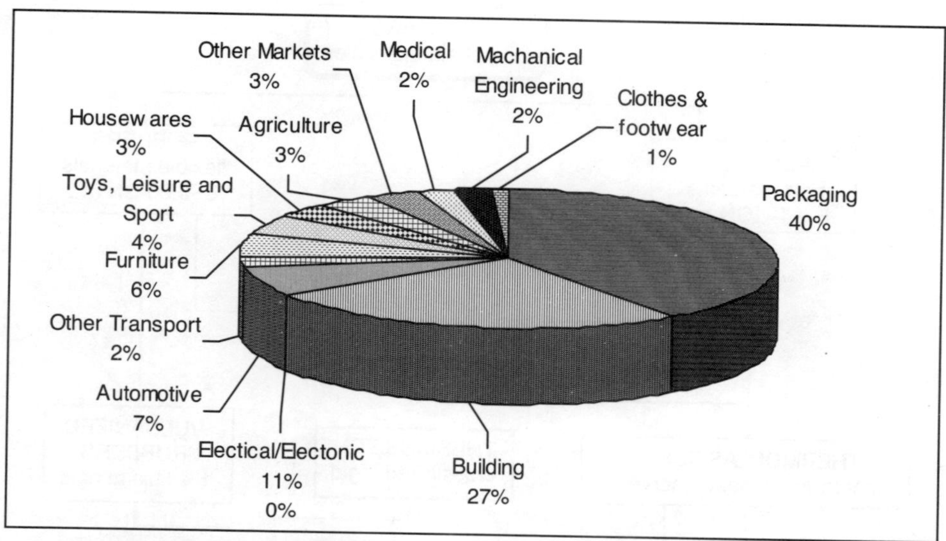

Source: data from BPF Statistics, 1995.

Fig. 10.2: UK plastics applications 1995.

10.3 Polymer Types

Traditionally, the industry has produced two main types of synthetic polymer—plastics and rubbers (Figure 10.3). The distinction is that plastics are, by and large, rigid materials at service temperatures while rubbers are flexible, low modulus materials which exhibit long-range elasticity. Plastics are further subdivided into thermoplastics and thermosets, the latter type being materials where the long chains are linked together by crosslinks, a feature they share with conventional vulcanized rubbers. As Figure 10.3 shows, however, the distinction in terms of stiffness has become blurred by the development of thermoplastic elastomers (TPEs). Moreover, all polymers, irrespective of their nature, can be reinforced by a very wide range of fillers to produce composite materials.

Another way of classifying polymers is in terms of their form or function, varying from additives to other bulk materials (e.g. viscosity modifiers in plaster), coatings to products (e.g. paints), film and membranes to fibres (e.g. textiles) and bulk products such as pipe, containers and mouldings (Figure 10.4). Some of these materials are of course used as products in their own right, or manipulated further into finished products. This does not always happen, however, some polymers being a disposable intermediary in certain industrial processes. Thus photoresists are used to create the circuit patterns on semiconductor chips through controlled degradation, and are entirely absent in the final product.

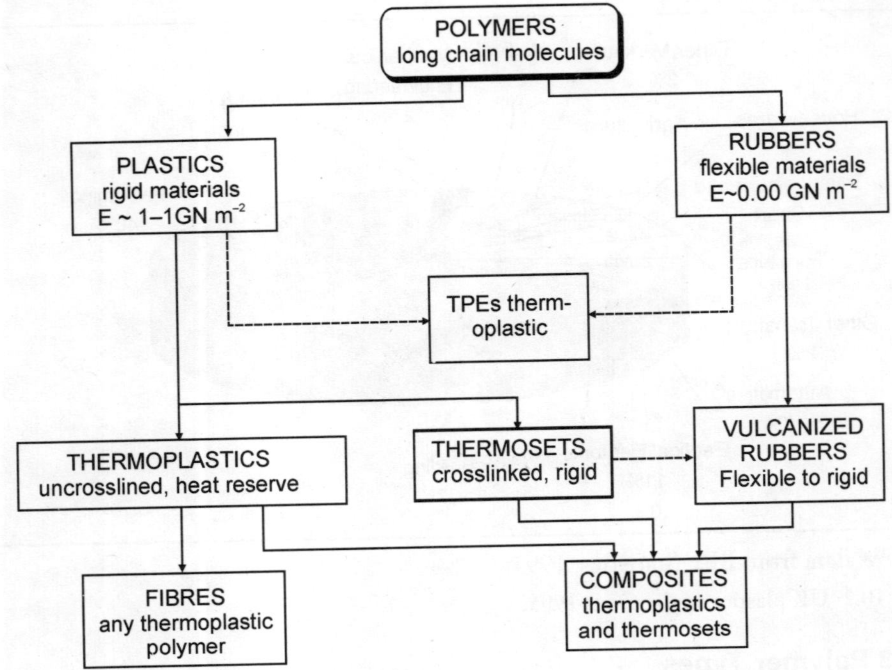

Fig. 10.3: Classification of polymers by property.

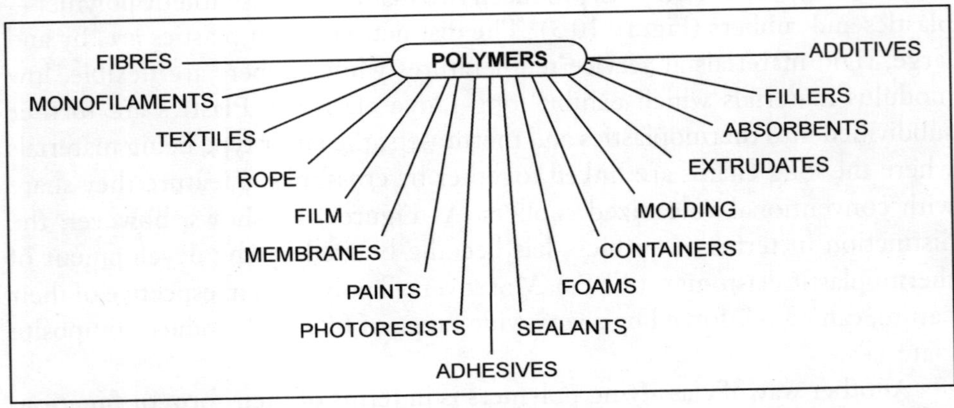

Fig. 10.4: Classification of polymers by design function.

Exercise 1

1. Identify the products in a typical modern house associated with the supply and disposal of water (or prevention of entry) which may be of polymeric origin, giving details of their generic type (Figure 10.3) and form or function (Figure 10.4).

Now Read the Answer

The first polymeric materials to be used were entirely natural in origin and required relatively little modification to be adapted for useful purposes. Such materials included wood from various species of tree, fibres for rope and textile fabrics, and amber adhesive for attaching stone and metal tools to wooden handles. Rubber was used by the early Americans for containers, shoes and balls. Ways of processing them to shape and improving their properties were developed during the Victorian era, but it was not until the growth of the organic chemical industry that the first synthetic polymers were made. The true molecular nature of materials like natural rubber and synthetics like Bakelite was not understood until about the 1920s when Hermann Staudinger recognised their chain-like structure. That period saw the growth of polymer chemistry, by which monomers could be synthesised and polymerized in a controlled way to give macromolecular materials. Some of today's major polymers were discovered in this period and were commercialised in the 1930s and 1940s. They included materials like polychloroprene (*Neoprene* rubber), nylon, polyester (*Terylene* or *Dacron*) and polyethylene (*Polythene*—note that trademarks for polymers are shown as proper names).

The raw materials for making the monomers had at first been based on coal tar derivatives but, with the rise of the petrochemical industry based on oil and natural gas, a much wider range of basic chemical building blocks became available. Fundamental advances in the understanding of catalysis led to the discovery of many new polymers in the post-1945 period—variations on simple polymers like polyethylene as well as entirely new stereoregular polymers like polypropylene. That progress has continued at an increasing rate up to the present. Novel polymers, like aromatic polyamides and polyimides which were discovered only in the 1960s, have been developed, while speciality, high temperature materials like polysulphones have penetrated new markets hitherto inaccessible to the traditional range of commercial polymers. The achievement has been a direct result of pioneering scientific research closely linked to development by industry.

Box 10.1: Plastics in Victorian times.

> The use of shellac as a moulding material was pioneered in the 1850s by Samuel Peck and Co. of the USA who added such refinements as the insertion of hinges during the moulding process. Metal inserts into thermoplastic mouldings are commonplace today. Casein made from skimmed milk was an early and reasonably successful protein plastic with a first patent in 1885 in Germany. The curds were separated from the whey, then after compounding

with plasticizers and colours, they were pressed into sheets, rods, tubes or discs. Finally the casein was hardened by immersion in formaldehyde. It made a tough material capable of accepting a high polish and hence was a popular substitute for horn, ivory and amber.

The first synthetic thermoplastic was developed in the 1860s when Parkes in England and Hyatt in the USA produced a mouldable cellulose nitrate by softening it (plasticizing it) through the addition of camphor. *Parkesine* has not survived as a product name or material, but Hyatt's *Celluloid* is still used commercially. Compounding polymers with additives to give a controlled range of properties is an essential step in the production of almost all the polymers used today.

The availability of phenol from cheap coal tar and formaldehyde from the oxidation of methanol led in 1877 to the development of phenol formaldehyde resin by Baekeland and Swinburne, working in the USA and Britain. These first condensation products of the controlled reactions between phenol and formaldehyde produced hard but relatively weak and brittle materials. Swinburne commercialised his resin as a range of varnishes but Baekeland mixed the resin with significant amounts of woodflour to produce the first polymer composite—*Bakelite*. It was the first synthetic thermoset, a material which becomes irreversibly hard (cures) either on heating as with Bakelite or by cold curing.

Natural and Synthetic Rubbers

Natural rubber was the first major polymer to be imported and used for commercial purposes. Long ago the natives of South America learned to tap the indigenous *Hevea Brasiliensis* trees to collect, dry and coagulate the latex. Today the main rubber plantations are in Malaysia and Indonesia. Natural rubber is well established as an important and versatile engineering material with an excellent balance of properties. However, almost two-thirds of the rubber now consumed world-wide is synthetic. The development of synthetic rubbers in Western Europe and the USA was accelerated by the demands of the Second World War and the associated loss of access to the natural rubber plantations in the Far East. Today's engineers have a complete spectrum of synthetic rubbers available to them, with properties ranging from the general-purpose to the highly specific. Hence the term 'rubber' or 'elastomer' is more properly the generic name for a class of polymeric materials of widely varying properties. The properties and common uses of a selection of both general-purpose and speciality rubbers are shown in Table 10.1.

Table 10.1: General-purpose and speciality rubbers: properties and uses.

Rubber		*General properties*	*Typical uses*
BR	butadiene (polybutadiene)	Special-purpose rubbers of density 0.93 Mg m?3. Good low-temperature properties and abrasion resistance. High resilience (low damping) and therefore low heat build-up at ordinary temperature. Poor resistance to oils and hydrocarbons.	Resilient mounts, tyre sidewalls (blended with NR)
CR	chloroprene (Neoprene)	Versatile special-purpose rubbers of density 1.20 Mg m?3 and good mechanical and electrical properties. Very good resistance to ozone oxidation, heat and flame.	Car radiator hose, gaskets, seals, conveyor belts, bridge bearings
EPM, EPDM	ethylene-propylene copolymer and terpolymer	The copolymer (EPM) and terpolymers (EPDM) are general-purpose rubbers of density about 0.85 Mg m?3 Good mechanical properties and resilience. Can accept very high loadings of oils and fillers. Very good resistance to ozone, oxidation, chemical, weathering and high and low temperatures.	Conveyor belts, hose, general goods
NR	natural rubber (cis-polyisoprene)	An excellent general-purpose rubber of density 0.93 Mg m?3 High resistance to tearing and abrasion. High resilience at 20 °C and thereforelow heat build-up under the action of dynamic stresses. Swells in mineral oils and degreasing solvents.	Tyres, suspension systems, bushes, bridge bearings
NBR	nitrile (acrylonitrile-butadiene copolymer)	Special-purpose rubbers of density 1.0 Mg m?3 and moderate mechanical properties. Poor cold resistance. Excellent resistance to swelling in hydrocarbons and alcohols. The greatest oil and alcohol resistance occurs in rubbers with a high acrylonitrile content.	Fuel lines and linings
SBR	styrene-butadiene copolymer	A good general-purpose rubber of density 0.94 Mg m?3, competitive in properties with NR when reinforced with carbon black. Very good abrasion resistance. Swelling and adhesion properties similar to NR, ageing resistance better than NR.	Tyres, often in direct competition with NR

Thermoplastics and Thermosets

As already stated, polymers including rigid plastics were first developed in the last century from natural precursors. The sealing wax employed by the

Victorians, for example, was usually based on the natural polymer shellac, an exudate of the Indian lac insect. Shellac is an early natural thermoplastic—defined as a material which softens and hardens reversibly on heating and cooling. In theory these reversible physical changes will take place without a corresponding change in the chemical structure of the material. This is why scrap thermoplastic can be re-used. In practice, some thermal and oxidative degradation occurs and recycling must be done only with an understanding of the effect that it has upon the properties of the final moulding.

Thermosets can be defined as those polymers which become irreversibly hard on heating or by addition of special chemicals. This hardening involves a chemical change (curing) and hence scrap thermoset cannot be recycled except as a filler material. The curing process invariably involves a chemical reaction which connects the linear molecules together to form a single macromolecule. These connections are known as crosslinks.

Scrap rubbers cannot be recycled easily, because of vulcanization which crosslinks the chains during moulding. It will be seen later that rubbers at the early stages of their processing can be considered to fit the definition of thermoplastics, but in their final moulded state they are properly defined as thermosets. As with rubbers, the impetus for the development of new and better synthetic plastics followed supply and demand. Initially the demand was for cheaper substitutes for traditional materials, but today no plastic is cheap and some are extremely expensive with unique properties designed to satisfy the stringent requirements of sophisticated products. Table 10.2 lists the names and acronyms of the bulk-use commodity plastics and some of the more specialised and expensive materials, and comments on their important properties and uses.

Consumption of Plastics

The consumption of plastics in the UK today is shown in Figure 10.5, with usage being dominated by PVC, closely followed by a range of polyolefins (polypropylene, various polyethylenes) and materials based on styrene monomer (PS and HIPS). These are the 'big five' bulk commodity polymers which dominate the market, and which have found application in almost every sphere of human activity. PVC is produced either in a flexible, elastomeric form (plasticized PVC) or as the rigid material familiar in pipes and profiles (unplasticized or uPVC). The ubiquitous plastic bag is usually made from any of the grades of polyethylene, and polypropylene is widely used for wrapping of consumer products. Polystyrene by itself is quite brittle, so is often used in its reinforced form (HIPS), or alternatively as a lightweight foam for insulation or protecting sensitive goods in transit.

Table 10.2: General-purpose and speciality plastics.

	Plastic	*Comments*
PMMA	acrylic, poly(methyl methacrylate)	Thermoplastic. A transparent rigid polymer.
ABS	acrylonitrile-butadiene-styrene	Based on SAN resin modified with polybutadiene rubber.
EP	epoxy	Thermoset. Resins used for encapsulation, adhesives, surface coatings and high-strength fibre-reinforced composites.
GRP	glass-reinforced plastic (mainly polyester)	Thermoset. Reinforced with glass fibre in various forms, such as chopped strand mat (CSM) and woven rovings (WR). Used for pipes, tanks, boat hulls, etc. May be applied as SMC or DMC.
HDPE	high density polyethylene	Thermoplastic. Linear polyolefin widely used in blow moulding.
HIPS	high impact polystyrene	Thermoplastic. A polystyrene modified by copolymerization with butadiene to improve its toughness.
LDPE	low density polyethylene	Thermoplastic. Branched polyolefin used for film and as electrical insulator, made at high pressures.
MF	melamine formaldehyde	Thermoset. Used in domestic ware, switches, plugs, etc.
PA	nylon, polyamide	Thermoplastic. Used in bearings, gears, mouldings, wall plugs, etc.
PF	phenolic, phenol formaldehyde	Thermoset. Moulding material and laminating resin. Sometimes known as Bakelite.
PAN	polyacrylonitrile	A fibre-forming thermoplastic polymer. One of the base polymers used to make carbon fibre.
UPR	polyester (unsaturated polyester resin)	Thermoset. A solution of polyester containing unsaturated groups in styrene or other polymerizable solvent. Matrix resin for GRP.
PET	poly(ethylene terephthalate)	Thermoplastic. A major fibre-forming polymer and a moulding material for beer bottles, etc. In competition with poly(butylene terephthalate) (PBT), a related thermoplastic polyester.
PVC	poly(vinyl chloride)	Thermoplastic. Can be plasticized to produce a leathery material. Unplasticized PVC (uPVC) used for rainwater goods, pipes, etc.
SAN	styrene-acrylonitrile	Thermoplastic. Rigid transparent material used for water jugs and beakers, etc.
SMC	sheet moulding compound	Thermoset. Sheets of glass fibre impregnated with polyester resin (td)

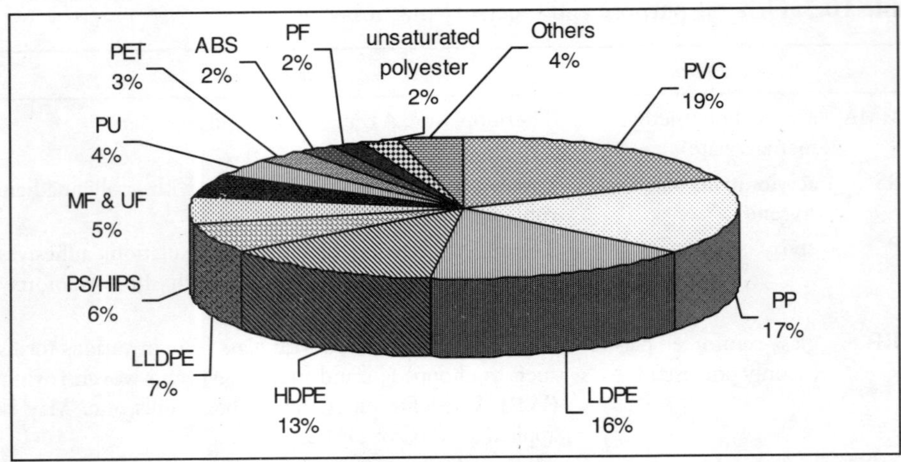

Fig. 10.5: UK consumption of plastics, 1995.

These five are followed by thermosetting melamine and urea formaldehyde, widely used for electrical insulation products as well as for reinforcing wood products. Polyurethanes are a very versatile group of polymers—they may be thermoplastic or thermosetting, and range in stiffness from flexible elastomers to stiff plastics. Stiff grades are used for car body panels, and in foam form for insulation, flexible foam being used in furniture. Another grade is widely used in paint varnishes. Polyester, PET, originally mainly used as a strong fibre, is often nowadays used in blends with cotton in textile fabrics. But its greatest application today is in tough lightweight bottles, displacing glass for its safer properties. ABS is a derivative of PS, being much tougher by blending with rubber particles, and so is widely used for enclosures. The oldest synthetic thermoset, PF, is used for reinforcing wood products, while unsaturated polyester forms the main matrix material for glass-fibre reinforced composites. Acrylics, which comprise PMMA and related polymers, first found use in transparent aircraft windows during the Second World War, especially since they could be easily formed into complex shapes. They should not be confused with acrylic fibre, actually based on a quite different chain structure, that of polyacrylonitrile or PAN.

Although consumption of the remaining named polymers in Figure 10.6 is relatively low, they represent a growing class of so-called engineering polymers, whose properties are often so unusual or interesting as to find unique application in addition to displacing traditional materials. The family of nylons centres around nylon 6,6 and 6, both invented in the 1930s, the former still popular as a fibre, although both forms are used for mouldings. A more spectacular example is aramid fibre, developed in the late 1970s in the USA, for its very high stiffness and resistance to high temperatures. These properties have resulted in aramid fibre being widely used in textiles, ropes and composite structures such as aircraft tailplanes and rotor blades (often in combination with epoxy resins). Acetal is a tough, crystalline polymer which until recently was widely used for domestic kettles. Polycarbonate also finds use as a tough material for consumer products, and displaces PMMA owing to its greater toughness. PBT is another kind of thermoplastic polyester mainly used as a moulding material. The fluoropolymers are a unique class of polymer exemplified by PTFE, the parent material used most frequently for its exceptionally low coefficient of friction and temperature resistance as in non-stick frying pans, plumbing tape and support pads for moving heavy equipment. A related elastomer, *Viton* rubber, is used for engine seals and aircraft hose.

10.4 Product Design and Manufacture

So what are the reasons for the continued growth in the use of polymers. It cannot be raw material cost, since the source of synthetic polymers is crude oil or natural gas, prices of which have risen over the same period of time. The comparative prices of polymers are considerably greater than traditional materials like mild steel, so we must look elsewhere for their success.

It is really necessary to focus on their end uses, as finished products, to even begin to attempt an answer. The largest area of consumption is that of packaging, where plastics are used to enclose perishable products such as food, drink, and fragile goods such as glassware and ceramics. Food preservation is a vital service today, where perishing by drying out or bacterial contamination has been reduced in Britain from between 30 to 50 per cent to 2–3 per cent in the last 50 years. Such packaging as shrink-wrap film, plastic bags and other containers, and foamed PS are now commonplace both for cooked and raw foodstuffs. Plastics are not alone here, because the 'tin' (actually steel or aluminium) can and glass bottle have also helped reduce wastage and harmful contamination (Figure 10.7). However, polymers do have a competitive edge in their low density and hence lower weight during transport. Thus vegetables are much more widely distributed in plastic packs, where deterioration is halted not by totally enclosing them in a metal can (so excluding air) but by freezing the vegetables and maintaining them at a low temperature in a film of low-air permeability. Carbonated soft drinks bottles are now almost universally made from PET, a tough transparent polymer, which cannot injure the consumer if it breaks and is substantially lighter than its glass equivalent. The barrier behaviour of polymers is not always as good as inorganic materials, but careful selection of impermeable polymers can reduce air penetration to a minimum. Thus PVDC is widely used to coat PET beer and cider bottles to prevent diffusion of air into these products and hence prolong their useful shelf-lives. Even with metal cans, plastic coatings are normal because they can be printed easily, and can contribute to product strength and barrier resistance.

Fig. 10.6: Various foods packaged in different materials including glass, PVC, PET, PP and LDPE.

So polymers do have certain intrinsic property advantages which have helped in designing new products. The most obvious one is low density for ease of transport and installation, but corrosion resistance is also important for products destined for a long life in a building (Figure 10.7). Ease of manufacture helps lower production costs, and also enables large sizes to be made. Most of these properties are also important in other areas of product design, especially in automotive, consumer, household products and office equipment.

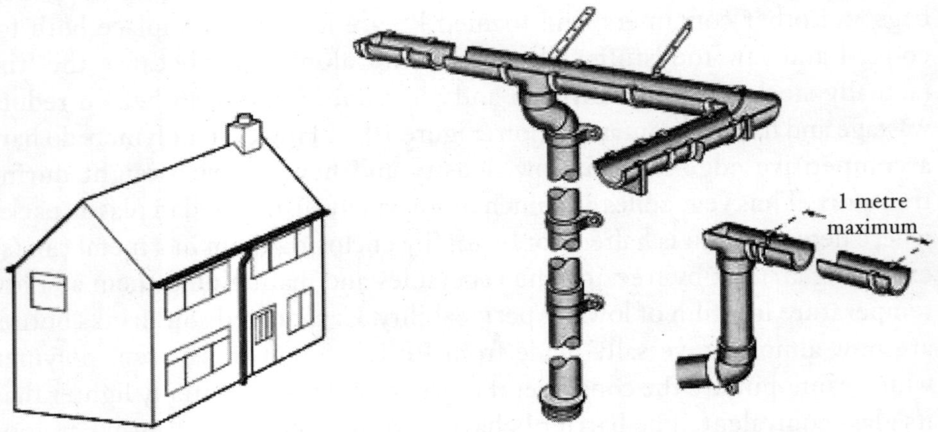

Fig. 10.7: uPVC rain water goods for a detached house.

Parts Consolidation

The most obvious use of polymers is for enclosures for working equipment, such as power and garden tools as well as cooking devices, and electronic products such as computers, video recorders and fax machines (as well as the products used in those machines). They are not just boxes for containment; such plastic enclosures can incorporate carefully designed ribs, webs and flanges on the hidden, inner sides to hold working components securely in place when in operation (Figure 10.8).

Fig. 10.8: Video cassette enclosure in polystyrene, separated to show internal working mechanism and supporting ribs, each part moulded in one piece. The spools are in transparent PS, the case parts in carbon black filled PS. The tape itself is PET loaded with soft magnetic particles.

Polymers offer several intrinsic advantages over conventional materials, and they include—

- ❖ electrical insulation since non-conducting,
- ❖ toughness to resist mechanical abuse such as impact,
- ❖ vibration attenuation,
- ❖ low weight, and
- ❖ one shot manufacture.

The final point in the listing may be difficult to appreciate fully since it involves an understanding of the nature of plastics moulding. Put briefly, it means that several separate functions can be incorporated in one (or two, since working parts must be inserted into the housing to make the finished product) component part, the enclosure.

Thus some of the internal ribs of a drill housing will support the electric motor, separate flanges will protect the trigger mechanism, while other inner ribs will form sockets to protect and support the wiring and leads. Even vents for cooling hot parts of the working mechanism can be incorporated into the form of the design during processing to shape without the need for a separate cutting operation. In early designs for such products, the support and protection were provided by separate parts which needed many assembly steps to make the finished product. This was especially true for sheet metal enclosures where complex internal projections are difficult if not impossible to incorporate in a single pressing operation. The underlying design philosophy with plastics materials is known as parts consolidation. The same philosophy can be seen at work in entirely new products where no conventional materials have been displaced, such as the videocassette cartridge.

Self Assessment Question

❖ Figure 10.10 shows a cross-section of a small lawnmower. It is a hovermower powered by electricity, provided by a cable attached to the mains supply. The 2 kW motor drives the rotor cutting blade in the base of the mower by a direct drive, to which a fan for the downdraft is fitted. It also creates an updraft by means of another fan to suck grass cuttings into a collecting bag. Why is the casing made using a tough plastic like ABS? Would alternative materials such as sheet steel be an appropriate choice for the hood? Give reasons for your answers in terms of:

- materials properties;
- suitability for function; and
- ease of manufacture including assembly.

There is another benefit to product function from parts consolidation, which might also escape superficial attention. All the internal projections improve the overall stiffness of the final product, which must achieve a minimum level to act as a safe and secure platform for the working mechanism. Very often, sufficient product stiffness can be achieved using a constant and relatively thin wall of uniform thickness throughout the housing. So product design, product function, manufacture and material of construction are actually intimately linked.

Human/Product Interaction

The balance of properties needed in a particular product varies enormously, depending on the exact duty that product will perform in service, the environment in which it will operate, and the way it will interact with the user or consumer. The last factor has assumed much greater influence in product design as competition between different manufacturers sharpens the perception of quality in users' eyes. The study of human-product interactions is variously known as human factors or ergonomics. The academic subject seeks to isolate those ways in which machine or product and the user interact, and how product design influences the effectiveness, or efficiency or safety of the product concerned.

So what polymer properties are important here? One argument might point towards the generally low stiffness of most polymers and their products, not dissimilar in fact to that of the users' hands or skin. Human skin is mainly composed of a natural polymer, the protein elastin, which is an elastomer and hence highly extensible and of low intrinsic stiffness. So external surfaces of products which must be handled are more compatible when the material is polymeric. Polymers also have low thermal conductivities, so such surfaces will not feel cold, and the user will be able to hold polymeric products longer and with greater ease than in the case of metals, for example. Plastic or rubber handles for tools are thus the preferred materials, and. in a way, the power tool enclosures examined above are really just an extension of the handle to encompass the whole tool.

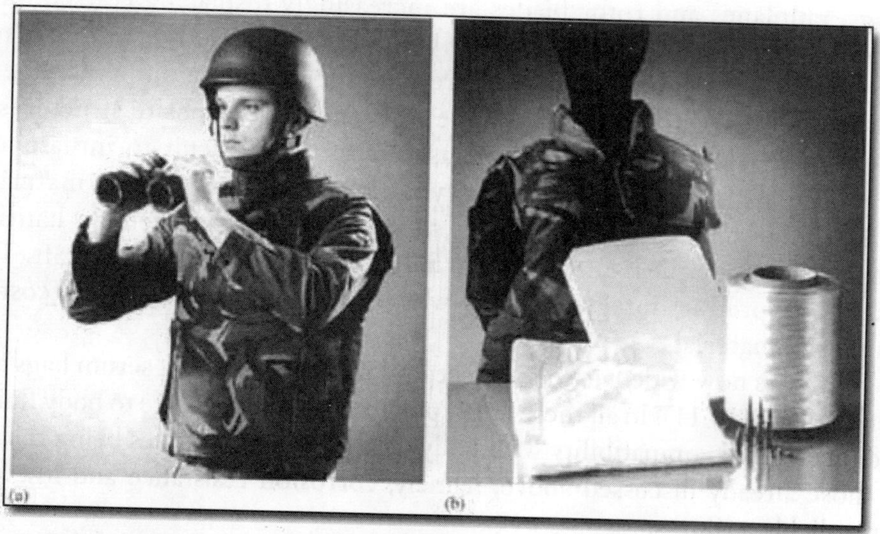

Fig. 10.9: Military applications of high performance PE fibre (*Dyneema*) (Dutch State Mines).

In a similar vein, polymeric textiles must be highly flexible to accommodate movement of the human body, so are composed of fibre assemblies where the stiffness is lowered further by the small diameter of the individual fibre. Low density and low thermal conductivity reinforce the selection of polymers for this product specification. Advances in polymer technology in the last few decades have further enhanced the profile of properties expected from polymeric materials however. Thus metal chain mail has in the past been the main way of protecting the human body from high external stresses, such as ballistic forces. Many polymers can now be processed into fibres of very high intrinsic stiffness, approaching if not exceeding that of inorganic materials. Such high-performance fibres can be woven into textile fabrics and they offer substantial ballistic protection, so can be used for bullet-proof jackets and rip-resistant clothing for sports competitors (Figure 10.9).

Speciality Sectors

There are many sectors of the market where polymers have made dramatic inroads, including medical/ethical, leisure and aerospace products. The influence of density as a property is most dramatic in the last area, simply because weight saving on aircraft fuselages creates large savings in fuel usage or, alternatively, means that more passengers and/or freight can be carried per aircraft. Although complete polymeric fuselages are rare, only being used to date in small aircraft (such as the 'Learjet' and 'Gossamer' man-powered plane) or military craft (such as the 'Stealth' fighter and bomber), composite wings, tailplanes and rotor blades are increasingly displacing conventional aluminium equivalents (Figure 10.10). And the reasons for this move are not just for the great weight savings offered. A crucial reason is the increase in mechanical integrity, since the fatigue resistance of composite structures is much greater than metal alloy because cracks are difficult to initiate and propagate owing to the multiplicity of interfaces within the material. Catastrophic propagation following slow and often undetectable hairline cracking from fatigue is a well-known failure mode for all metal aircraft. The increase in product integrity is well worth the extra manufacturing cost of composite materials.

Products now widely used and accepted include catheters, serum bags and stents (Figure 10.11). In all these areas, polymers offer resistance to body fluids, and mechanical compatibility with body tissues, such properties being related to those already discussed above, namely, corrosion resistance and low and controllable stiffness.

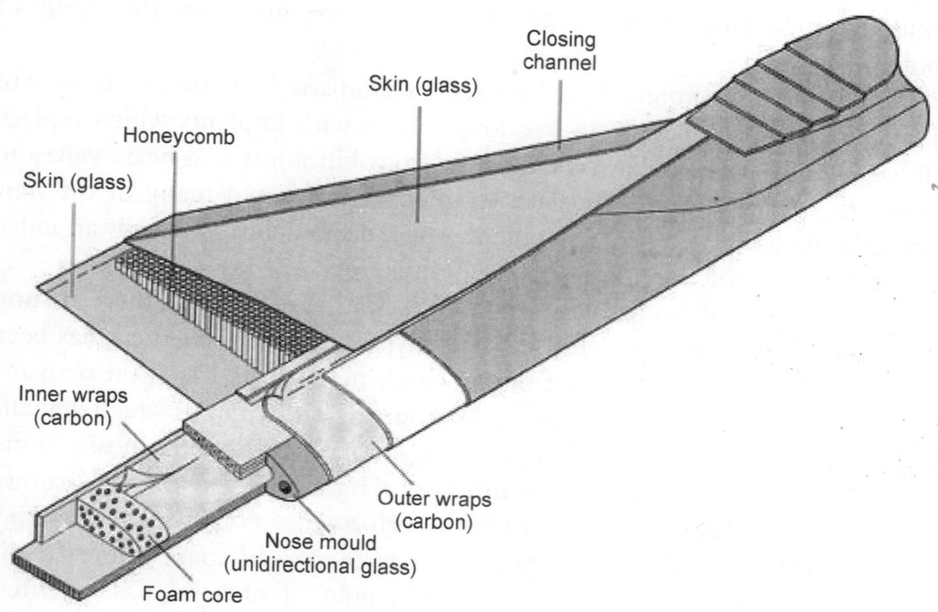

Fig. 10.10: Sea King rotor blade composite blade construction.

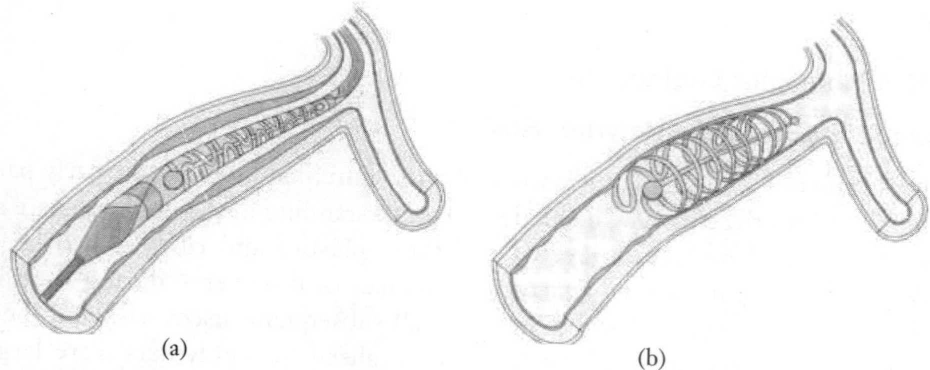

Fig. 10.11: Medical applications of polymers in balloon catheter used to expand stent in coronary artery: (a) catheter inserted into stent over un-inflated baloon. (b) balloon inflated to expand stent which is left in place to support artery after catheter has been removed.

Other materials are used in combination with polymers, where specially high stiffness may be required (e.g. needles) or where shaping is easier (e.g. the ball part of the joint in replacement hip joint. Similarly, stents for supporting weakened or collapsed veins and arteries are more often made using special metals such as tantalum and platinum rather than polymers. The balloon catheters which inflate them in place, however, are universally elastomeric since the long-range elasticity needed is a property unique to this group of polymers.

Medicine and surgery have also seen a dramatic increase in the uses of polymers, enabling people to live longer lives with implants which replace diseased body parts. Such devices range from hip joints and heart valves to synthetic skin and bone. Invasive surgery has exploited many of the new polymers for a wide range of equipment, where disposability is actually an added bonus for the reduction in the chance of infection.

With improved productivity across wide swathes of industry, there are now much greater opportunities for leisure activities. In this area, there has been much inventiveness in product design, with new sports based on or using polymer products. Surfboarding and windsurfing are activities which demand lightweight products, often composed of thermoplastic or composite skins, sometimes reinforced with foam interiors. Highspeed yachting has shown similar developments in the use of high-performance polymers. In athletics, the Olympic record for polevaulting has increased, partly due to the displacement of wood or metal by composite poles. Tennis too, has benefited, especially with composite rackets which have increased product lifetime by reducing fatigue, and increased performance through the use of light-weight materials.

10.5 Molecular Engineering

Understanding the Polymer State

It was the pioneering scientific work of Hermann Staudinger in the early part of the twentieth century which led to an understanding of the polymer state at an atomic and molecular level. Until then, plastics and rubbers had been developed from naturally occurring substances or discovered during routine synthesis. His research laid the basis for all subsequent discoveries and their commercial development. In essence, he realised that polymers were large molecules built up by the repetition of small chemical units, known as repeat units, to create linear chains. More complicated structures consisted of linear chains with branches, and crosslinked molecules were effectively a single macromolecule of almost infinite molecular mass.

The development of entirely new polymer structures relies on a thorough understanding of the various ways in which non-metal atoms can be manipulated into chain structures, limited only by their valency and bonding properties. This subject is known as molecular engineering, and a glossary of terms in the subject is shown in Box 10.2. Owing to the versatility of carbon, it is this element more than any other which has been exploited in molecular engineering. Even now, fundamental discoveries about carbon are still being made by research scientists. Witness the discovery and development of fullerenes, molecules in which carbon atoms are linked together into closed three-dimensional structures such as balls and cylinders (Figure 10.13).

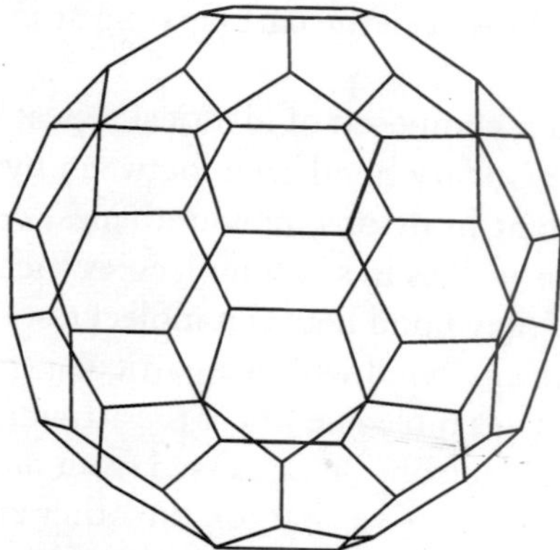

Fig. 10.12: Fullerene Structure in Elemental Carbon (C_{60}).

Box 10.2 Structural and bonding terms.

- *aliphatic:* a term used to describe non-aromatic carbon compounds, so includes both saturated and unsaturated linear, branched or cyclic compounds or structures.
- *alkane:* a carbon compound where all the carbon-carbon bonds are single and saturated (C—C), like ethane and propane (also *paraffins*).
- *alkene:* any organic compound containing a carbon-carbon double bond (C = C), like ethene (ethylene) or propene (propylene) (also *olefins*).
- *alkyne:* any compound with the carbon-carbon triple bond (C•ßC), like ethyne (acetylene).
- *aromatic:* any structure possessing a benzene ring or higher derivatives such as naphthalene, and also heterocyclic rings as in polyimides.
- *asymmetric carbon atom:* a saturated carbon atom where the four substituents are all dissimilar, so giving rise to stereoisomerism.
- *branched polymer:* a linear chain to which are attached side chains.
- *catenation:* the tendency of atoms of an element to link together to form chains.
- *cis:* an isomer of a double bond in which two similar substituents are on the same side of the bond (inverse of *trans* isomer).
- *crosslink:* any kind of tie between chains, whether covalent (as in vulcanized rubber) or based on secondary bonds (as in TPEs).
- *copolymer:* chains in which there are two or more different types of repeat unit.

- *configuration:* the structure of a chain fixed by covalent bonds.
- *conformation:* the structure of a chain determined by intramolecular rotation.
- *covalent bond:* the strongest bond between atoms, where the outer electrons are shared between the participating atoms (also called *primary bond, chemical bond*).
- *homopolymer:* a chain composed of identical repeat units.
- *hydrogen bond:* a secondary bond type, between hydrogen and nitrogen or oxygen; important in nylons, polyurethanes, cellulosics and nucleic acids (e.g. DNA) as well as in small molecules such as water.
- *intermolecular bond:* any bond between molecules.
- *intramolecular bond:* any bond within a particular molecule.
- *ionic bond:* any bond in a molecule where the attraction between the atoms is electrostatic, the electrons being passed from one atom to another, so that one part is cationic (+ ve charge), the other anionic (-ve charge).
- *isomerism:* phenomenon where polymers or small molecules have the same chemical formula, but different structures.
- *random coil:* conformation adopted by single polymer chain where each repeat unit is randomly placed relative to its predecessor or successor.
- *stereoisomers:* polymers or small molecules with the same formula but which differ in the spatial arrangement of the substituent atoms, as in *tactic* polymers; caused by presence of an *asymmetric carbon atom*.
- *tacticity:* type of isomerism found in vinyl polymers, and caused by an asymmetric carbon atom in the main chain; as in isotactic, syndiotactic and atactic polymers.
- *van der Waals bond:* type of weak, secondary bond between atoms and molecules; occurs widely in organic compounds or polymers, especially between distinct molecules.

Realisation of theoretical ideas on the structural possibilities, however, has always presented practical problems of synthesis, availability and cost of raw materials, and ingenuity in finding application in real products which will sell in the open market. Thus, as of 1997, fullerenes remain a laboratory curiosity in terms of commercial application. Nevertheless, it is only a matter of time and development before such compounds and their isomers are of practical benefit and use.

Chain Repeat Units

The repeat units of a range of polymers together with the monomer units from which they are derived are shown in Table 10.3. The simplest repeat unit is that for polyethylene, and consists of two carbon atoms linked to four hydrogen atoms. The difference between the monomer and the repeat unit is the loss of the double bond in the former to give the chain-linked repeating group. Thus the molecular masses of both monomer and unit are identical at 28. The molecular mass of the repeat unit is usually designated M_R and is simply the sum of the atomic masses A_i of the component atoms (i) of the repeat unit:

$$M_R = \sum_i A_i \qquad (1)$$

Table 10.3: Repeat Units of Some Polymers.

Polymer	*Monomer*	*Repeat unit*	*MR*
PE	$CH_2=CH_2$	$-[CH_2-CH_2]-$	28
PP	$CH_2=CHCH_3$	$-[CH_2-CHCH_3]-$	42
PVC	$CH_2=CHCl$	$-[CH_2-CHCl]-$	62.5
PS	$CH_2=CHC_6H_5$	$-[CH_2-CHC_6H_5]-$	104
BR	$CH_2=CH-CH=CH_2$	$-[CH_2-CH=CH-CH_2]-$	54
NR	$CH_2=CH-C(CH_3)=CH_2$	$-[CH_2-CH=C(CH_3)-CH_2]-$	68
CR	$CH_2=CCl-CH=CH_2$	$-[CH_2-CCl=CH-CH_2]-$	88.5
PA6	$(CH_2)_5$ ring with CO and NH	$-[NH-(CH2)5-CO]-$	113
PA 6,6	$HO_2C-(CH_2)_4-CO_2H$ and $H_2N-(CH_2)_6-NH_2$	$-[NH-(CH_2)_6-NH-CO-(CH_2)_4-CO]-$	226

The M_R of polypropylene (PP) is (3 × 12) + (6 × 1) = 42. The number of repeat units in a chain is specified by the degree of polymerization, n, but a much more commonly used measure is the chain molecular mass M

$$M = nM_R \qquad (2)$$

Various types of polymer can be generated from similar monomer units, and Table 10.3 shows three families of closely related polymers. Replacing a single hydrogen atom in ethylene by a methyl group ($-CH_3$) yields propylene, which when polymerized forms polypropylene (PP). When a chlorine atom (Cl) is substituted, the polymer is PVC and when a benzene ring ($-C_6H_5$) is substituted, the polymer is polystyrene. M_R increases in this series, so that a polystyrene sample of molecular mass 10 000 will possess only $10^4/104$ or roughly 100 repeat units compared to nearly 360 for a linear polyethylene of the same molecular mass. When just one hydrogen atom is substituted in the ethylene molecule, so-called vinyl polymers are created. When *both* hydrogen atoms on one of the two carbon atoms in the repeat unit are substituted, vinylidene polymers are formed. Thus poly(vinyl fluoride) has the repeat unit

$[CH_2—CHF]_n$ while poly(vinylidene fluoride) or PVDF has the repeat unit $[CH_2—CF_2]_n$. Polymerization of tetrafluoroethylene monomer of structure $CF_2 = CF_2$ yields PTFE (*Teflon*).

While most vinyl polymers are rigid thermoplastics at room temperature, introduction of a double bond into the repeat unit creates some of the common rubbers already described. The precursor monomer units are dienes, that is, they possess two double bonds as shown by BR and NR in Table 10.2.

A quite different way of constructing polymer chains is shown by the repeat units of polyamide 6 and 6,6 (nylon 6 and nylon 6,6). Here the monomer molecules are linked together by acid and amine groups at their ends. They are known as functional groups and play a significant role in controlling the physical properties of the final polymer. An important way of studying such functional groups, and other ways in which atoms are linked together in polymer chains, is spectroscopy.

10.6 Chain Configuration

The structure of repeat units is fixed by the chemical bonds between adjacent atoms. The shape or shapes thus created is known as the configuration, and for chains will be the chain configuration. Like children's plastic building blocks, however, there can be many different configurations for a given set of atoms of a particular type. The different structures which have identical chemical formulae are known as isomers, and such isomers can have quite different properties. Isomerism becomes an important structural feature as the complexity of molecules increases, and is thus very important for long polymer chains. There are various types of isomerism of especial interest for polymers, all essentially based on the nature of the carbon bond and the way carbon bonds are oriented in space.

Structural Isomerism

In the saturated hydrocarbons, whose structural formulae are shown in Figure 10.16, it is not possible to form distinct isomers with just three or less carbon atoms linked together. There is only one way in which one carbon and four hydrogen atoms can be linked together, the single compound being methane, CH_4. A similar situation holds for ethane, C_2H_6 and propane, C_3H_8. But with butane, two possible structures can be formed: *n*-butane, which has a linear structure, and *iso*-butane, where the central carbon atom is linked to three adjacent carbons rather than one (Figure 10.13). Their physical properties are slightly different, for example their boiling points differ by 10 °C, but otherwise they are very similar compounds. Drawing the possible structures (or using structural models) of the isomers gives 3 isomers for pentane, 5 for hexane, 9 for heptane, and so on. As the number of possible structures increases, their properties diverge: the boiling points of the heptanes range over 20 °C and the melting points by no less than 110 °C. The chemical properties differ too: witness the different combustion behaviour of *n*- and *iso*-octane in petrol engines. The straight chain isomer causes 'pinking' during combustion, while the branched chain burns smoothly in the engine chambers (hence the familiar 'octane rating' for petrols).

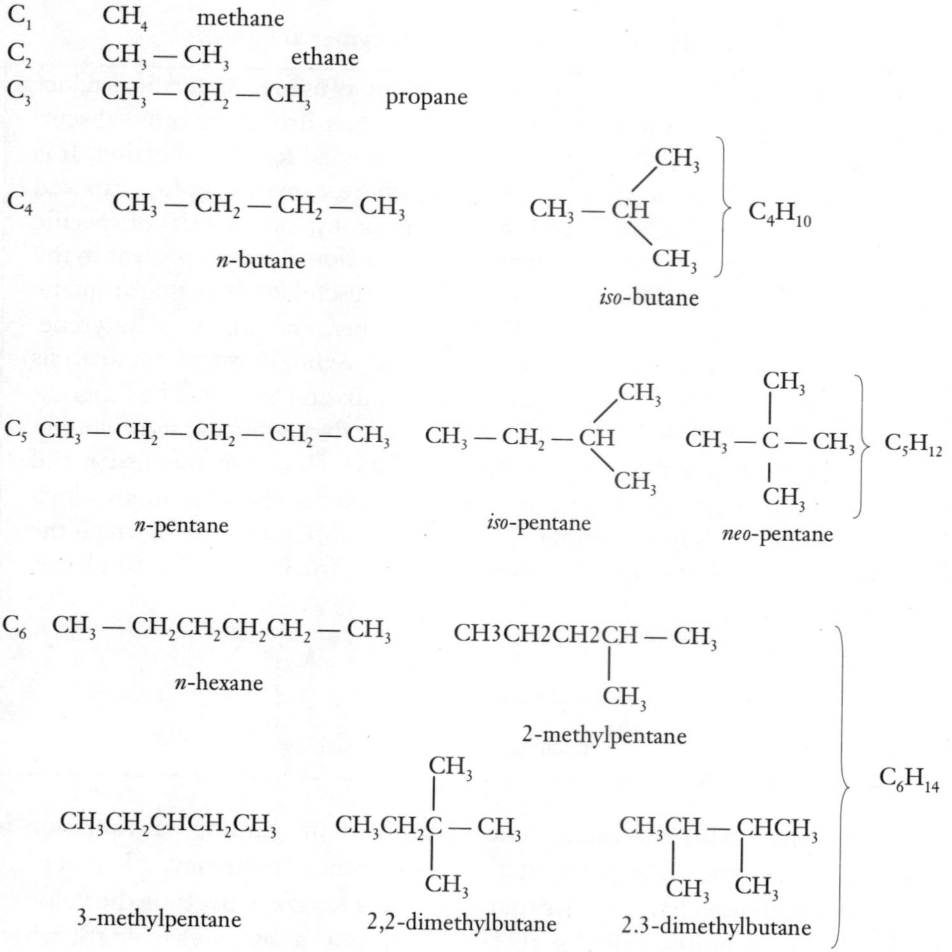

C_1 CH_4 methane

C_2 CH_3-CH_3 ethane

C_3 $CH_3-CH_2-CH_3$ propane

C_4 $CH_3-CH_2-CH_2-CH_3$ *n*-butane; $CH_3-CH(CH_3)-CH_3$ *iso*-butane } C_4H_{10}

C_5 $CH_3-CH_2-CH_2-CH_2-CH_3$ *n*-pentane; $CH_3-CH_2-CH(CH_3)-CH_3$ *iso*-pentane; $CH_3-C(CH_3)_2-CH_3$ *neo*-pentane } C_5H_{12}

C_6 $CH_3-CH_2CH_2CH_2CH_2-CH_3$ *n*-hexane; CH3CH2CH2CH(CH₃)—CH₃ 2-methylpentane; $CH_3CH_2CHCH_2CH_3$ 3-methylpentane; $CH_3CH_2C(CH_3)_2-CH_3$ 2,2-dimethylbutane; $CH_3CH(CH_3)-CH(CH_3)CH_3$ 2.3-dimethylbutane } C_6H_{14}

Fig. 10.13: Configurational isomers of saturated hydrocarbons (alkanes). The number of isomers increases rapidly with the number of carbon atoms in the structure. There are no isomers for the first three members of the series, two for butane (C_4), three for pentane (C_5) and five for hexane (C_6).

The increase in isomeric structures is so rapid that at C_{30}, there are no less than 4 111 846 763 theoretically possible compounds! So with even the lowest molecular mass polyethylene, there is an almost infinite number of isomers. Fortunately, the situation is simplified enormously by the way polymerization occurs and in fact there are relatively few chemically distinct polyethylenes. As we shall see later, the concept of a distinct molecular formula is redundant with most commercial polymers since chain lengths are very variable even within a single sample, but the idea of branching is important for polyethylene in particular.

Box 10.3: Spectroscopy and polymer analysis.

Although information on the polymer or polymers used in a specific product may be provided on the packaging or the product itself, it is often absent, and so some way of analysing the material is needed for identification. It is usually provided by spectroscopic analysis, where a small sample is exposed to electromagnetic radiation and the absorption by the material of specific frequency bands is usually diagnostic of the functional groups present in the polymer. UV or ultraviolet spectroscopy is most useful for detecting aromatic groups in polymer chains, such as the side chain benzene rings in polystyrene. It is infrared or IR spectroscopy which is most useful, however, for analysis of polymers. The method essentially detects radiation absorbed by different bonds vibrating within the chains. Real chains are always moving and vibrating in polymers (unlike the static models for chains shown in this text!), and their vibration frequencies depend on the sizes and masses of the atoms which the covalent bonds link together. There are also different ways in which the bonds can vibrate: they can stretch along or bend (rock) about the bond axis, for example:

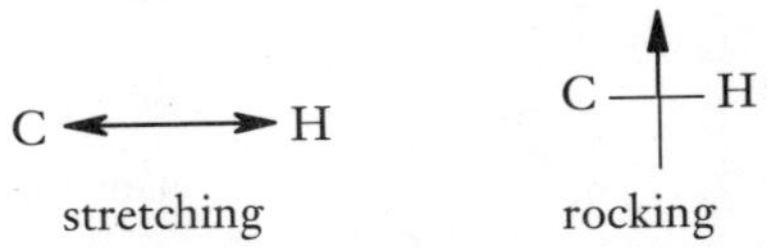

C ⟷ H
stretching

C —|— H
rocking

Stretching involves a higher energy input than bending, so radiation is absorbed at a higher frequency than rocking. Since frequency, v is inversely proportional to wavelength, λ (from the formula $\lambda = c/v$, where c is the velocity of light), C—H bonds will absorb IR radiation at a *lower* wavelength when stretching compared to rocking. Of functional groups, the carbonyl group (C = O) is especially important for diagnosis of the polymer since it occurs in nylons and polyesters. Bond stretching occurs in a very narrow band centred at about 6 ìm wavelength, so when a strong peak is detected at this position in the IR spectrum, it is likely that the carbonyl group is present in the polymer. It doesn't necessarily follow that the polymer is a nylon or polyester, however, because such groups can be present even in polyethylene chains as a result of oxidation. Band 3 of the AV cassette gives an account of a forensic investigation concerning a PE product which fractured as a result of oxidation and chain degradation. IR and UV spectroscopy were critical tools in the determination of the causes of the failure of the product.

Chain Branching

A germ of the idea is shown by the formulae for 2- and 3-methylpentane in Figure 10.14. A single methyl group (CH_3—) can occur in two different positions along an essentially linear carbon-carbon chain. The methyl group is a very simple kind of branch along the chain, and it is easy to extend the idea to much larger molecules. Thus LDPE is a polymer based on a linear backbone chain with the repeat unit [CH_2CH_2], but is in addition branched with very long chains at infrequent points along the main chain (about 1 in 1000), as shown in Figure 10.14. Branching is caused during polymerization at high pressure by growth sometimes starting from an initiation point *in* a chain rather than at the end. An alternative way of making polyethylene is at low pressure using a special catalyst, and this usually results in a highly linear chain without branching (HDPE).

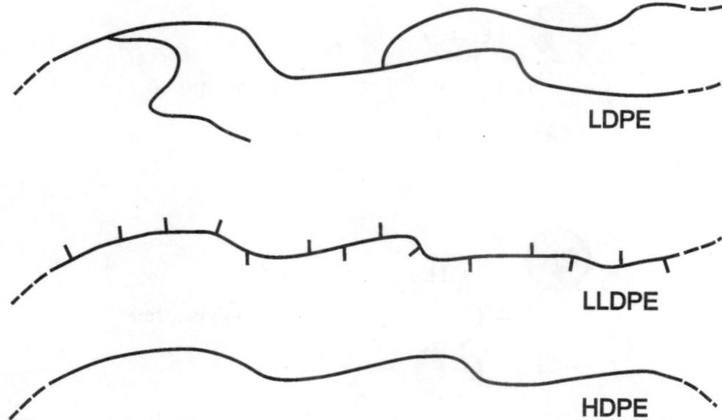

Fig. 10.14: Different chain configurations caused by branching of main back bone chain in HDPE (lower). LDPE has a small number of long chain branches, LLDPE a large number of short branches.

However, it is easy to polymerize a mixture of ethylene with a higher alkene such as hex-1-ene, so that the new units co-polymerize together to form a chain where a certain proportion of the chains have tails or short branches along the linear sequence (Figure 10.15). This and similar copolymers are generically part of the polyethylene family, and are known as LLDPE, short for linear low density polyethylene.

This kind of structural variation is important because it affects the properties of the polymers, as their names indicate. Thus branches along the chain hinder crystallisation of the chains, resulting in a less dense and lower modulus material. LDPE typically has a density of 0.92 Mg m^{-3} while HDPE has a higher density of 0.96 Mg m^{-3}. (Note that these density values are numerically identical to those expressed in g cm^{-3}.) Intermediate grades, MDPE, are other important relatives, finding wide application in gas pipes, for example.

Geometrical Isomerism

A second type of isomerism occurs with diene monomers, and is present in both NR and butadiene rubbers (BR). It occurs because the single double bond in the final polymer can exist in two ways: a *cis* form and a *trans* form. The repeat unit shown in Table 10.3 for NR does less than justice to the two-dimensional structure of this material. In this planar formula where the bonds are shown in their correct orientation to one another, the base polymer in natural rubber can be seen to be m-polyisoprene, where the pendant methyl group appears on the same side as the lone hydrogen atom. The two parts of the chain in which this single repeat unit sits, lie on the opposite side of the double bond.

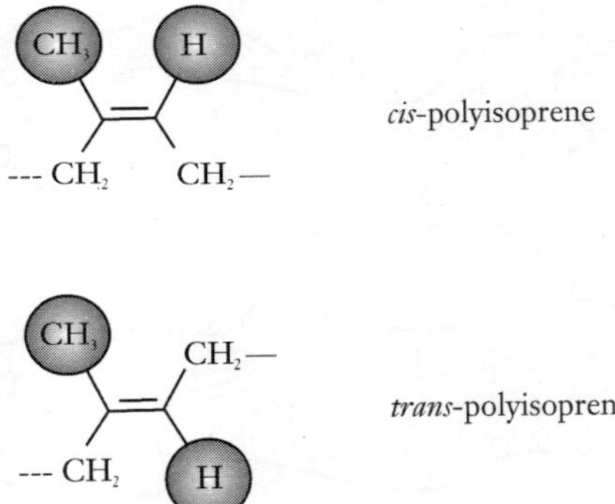

Fig. 10.15: Geometrical isomerism in polyisoprene. *Cis*-polyisoprene in the main polymer occurs in natural rubber, *trans*-polyisoprene in gutta percha.

But there is an alternative structure with exactly the same formula: the polymer is gutta percha, and its structural formula is shown below that of NR in Figure 10.15. Here, the two parts of the chain are on opposite sides of the double bond, giving the overall chain a zig-zag appearance. It is also a naturally occurring polymer, but has quite different properties to natural rubber. It is a highly crystalline and rigid material without the long-range elasticity characteristic of an elastomer. It was used formerly as an electrical insulator, but has now been largely superseded by synthetic plastics.

Stereoisomerism

A final type of isomeric variation occurs as a result of the three-dimensional structure of some polymers. It is possible because a four-valent atom like carbon can exist in two different forms when the subsidiary groups or atoms attached to the carbon are all different. The carbon atom is then known as an asymmetric carbon atom. A very simple example of the phenomenon is the structure of a small molecule, lactic acid. As Figure 10.16 shows, it can exist in two forms which are mirror images of one another. One of the two possible compounds, *laevo-* (standing for left-handed), or *l*-lactic acid occurs in muscle after vigorous, anaerobic exercise and causes muscle cramp. It is a good example of stereoisomerism in a small molecule, a feature it also shares with a large number of biological molecules, as well as some polymers.

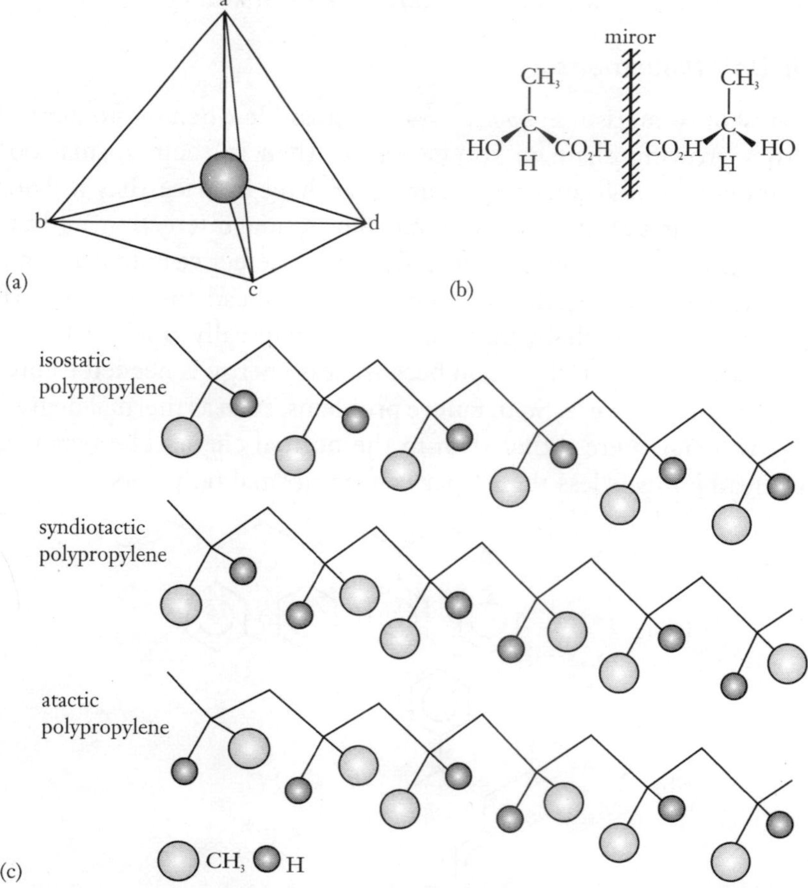

Fig. 10.16: (a) Tetrahedral configuration of single carbon atom; (b) left and right-handed forms of lactic acid; (c) tacticity in polypropylene.

The carbon atom in a vinyl polymer to which is attached the pendant side group (i.e. every alternate carbon atom in the main chain) is another example of an asymmetric carbon atom. It gives rise to tacticity. When the zig-zag chain is written with the plane of the chain in the plane of the paper, there are three ways in which the position of the pendant side group can exist. The methyl groups in polypropylene, for example, can occur all on one side (isotactic), on alternate sides (syndiotactic) or placed at random (atactic). These possibilities are shown in Figure 10.16(c). The properties of each type of polymer are quite different to one another, primarily because isotactic and syndiotactic PP have ordered chains and so can crystallise, but atactic chains are quite irregular and cannot crystallise. Isotactic PP is the common form of the commercial material, although atactic PP is used as a binder for paper for example. Syndiotactic PP has recently become available commercially (1997) being made using a new family of catalysts, known as metallocenes.

Repeat Unit Placement

A final kind of isomerism in *homopolymers* is possible when monomer units are added to a growing chain in reverse rather than in their normal position. Because monomer molecules have a particular shape in space, they will normally approach a growing chain end to minimise any spatial interaction, and a regular chain structure results from head-to-tail joints. A defective joint can sometimes occur, however, when heads combine to form a head-to-head joint (Figure 10.17). Although the chain units are fully chemically bonded together, it represents a weak link in the chain because less energy is needed to break the chain here than elsewhere. So in failure problems, such as thermal degradation, breakage will start here rather than in the normal chain. The occurrence of head-to-head joints is less than 1 per cent in normal polymers.

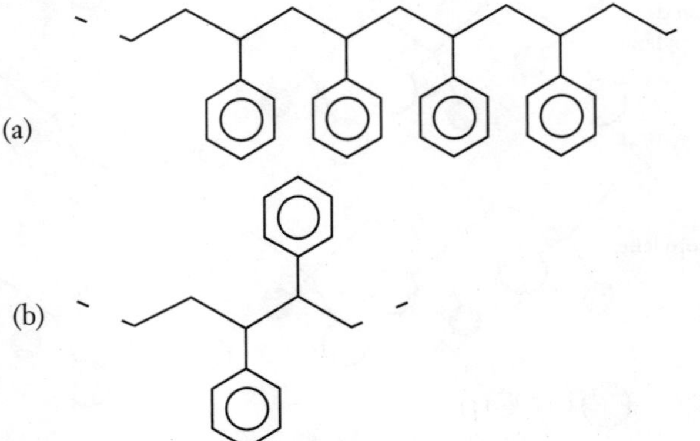

Fig. 10.17: (a) Regular head-to-tail repeat units in polystyrene; (b) irregular head-to-head joint in polystyrene chain.

Copolymers

So far, the discussion has been confined to polymers with only a single type of repeat unit, but in reality, a large and growing number of commercial polymers are actually composed of different types of unit attached together by chemical covalent bonds. They are known as copolymers, and can comprise just two different units (binary copolymers) or three (ternary), and so on. It is one of the common strategies used by molecular engineers to manipulate the properties of polymers to gain just the right combination of properties for a specific application.

One of the best known examples involves polystyrene. In its homopolymer form, it is a rigid, transparent thermoplastic which is also very brittle. It thus finds little application for stressed applications in its original state. It also shows a glass transition temperature of about 97 °C, so is useless for containers which could hold boiling water (like very hot coffee). The glass transition temperature (T_g) is the temperature at which an amorphous thermoplastic becomes flexible and rubbery (Box 10.4). This problem can be solved by copolymerizing styrene with acrylonitrile to produce SAN polymer, where the styrene and acrylonitrile units alternate along the backbone chain of the material (Figure 10.20). SAN is commonly used for transparent drinks containers since the acrylonitrile units raise the T_g to about 107 °C.

The problem of brittleness can be solved in a quite different way. Paradoxically, if rubber (polybutadiene) chains are grafted onto the main backbone polystyrene chain, the graft copolymer so formed (Figure 10.20) is much tougher owing to molecular segregation of the rubber chains into tiny particles. Although they reduce the stiffness of the copolymer compared with the parent PS, the particles act as nuclei for minute crazes. Such crazes are so plentiful when the solid is stressed, that a great deal of energy is absorbed and so the bulk material appears ductile and tough. The material is HIPS or high-impact polystyrene. The benefits of both high T_g and toughness are achieved with ABS, a terpolymer of the three component repeat units, with butadiene present to about 25 weight per cent (Figure 10.20). The strategy of adding rubber particles to toughen a brittle polymer is known as rubber-toughening.

Because the different repeat units added to the original polymer are always covalently bonded, they are in effect locked into the structure, so in theory at least, the composition is infinitely variable. This is quite unlike metal alloys or mixed glasses, where the composition is only possible between certain, fixed limits. Thus mild steel is an alloy with 0.1–0.4 per cent carbon and small deviations above or below cause large changes in properties. So what happens if the second rubbery component in a copolymer is increased? Not surprisingly, the properties change from those of a plastic to that of a reinforced rubber. In fact, copolymerization of butadiene and styrene was employed at a very early stage in the development of synthetic rubber during the last World War particularly. It was found that the stiffness of polybutadiene rubber could be improved by copolymerization with about 24 weight per cent styrene, to give a random copolymer of the two units, known as SBR (Figure 10.18).

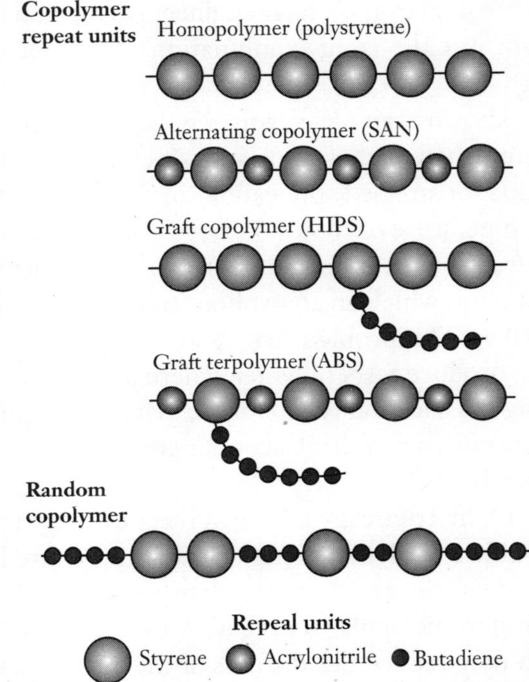

Fig. 10.18: Various configurations of styrene, butadiene and acrylonitrile in different copolymer structures.

Box 10.4: Thermal transitions in polymers.

The two most important thermal transitions exhibited by polymers are the glass transition temperature, T_g and the crystalline melting temperature, T_m. The glass point is the temperature at which amorphous polymer becomes elastomeric and flexible as the temperature is raised. The crystalline melting point is that point when the crystalline component loses coherence and long-range order. Since most polymers are rarely completely crystalline owing to chain entanglements, most crystalline polymers will show both a T_g and a T_m. This is illustrated below by the thermogram for polyethylene terephthalate) or PET (Figure 10.23), where the T_g is shown at about 84 °C by the inflection in the curve. The material used was sampled from a soft drinks bottle.

The large dip at about 248 °C represents the melting point. The thermogram was obtained using a technique known as differential scanning calorimetry or DSC for short. It is a very accurate way of evaluating the thermal properties of polymers, the apparatus needing only a few milligrams of the polymer for analysis. The temperature of the sample is raised in a controlled and regular way (horizontal scale of Figure 10.19). The device measures heat flow into or from the sample, as shown by the vertical scale in Figure 10.19.

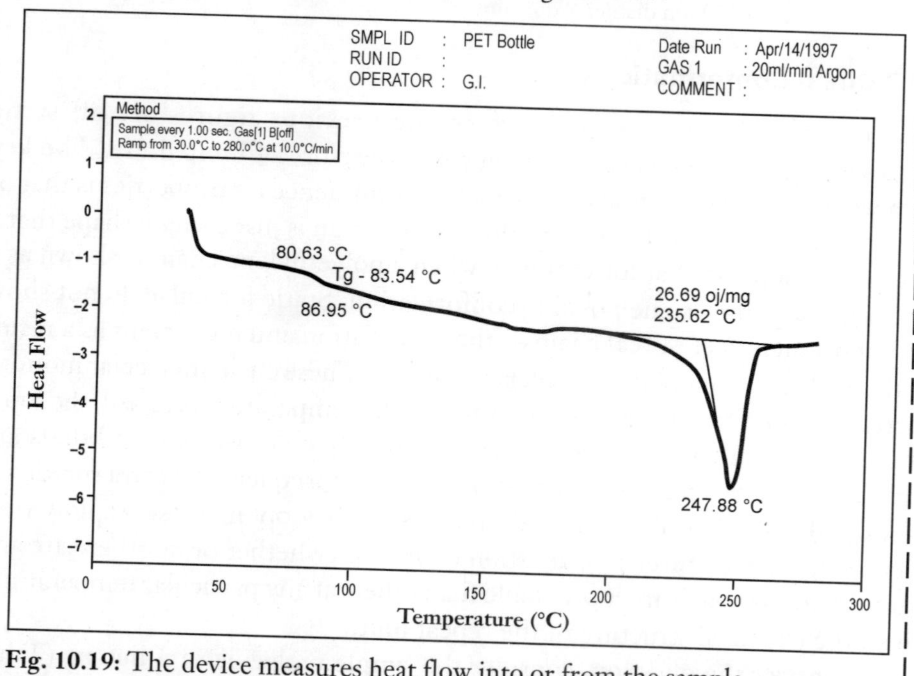

Fig. 10.19: The device measures heat flow into or from the sample.

Alternatively, it was found in the 1960s that another way of putting the units together was possible. If, rather than using a mixture of monomers, they were added sequentially, then a block copolymer resulted (Figure 10.22). The properties are different again to those of the random copolymer because each type of chain segregates together to form minute domains, as shown by the microstructure of Figure 10.20. Such materials retain thermoplastic behaviour yet behave as crosslinked rubbers, and the SBS block copolymer was the first commercial thermoplastic elastomer or TPE. Although most polymer chains are incompatible with one another, there are some exceptions to the rule. One in particular has gained commercial success, and is a blend of polystyrene and poly(phenylene oxide) (PPO). The T_g is increased as is the toughness of the resulting physical mixture of different chains, and the polymer mixture is known by the trade name *Noryl*. It is used widely for enclosure of consumer products.

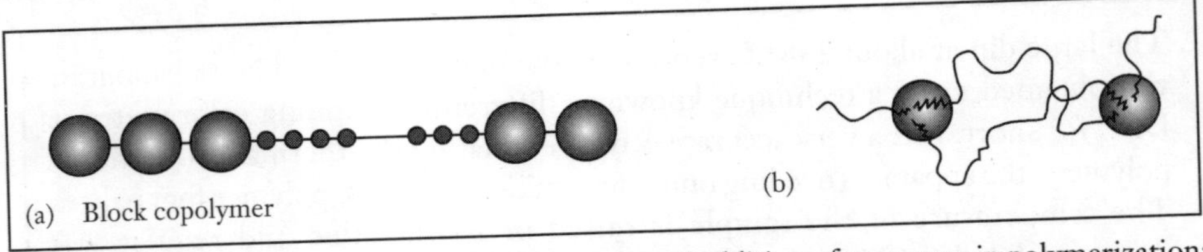

Fig. 10.20: Block copolymers are formed by sequential addition of monomer in polymerization: (a) SBS block copolymer; (b) microstructure with polystyrene and chains segregated to form 10–20 nm diameter domains.

10.7 Chain Conformations

The repeat units or chains shown in the previous figures are all static representations of real chains and they are therefore of limited use. The key idea we need to explore real chains, and their influence on properties is that of the conformation of a chain. A single conformation is just a single shape that a chain can adopt, so that for example, when a polyethylene chain is shown as a linear zig-zag, this is one possible conformation. Static formulae do not show an important aspect of real chains—their oscillation and movement as a result of thermal vibrations of the molecular structure. These chain molecular motions increase in both frequency and amplitude as the temperature is raised, the most important being rotation about single bonds. By contrast, double bonds are rigid and the adjacent atoms are immobile. As a consequence of extra rotational motion, the number of conformations a chain can adopt increases rapidly with increasing temperature. The question then arises whether or not there are any conformations which are more stable than others at any particular temperature, given the chemical structure of the repeat unit.

To answer the question, consider the simple molecule of *n*-butane (Table 10.4), which consists of four carbon atoms linked together in a linear chain together with hydrogen atoms along the periphery. There will be considerable interference between adjacent hydrogen atoms when they are aligned. The interference will be lower when they rotate about their own axis to give a staggered conformation [Figure 10.21(a)]. Concentrating on the two central carbon atoms (C_2 and C_3), and looking *along* the C_2—C_3 axis, it is also clear that there are two different ways in which the outer carbon atoms (C_1 or C_4) can be situated with respect to one another. One position where C_1 and C_4 are in opposite positions is known as the *trans* conformation, the other where they are adjacent is known as the *gauche* conformation. It will be readily apparent from the butane model that interference occurs between the hydrogen atoms on C*1* and C_4 carbon atoms in the *gauche* conformation. This so-called steric hindrance makes the gauche conformation of higher energy than the *trans* conformation, as shown in Figure 10.21(b). The net effect then is that the *trans* conformation is favoured and this also applies in longer chain hydrocarbons.

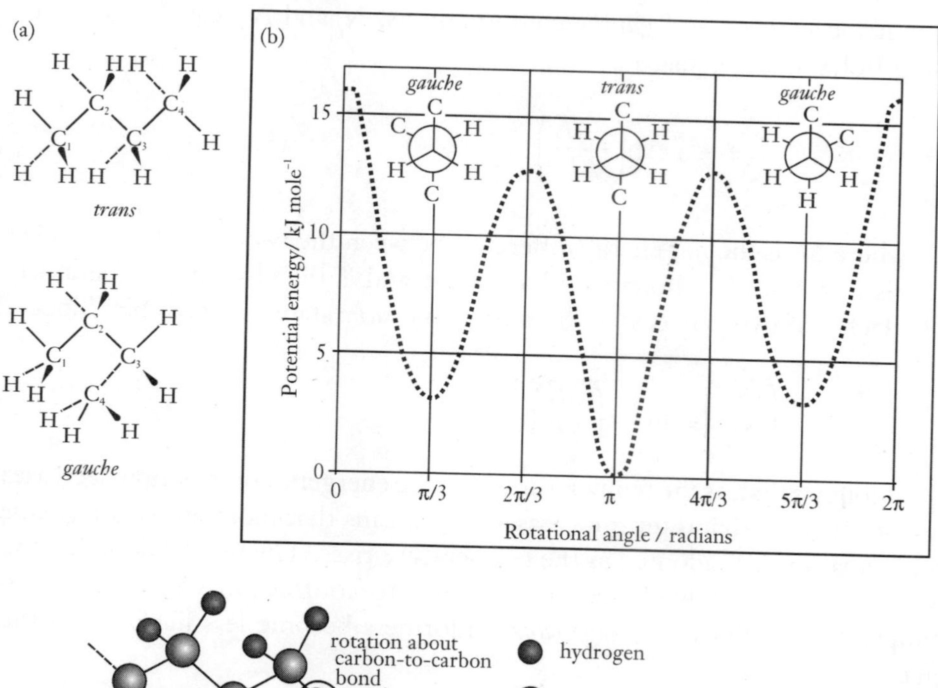

Fig. 10.21: (a) Rotational isomerism in *n*-butane; (b) potential energy wells for *gauche* and *trans* conformers in butane (views along C—C bond); (c) rotational isomerism in polyethylene.

Table 10.4: Properties of linear saturated hydrocarbons.

Name	*Formula*	*n*	T_m/°C	*Density of liquid/Mg m^{-3}*	*State at 20 °C*
ethane	C_2H_6	1	-172	0.5462	gas
n-butane	C_4H_{10}	2	-135	0.5788	gas
n-octane	C_8H_{18}	4	-57	0.7028	liquid
n-hexadecane	$C_{16}H_{34}$	8	18	0.7749	liquid
n-triacontane	$C_{30}H_{62}$	15	66	0.7786	waxy solid
n-heptacontane	$C_{70}H_{142}$	35	105	0.7940	waxy solid
polyethylene	$C_{700}H_{1402}$	350	130–140	ca. 0.08	solid polymer

When polyethylene crystallises it adopts an overall linear conformation in which the carbon atoms form a planar zig-zag, where they are all *trans* with respect to their near neighbours. In the non-crystalline state, the relative proportion of *trans* and *gauche* conformations, N_t and N_g, can be predicted using Boltzmann's equation:

$$\frac{N_g}{N_t} = 2\exp\left(\frac{-\Delta E}{kT}\right) \qquad (3)$$

where ΔE is the net energy difference between the two states ($\approx$3 kJ mole^{-1} in this instance), k is Boltzmann's constant and T the absolute temperature. The factor of two is necessary because two *gauche* states are possible. Since

$$\ln\left(\frac{N_g}{N_t}\right) = \ln 2 - \left(\frac{\Delta E}{kT}\right)$$

it follows that, as the temperature rises, the energetic contribution decreases and so the right-hand term increases; this means that more and more *gauche* conformations are adopted as the temperature rises. The net result is that the randomly coiled polyethylene chain tends to contract in size with rise in temperature as linear zig-zag *trans* conformers become less likely along the chain.

10.8 Structure-Property Relationships

Given the large number of possible configurations in polymers, what guides to likely properties are available? We have already seen some of the effects on properties of changing tacticity, for example, which can affect crystallinity. Control of copolymer structure, too, can have substantial effects on their thermal properties.

Homologous Series

Another approach to the problem is to consider what happens to the properties of a related series of compounds of increasing chain length. The simplest precedent is the sequence of properties for the saturated paraffin hydrocarbons (normal alkanes). Such a series of compounds is known as an homologous series. The lowest members are gases like methane and ethane, but as the length of the chain increases the hydrocarbons become first liquids and then waxy solids. Both the melting points and densities increase in a regular manner. When many hundreds of carbon atoms are linked together the properties reach a plateau, and it is only here that one can talk of true polymeric properties (Table 10.4).

During investigations of other types of repeat unit, considerable efforts were and still are made to determine how properties vary with increasing chain length for the lower members (oligomers) of the family, and how it may affect the final properties of the true polymers.

Polymer Families

But how do small changes in chain configuration for a given family of polymers affect their properties? A very clear example of slight changes in the repeat unit structure is exhibited by polyamides, polyesters and polyurethanes. They are all polymers linked together by a particular kind of functional group, which gives the name to each family. Their backbone chain may either be aliphatic or aromatic in nature, although here we'll only be examining trends for the aliphatic polymers. One way of looking at their structure is to regard them as comprising a polyethylene chain into which is inserted the functional group of interest.

They are crystalline polymers which show a distinct melting point (T_m) and we are interested in the effect on T_m of changing the chain length of the sectors between functional groups. Since such polymers are made from two separate monomers (cf. the repeat unit in nylon 6,6 of Table 10.3), only one of the sectors will be varied. The melting temperatures are shown in Figure 10.11 as a function of increasing sector chain length. The constant melting point of HDPE is shown on the same figure for comparison.

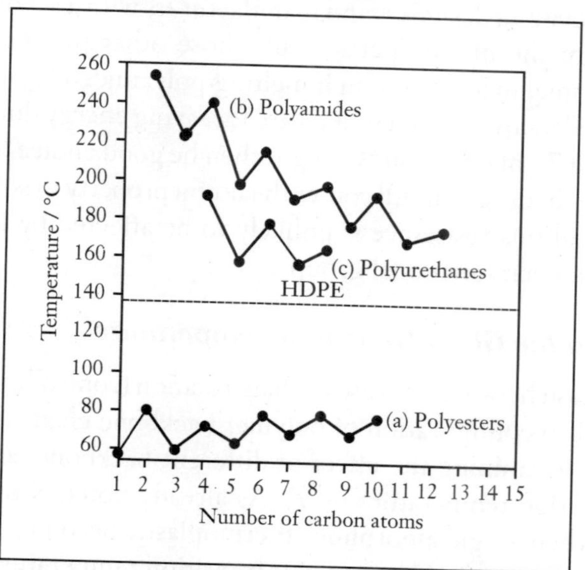

Fig. 10.22: Dependence of T_m on polar group spacing. Number of carbon atoms refers to (a) acid for polyesters made with decamethylene glycol, $HO(CH_2)_{10}OH$; (b) diamine for polyamides made with sebacic acid, $HOOC(CH_2)_{10}COOH$; (c) diisocyanate for polyurethanes made with fetramethylene glycol $HO(CH_2)_4OH$.

The first point of interest is that the polyamides and polyurethanes always have melting points above that of HDPE, an effect due to hydrogen-bonding between adjacent chains. This is a secondary kind of bonding between hydrogen and oxygen or nitrogen, which occurs in water and ice as well as many natural polymers. Hydrogen bonds occur between the functional groups on neighbouring chains and effectively tie the chains together in a weak kind of crosslink. This means that the crystalline chains are held together in an energetically more stable conformation than in polyethylene itself, so a higher temperature is needed to melt or decompose the crystals. Hydrogen bonding is absent from polyesters, so the crystals actually show a lower melting temperature than PE because the functional groups make for a less well packed, and hence less stable, structure.

The effect of increasing the number of carbon atoms in the intervening PE chains in the amides and urethanes is similar, lowering T_m in a regular way and approaching the limiting value for HDPE. The decrease is not smooth, however, values oscillating about the line. This effect is caused by packing effects in the crystal structure, odd numbers of carbon atoms fitting together less easily than even numbered chains. With the polyesters, the packing effect still creates oscillations, but increasing chain length has a minimal overall effect on melting temperature.

Such trends are important in choosing specific grades for specific end products, or providing designers and manufacturers with a set of polymers with slightly different thermal properties but whose other properties remain the same. For example, nylon 6 is a high melting polyamide (T_m = 225 °C), and it might be desired to lower this temperature for saving energy during processing. Nylon 11 with a T_m of circa 180 °C might then be good choice of material. The hydrogen bonds in the chain still confer the useful property of solvent resistance, for example, and this resistance is unlikely to be affected by lengthening the chain sector between functional groups.

Structure and the Glass Transition Temperature

There is a relation between the ease of chain rotation (controlling conformation) and the locked-in configuration of polymer backbone chains. It is most easily appreciated by examining the effect of different backbone configurations on the glass-transition temperature or T_g. As already noted above, the T_g is the temperature when a rigid amorphous thermoplastic becomes elastomeric, and its stiffness drops steeply. How can this transition temperature be interpreted at a molecular level?

Table 10.5: Glass transition temperatures and repeat unit structures.

Polymer	*Tg/°C*	*Repeat unit*
silicone rubber	–125	$[-Si(CH_3)_2-O-]$
polyethylene	–90	$[-CH_2-CH_2-]$
cis-polybutadiene	–85	$[-CH_2-CH=CH-CH_2-]$ (cis, H on C=C)
poly(methylene oxide)	–75	$[-CH_2-O-]$
poly(ethylene oxide)	–40	$[-CH_2-CH_2-O-]$
polypropylene	5	$[-CH_2-CH(CH_3)-]$
poly(vinyl acetate)	30	$[-CH_2-CH(CO_2CH_3)-]$
PVC	80	$[-CH_2-CH(Cl)-]$
polystyrene	97	$[-CH_2-CH(C_6H_5)-]$
SAN PS/ANcopolymer	107	$[-CH_2-CH(C_6H_5)-]_m[-CH_2-CH(CN)-]_n$
ET	65	$[-O-CH_2CH_2-O-C(=O)-C_6H_4-C(=O)-]_n$
PC	149	$[-O-C-O-C_6H_4-C(CH_3)_2-C_6H_4-]_n$

The simplest way at looking at the problem is in terms of the chain rotational model of Section 2.3, where the effect of raising temperature on the conformation of polyethylene was considered. It was of course implicitly assumed that the chain was flexible, and that rotation about carbon-carbon bonds created chain flexibility. And that is true at ambient temperatures of 25 °C, say, provided the level of crystallinity is low or absent. But what happens if the temperature is lowered? As the temperature decreases, there must come a point when all rotation about chain bonds ceases entirely; in other words, the energy available locally in the form of thermal vibration is insufficient to cause neighbouring atoms to twist around one another. The energy needed to achieve rotation is actually shown by the potential energy banners in Figure 10.22(b). They lie at about 13 kJ mole^{-1} and 16 kJ mole$_{-1}$ above the energy minimum for the *trans* conformer in *n*-butane.

The temperature at which chain molecular rotation ceases is the glass transition temperature, because the chains can no longer respond to external strain by uncoiling and lengthening by chain rotation. In other words, the polymer becomes glassy and rigid. For polyethylene, the T_g is very low and occurs at about -90 °C. However, it is important to mention that the T_g is not necessarily a sharp transition, like the melting point, for example. It can be very broad indeed, with a progressive stiffening effect as temperature is lowered. Indeed with PE, substantial stiffening is already present by -20 °C.

So how do changes in chain structure affect T_g, if the transition is largely controlled by rotation about chain bonds? Introduction of atoms like oxygen where there are no hydrogens to create steric hindrance in the chain would be expected to lower T_g, and this is found to be in general the case (Table 10.5). POM or acetal resin and PEO also have low T_gS, and silicone polymer is exceptional in having one of the lowest T_gS of any material, at -125 °C. This is why the (crosslinked) rubber is widely used in gaskets and fuel hose for aircraft, where low temperatures will be encountered when flying at height.

It would also be expected that chains having double bonds, such as BR and NR might have low T_gs since there is little steric hindrance adjacent to this bond. Again, this is found to be the case (Table 10.5).

On the other hand, if ways of hindering chain rotation are used, for example by increasing the physical size of pendant groups, then the T_g would be expected to increase. Thus polypropylene with a large methyl group on every alternate chain atom has a T_g of about 5 °C. a value that implies PP milk bottle crates could crack on a frosty morning! This problem was overcome by using a copolymer grade with ethylene to lower the T_g to below 0 °C. Increasing the size of the side group in vinyl polymers shows a reasonably regular increase in T_g, with PVA having a value of about 30 °C, PVC a value of about 80 °C and polystyrene with a very large benzene pendant ring hindering rotation has a T_g of 97 °C.

Larger side groups than this are uncommon, so it is worth returning to the structure of the main chain. Benzene rings trapped within the backbone should increase steric hindrance, and hence T_g. The effect is shown in many polymers and was indeed a strategy used for developing polymers stable to high-temperatures. PET for example, is an aromatic polyester and contains such a ring (C_6H_4) in its repeat unit together with a short aliphatic portion (Table 10.5). It has a T_g of about 65 °C, but a polymer like polycarbonate is much more hindered by its repeat unit, since the short chain is absent. Here, the large group which forms the bulk of the unit is a bisphenol A group consisting of two benzene rings connected by a carbon atom with two pendant methyl groups, so there is considerable resistance to rotation. Its T_g is about 149 °C. Finally, there are materials like aramids and polyimides where the chain is either completed prevented from rotation at all (PI), or degrades by chain breakage before rotation is possible (aramids). The concept of T_g in these cases becomes redundant.

Melting and Structure

For those polymers which can crystallise, one would expect some relation between chain rotation and melting. Since all crystallisation demands that chains form an ordered conformation (e.g. the PE planar zig-zag) before they can pack together, the chance of this happening should be related to the ease of twisting into the required conformation.

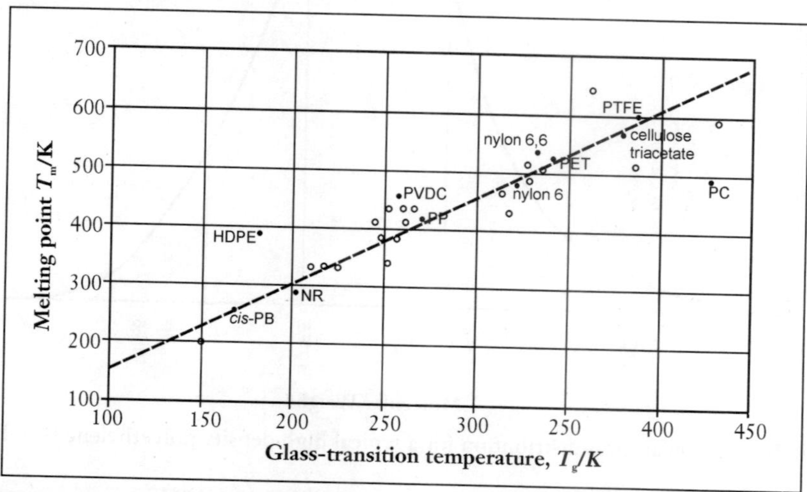

Fig. 10.23: The melting points T_m of crystallisable polymers plotted against their glass transition temperatures T_g. The dotted line shows T_g ~2/3 T_m.

That there is a rough correlation between T_g and T_m can be judged from Figure 10.23, where the two transition temperatures are plotted against one another. Averaging over the scatter of points shows that when the transitions are plotted in kelvins, then

$$T_g \approx 2/3\ T_m$$

Some care is needed in interpreting such a rule of thumb, and there are known to be other factors at work, such as hydrogen bonding as we've already seen above in polyamides. But overall, there does indeed appear to be a correlation between ease of chain rotation and the most important thermal transitions observed in polymers. Such relations can be a guide to synthesis of new repeat units.

10.9 Molecular Mass Distribution

In Figure 10.24 the peak (M_p) is at about 40 000. Number average $(\bar{M}_n)$ and weight average $(\bar{M}_w)$ molecular masses fall below and above the peak value. Determined using high temperature GPC.

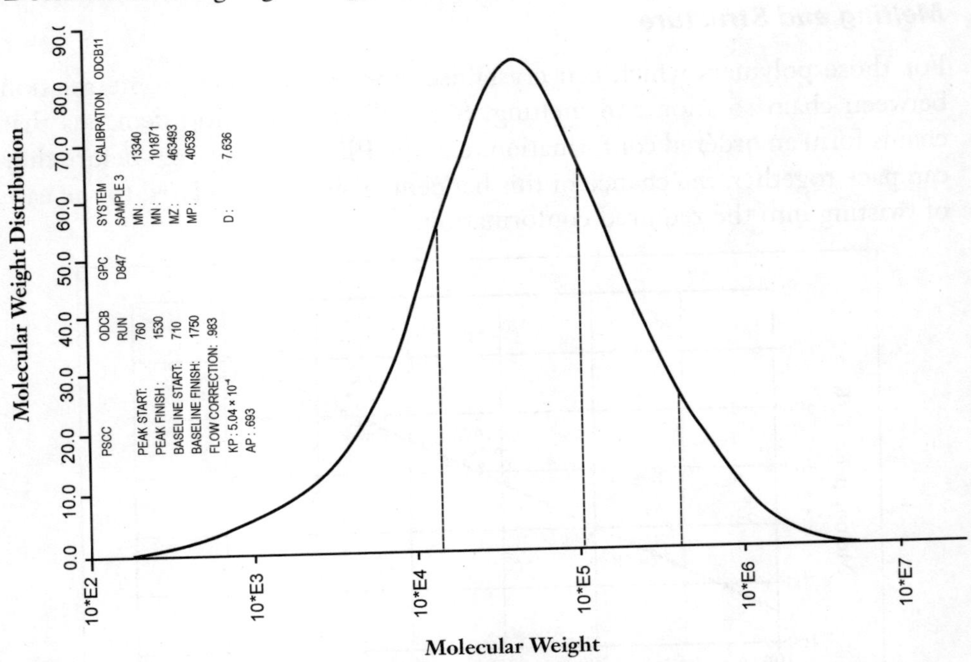

Fig. 10.24: Molecular mass distribution for a typical high density polyethylene (HDPE).

Chain molecules give rise to types of structure which are not formed in small molecules, and they exhibit molecular mass variations which are critical not only for their properties but also for processing into shape. In general, commercial polymers do not consist of assemblies of chains of constant length but rather a distribution of lengths, so that statistical techniques of analysis are needed to characterize them. The spectrum of molecular masses for a typical HDPE is shown in Figure 10.28, with M varying from $M = 200$ through a peak at about 40 000 to over a million. Such a distribution can be described by average molecular masses, of which the most important are the *number-average molecular mass.*

$$\left(\overline{M}_n\right) = \frac{\sum_i N_i M_i}{\sum_i N_i} \tag{4}$$

the weight-average molecular mass

$$\left(\overline{M}_w\right) = \frac{\sum_i W_i M_i}{\sum_i W_i} \tag{5}$$

and the z-average molecular mass

$$\overline{M}_z = \frac{\sum_i W_i M_i^2}{\sum_i W_i M_i} \tag{6}$$

where N_i and W_i are the number and weight of chains of molecular mass M respectively. Since $W_i = N_i M_i$, Equation (5) can be rewritten as

$$\overline{M}_w = \frac{\sum_i N_i M_i^2}{\sum_i N_i M_i} \tag{7}$$

and Equation (6) becomes

$$\overline{M}_z = \frac{\sum_i N_i M_i^3}{\sum_i N_i M_i^2} \tag{8}$$

Each measure of molecular mass gives a different emphasis. $\overline{M}_n$ emphasises the smaller molecules present whereas $\overline{M}_w$ and $\overline{M}_z$ emphasise the larger molecules. It is clear therefore that $\overline{M}_z > \overline{M}_w > \overline{M}_n$ as shown in Figure 10.28. A simple example suffices to demonstrate the point. An equimolecular (i.e. equimolar) mixture of samples of molecular mass 50 000 and 100 000 will have the following average values:

$$\overline{M}_n = \frac{(0.5 \times 50000) + (0.5 \times 100000)}{1.0}$$

$$= 75\ 000$$

$$\overline{M}_w = \frac{(0.5 \times 50000^2) + (0.5 \times 100000^2)}{75000}$$

$$= 83\ 000$$

$$\overline{M}_z = \frac{(0.5 \times 50000^3) + (0.5 \times 100000^3)}{62\ 500 \times 10^5}$$

$$= 90\ 000$$

The breadth of the distribution is usually defined by the dispersion, which is the simple ratio of weight- to number-average molecular masses

$$\text{dispersion} = \frac{\overline{M}_w}{\overline{M}_n} \qquad (9)$$

which is unity for a monodisperse polymer but is usually greater than unity. High density polyethylenes have values from 6 to 12 while LDPE is usually much broader in distribution ($\overline{M}_w / \overline{M}_n$ up to ca. 30). New metallocene polyolefins, however, have low dispersions of about 2.5. In polymers like nylon it is also much sharper, with values in the region of 2.0.

The molecular mass distribution (MMD) is important for many physical properties of the polymer and is best expressed by a complete curve as shown in Figure 10.26, obtained directly by a method known as gel permeation chromatography (GPC) (Box 10.5).

Single-point averages can be obtained by other means such as osmometry (M_n), light scattering (M_w) or ultracentrifugation (M_z). Dilute solution viscometry gives a direct measure of the viscosity-average molecular mass M_v, which usually lies between M_n and M_w. It is defined by the equation

$$\overline{M}_v = \left[\frac{\sum_i N_i M_i^{1+a}}{\sum_i N_i M_i} \right]^{1/a} \qquad (10)$$

where a is a constant in the equation

$$[\eta] = KM^a \qquad (11)$$

$[\eta]$ is the intrinsic viscosity of a polymer in solution and K is another constant characteristic of the solution. K usually lies between 0.5×10^{-4} and 5×10^{-4}, while a typically lies in the range 0.6–0.8.

Self Assessment Question

- To improve processing properties, an equal weight of low molecular mass nylon 6 (degree of polymerization $n = 100$) is blended with a moulding grade of nylon 6 ($n = 500$). Assuming both materials are monodisperse, what will be
  - the weight-average molecular mass,
  - the number-average molecular mass,
  - the dispersion of the blend?

Box 10.5: Gel permeation chromatography (GPC).

GPC is a technique for determining the molecular mass distribution of a polymer by analysis of a solution of the polymer concerned. It is normally a rapid and accurate way of determining the MMD compared with older methods such as osmometry or light scattering. Moreover, it gives the *whole* distribution in one pass, while the other methods give only single point averages. Light scattering, for example, only yields the weight-average molecular mass. When the complete MMD is available from GPC, *any* average molecular mass can be computed simply and easily.

But since a solution is the basis for analysis using GPC, the technique is limited to thermoplastic, uncrosslinked polymers. Polymers vary widely in their ability to form solutions. Some, such as polystyrene and polycarbonate are soluble in a wide selection of organic media. Tetrahydrofuran or THF is a useful solvent for their GPC analysis. Many other plastics materials, however, are quite insoluble in the range of solvents commonly available. They include the polyolefins, PE and PP, as well as PET and other engineering polymers with high T_gS. GPC in a high boiling point solvent such as decalin (*T*b = 169 °C) is often used for such intractable materials.

So what is the basis of GPC? The method involves passing the polymer solution down a column containing gel spheres into which the smallest chains will diffuse most easily, the largest with extreme difficulty. The latter are thus the first to be passed out at the bottom of the column, at low elution counts, while the smallest chains diffuse out last of all. The polymer concentration is measured in the eluate, so giving an estimate of the MMD. The method demands calibration with standard samples, usually monodisperse polystyrene of various molecular masses (Figure 10.25).

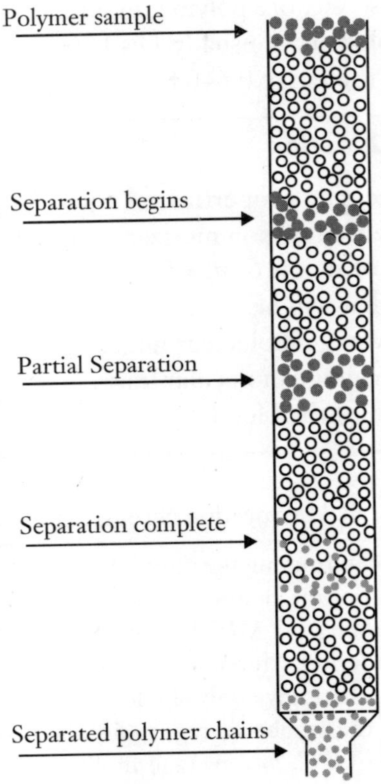

Fig. 10.25: Method demands calibration with standard samples.

When is GPC used? Basically, whenever a user wants to check the MMD of a polymer specimen against, say, a virgin or unused sample. The user may be interested in checking whether the correct grade of polymer has been used, for example, in a plastic moulding, or alternatively, the effects of chain degradation.

10.10 Commercial Polymers

The increasing control of polymer structure by fine-tuned catalysis of polymerization opened up an enormous area for commercial exploitation, and new polymers are still being produced in this way (such as the metallocene polymers). A revolution of equal magnitude has occurred with polymers containing functional groups, for example, the nylons, polyesters and polyurethanes, resulting in polymers ranging from quite simple structures like aramid fibre to relatively complex repeat units like those in polyimide or polycarbonate. These polymers are mainly based on carbon-carbon, carbon-oxygen or carbon-nitrogen bonds, but a significant amount of research is also being put into polymers like the silicones (Table 10.6) which are termed 'inorganic' (Box 10.6). Silicone polymers have been known since the 1940s and were developed for both low and high temperature resistant seals for aircraft engines, although many other uses were found for their water-repelling properties. A number of speciality silicones now exist, mainly varying in molecular mass and the pendant organic side groups.

Polymeric adhesives were traditionally based on concentrated solutions of latex rubber, PVA or epoxy resins, but a novel approach developed in the sixties involved direct polymerization of cyanoacrylate monomer at the surfaces to be bonded. The main advantage is ease of application of the liquid monomer to the surfaces, where its low viscosity enables it to penetrate into the finest crevices. The monomer polymerizes very rapidly by contact with traces of water and gives a very strong bond. It is now finding widespread use in industrial assembly as well as for domestic purposes. Poly(vinyl butyral) is another speciality adhesive primarily used for bonding glass sheets together to form laminated or safety glass. PVA is partially reacted with butyraldehyde to make the adhesive which is then bonded to the glass surfaces. In service, it acts to stop crack growth penetrating from one glass sheet to the next by absorbing energy like a rubber.

Silicone rubber is an example of an inorganic polymer, one member of a very large family of inorganic materials which have a chain or sheet molecular structure. Elements which can link together with themselves ('catenation') are not limited to carbon, with nitrogen forming a dimer (N_2 gas), phosphorus in its coloured forms (P_n), plastic sulphur (Se_n) and amorphous selenium (Se_n). Generally, it is those non-metals or metalloids lying near carbon which catenate, as highlighted in Figure 10.28.

Table 10.6: Some Speciality Polymers.

Polymer	*Repeat unit*	M_R	*Comments*
silicone rubber	$-[Si(CH_3)_2-O]-$	74	Special-purpose rubber with very low glass T_g and high temperature mobility
poly(methyl cyanoacrylate)	$-[CH_2-C(CN)(CO_2CH_3)]-$	111	'Superglue' adhesive polymerized *in situ* from liquid monomer
poly(vinyl butyl)	$-[CH_2-CH-CH_2-CH]-$ with O–CH(C_3H_7)–O bridge	138	Modified PVA for laminated glass interlayer, tailored at a molecular level to give high adhesion to glass surfaces
poly(ethylene terephthalate) PET	$-[OCH_2CH_2-O-C(=O)-C_6H_4-C(=O)]-$	192	Fiber-forming polymer also used for moulding bottles
poly(butylene terephthalate) PBT	$-[O(CH_2)_4OC(=O)-C_6H_4-C(=O)]-$	220	Moulding polymer with lower melting point than PET
poly(phenylen e sulptide) PPS	$-[C_6H_4-S]-$	108	Heat-and flame-resistant polymer of particular value in electrical applications
polyethersulph one PES	$-[C_6H_4-O-C_6H_4-S(=O)_2]-$	232	Transparent, heat-resistant polymer of particular value in 'under-the bonnet' and aerospace applications; commercialized in 1972
Tragamid nylon	$-[NHCH_2C(CH_3)_2CH_2CH(CH_3)CH_2CH_2NHOC-C_6H_4-CO]-$	288	Transparent, amorphous nylon with better resistance to water than nylons 6 or 6,6
polyetherether ketone PEEK	$-[C_6H_4-C(=O)-C_6H_4-O-C_6H_4-O]-$	276	Commercialized in 1977, and used for high temperature resistant insulation. One of the most beat-resistant commercial polymers

Box 10.6: Inorganic Polymers.

The polymers formed by linkage with another element are much more common, however, silicon being an element which forms extremely strong bonds with oxygen (Si-O) in many silicate minerals, for example. Chain formation in inorganic silica-based glasses is shown by the amorphous, non-crystalline nature of the material (transparent window glass, for example) and the high viscosity of molten glasses. The strength of Portland cement derives from the formation of a polymerized calcium silicate formed during setting.

Phosphorus also forms very strong chemical bonds with oxygen, frequently forming long chain polymers. Indeed. DNA, the stuff of life itself, possesses a backbone chain which is a copolymer of phosphate groups (PO4) with sugar molecules.

				1 H	
5 B	6 C	7 N	8 O	9 F	10 Ne
13 Al	14 Si	15 P	16 S	17 Cl	18 A
31 Ga	32 Ge	33 As	34 Se	35 Br	36 Kr
49 In	50 Sn	51 Sb	52 Te	53 I	54 Xe
81 Tl	82 Pb	83 Bi	84 Po	85 At	86 Rn

Functional groups which form part of the main chain offer very great flexibility in synthesis not only because of the variety which exists but because of the very large number of possible structures that can be linked together. The selection shown in Table 10.6 includes staple polymers like PET, which has recently received a boost through development of blow moulding grades, but also high technology materials like PEEK. Traditional bulk polymers like the polyolefins, PVC and PS are limited in application particularly for engineering use by virtue of their low melting or glass transition temperatures. The concept of rotational isomerism stimulated research into ways of stiffening chains by the introduction of aromatic groups. Outstanding success stories include polycarbonate, a tough transparent thermoplastic discovered in 1954 and available commercially in the early sixties, *Kevlar* aramid fibre (1968) and the polysulphones. They show progressively increasing temperature resistance: PC has a T_g of 145 °C, but polyethersulphone a T_g of about 230 °C. Polyetheretherketone (PEEK) exhibits a T_g of 144 °C but melts at 335 °C. These polymers are also highly resistant to burning, a feature which reflects the chemical stability of the aromatic benzene rings. It is a characteristic which reaches its ultimate in refractory graphite, which can resist temperatures above 3000 °C. Thus PPS has an oxygen index of 0.53 (the fraction of oxygen in air needed before burning will occur) compared to most common polymers which have indices less than 0.21, which is the normal oxygen content of the atmosphere. Introduction of aromatic rings into the nylons has similar effects, stiffening the chain and providing chemical stability. It is seen most dramatically with aramid fibre, and to a lesser extent with *Trogamid* nylon (Table 10.6). This polymer has a high T_g of about 150 °C yet is completely non-crystalline, and thus transparent, owing to the disruption of the crystal lattice by the irregular aliphatic diamine chain. The T_g of nylon 6 or 6,6 is about 0 °C but varies with the amount of water absorbed in the material.

Thermosets

There are some limitations to the concept of the repeat unit when applied to crosslinked polymers, the thermosets. This is because of the complexity of the crosslinking reactions, the way molecules link together chemically during thermoset processing. For example, phenolic resins (the basis for materials like *Bakelite*) are prepared initially as prepolymers, i.e. polymers of low molecular mass (ca. 1000) by reaction between phenol and formaldehyde (Figure 10.26). Reaction can occur in several ways depending on the catalyst used and the ratio of phenol to formaldehyde: acid catalysis gives novolaks [Figure 10.26(b)] and alkaline catalysis produces resols [Figure 10.26(c)]. The latter always possess free alcohol groups which are not present in novolaks. The prepolymers are blended with fillers before moulding to shape. With resols, no extra crosslinking agent is needed whereas multi-functional agents such as hexamine are added to novolaks to ensure crosslinking. Since the reaction occurs at three possible sites on the phenol molecule [Figure 10.26(a)], it is clearly not possible to specify a repeat unit in a simple way. The molecular mass is practically infinite since all parts of the polymer are linked together by chemical covalent bonds.

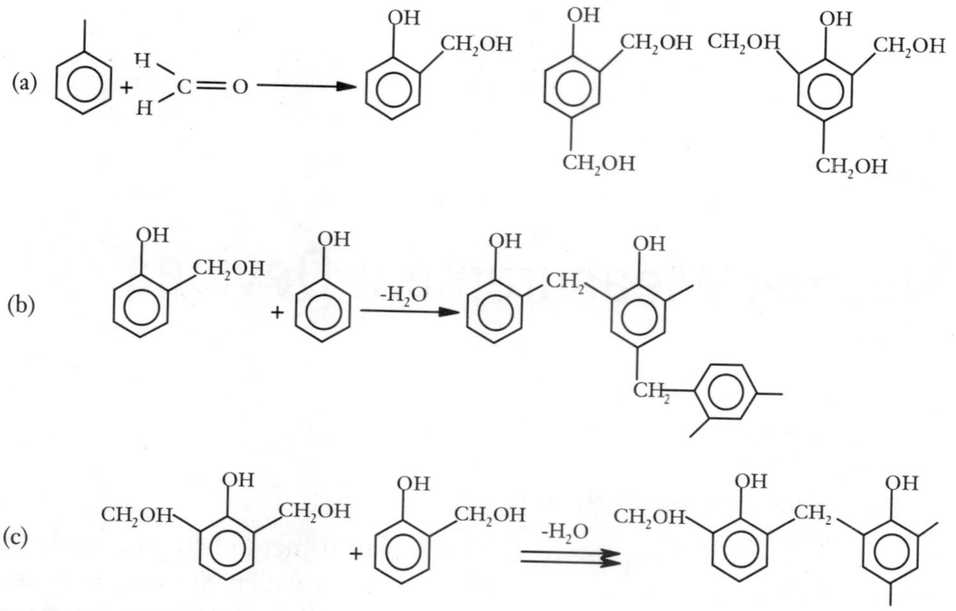

(a)

(b)

(c)

Fig. 10.26: Preparation of phenolic prepolymers from aromatic phenol and formaldeyde; (a) several possible structures can be formed; (b) novolaks result during acid catalysis; (c) resols result from an excess of formaldehyde and alkaline catalysis.

11

Polymer Measurement Devices

11.1 Time for Polymer Diffusion

Neutron reflectivity has proved to be a powerful method for the study of polymer diffusion, normally using a technique in which the polymers are annealed followed by rapidly frozen to suspend diffusion whilst a measurement is made. However, recent experiments on the CRISP and SURF reflectometers have shown that *in-situ*, real-time measurements of polymer diffusion are possible owing to the rapid data collection times at ISIS—of the order of a minute in some cases.

Reptation theory predicts that for the polymer to move in an entangled melt it can only do so along a tube formed by the constraints provided by the surrounding polymer chains. The time taken for a polymer chain to diffuse out of this tube is known as the reptation time (τ_r). Motions within the tube and those along its contour are predicted to occur with different time dependencies. In particular, above τ_r movement of the molecules is proportional to the square root of time ($t^{1/2}$), whereas below τ_r diffusion is proportional to $t^{1/4}$. Studies to test the reptation theories have been most successful when measuring the change in interfacial width developing between two polymer layers as they interdiffuse. Although the early stages of interdiffusion at polymer-polymer interfaces have been studied by a number of techniques neutron reflectivity (NR) has proved to be an extremely powerful method, especially for the very early stages of diffusion. Not only does it provide an accurate measure of the width of the interface but also the interfacial concentration profile between the interdiffusing polymers.

The normal way of studying the diffusion process between thin films is to use an 'anneal-quench' cycle. Here, the sample is heated for a specified time above the glass transition temperature, T_g, of both polymers and then rapidly quenched to room temperature where the polymer molecules are frozen in place and a full reflection profile can be collected. Despite the success of this procedure, it is not applicable for all systems and another approach—*in-situ* real time reflectivity measurements—has been developed.

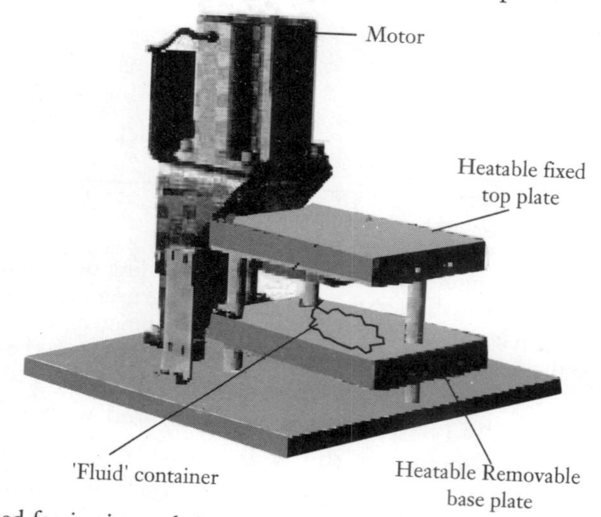

Fig. 11.1: Cell used for in situ real time reflectivity measurements performed on the SURF reflectometer.

Such measurements have been attempted in the past with very limited success using point-by-point θ-2θ fixed wavelength NR measurements by taking a partial reflectivity curve of a limited number of points. However, the technique is limited due to the difficulties presented by fitting models to a sparse data set. In contrast, NR profiles collected using a white beam on a pulsed neutron source give a full reflectivity profile at a fixed incident angle. Data collection times are then limited by the time required to gain adequate statistics. Using the CRISP and SURF reflectometers at ISIS, a full reflectivity profile of limited Q range for a single incident angle can be collected in a few minutes or less and provides sufficient statistics for extraction of interfacial widths and profiles from the data.

In order to be able to do these *in situ* measurements specialised sample cells (Figure 11.1) have to be employed to allow for neutron alignment before the two diffusing species come into contact. The cell consists of an inverted silicon wafer coated with a thin film of the deuterated polymer, which is held in a top heating plate. Below this a second heating plate holds a trough containing the low molecular weight material. Once the lower plate is raised into contact data collection is started, and reflectivity profiles are collected every few minutes.

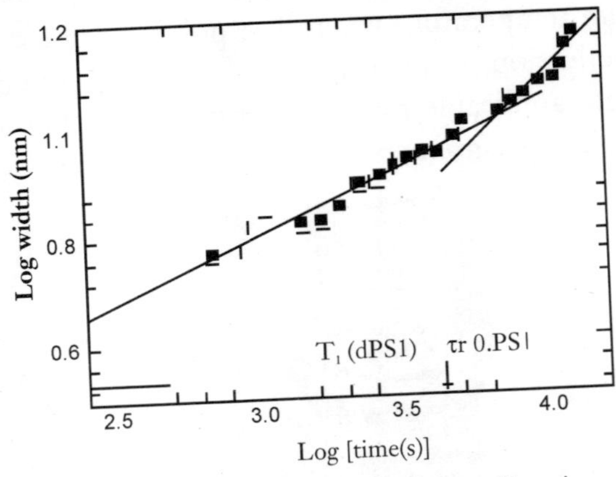

Fig. 11.2: A log-log plot of interfacial width, w, versus annealing time, obtained from fits to reflectivity. The lines through the data are calculated from reptation theory assuming the time dependencies shown. The values of the reptation times for the two polymers based on the calculated diffusion coefficient are indicated on the time axis.

The potential of the technique can be illustrated with reference to two recent experiments. The first is the well known example of polystyrene-polystyrene inter-diffusion, which has been successfully studied by using the 'anneal-quench' procedure. Investigating the system while maintaining it above the T_g of the polymers necessitates the use of real-time measurements. We have measured the interdiffusion between hydrogenated and deuterated polystyrene at 115°C, approximately 10-15° above T_g in real time taking data every six minutes over a 4 hour period. Careful data fitting enabled the interfacial width of the diffusing polymers to be determined and the results are shown in Figure 11.2. Clearly the accuracy of the measurements even with only 6 minute runs is enough to confirm the potential power of the technique.

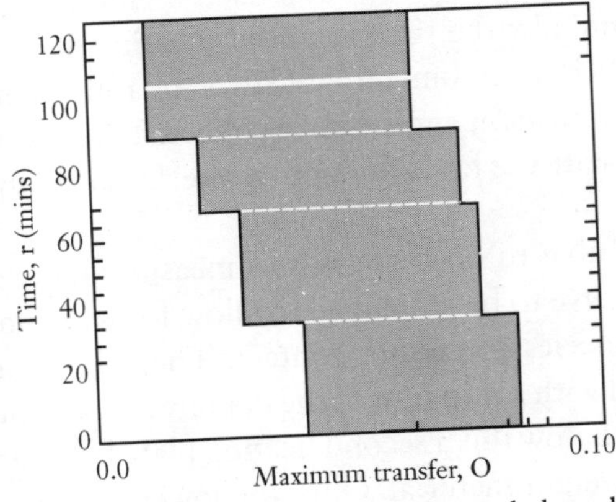

Fig. 11.3: Contour plot reflectivity obtained from poly(methyl methacrylate)-oligo(methyl methacrylate) diffusion studied in real time. The blocks of data illustrate where data is taken at fixed angles, which are incrementally decreased.

When studying smaller molecules, such as plasticisers or oligomers, diffusing into polymers, it is possible that diffusion continues at room temperature. In addition, the time scale over which the early stages of diffusion occur may be much quicker than polymer-polymer diffusion. Such systems have been tackled by reducing the data collection time in some cases to as little as 20 seconds per reflection profile. A plot of a series of data taken from a polymer-plasticiser example is shown in figure 11.3. Careful analysis of these data indicates that the interface has a complex asymmetric profile qualitatively similar to theoretical predictions.

The potential of these in-situ real time measurements is clear, and there is now a whole spectrum of new and exciting experiments waiting to be attempted which could not be performed before the advent of higher flux white beam reflectometers.

Molar Mass Distribution

The Molar mass distribution (also known as the molecular weight distribution) in a polymer describes the relationship between a polymer fraction and the molar mass of that polymer fraction. In linear polymers the individual polymer chains rarely have the exact same degree of polymerization and there is always a distribution around an average value.

Definition

Different average values can be defined depending on the statistical method that is applied. The weighted mean can be taken with the weight fraction, the mole fraction or the volume fraction:

- Weight average molar mass or M_w
- Number average molar mass or M_n
- Viscosity average molar mass or $M_{\acute{i}}$
- Z average molar mass or M_z

$$M_n = \frac{\sum M_i N_i}{\sum N_i},\ M_w = \frac{\sum M_i^2 N_i}{\sum M_i N_i}, M_z = \frac{\sum M_i^3 N_i}{\sum M_i^2 N_i},\ M_\upsilon = \left[\frac{\sum M_i^{1+a} N_i}{\sum M_i N_i}\right]^{\frac{1}{a}}$$

Measurement

These different definitions have true physical meaning because different techniques in physical polymer chemistry often measure just one of them. For instance osmometry measures number average molar mass and small angle laser light scattering measures weight average molar mass. M_v is obtained from viscosimetry and M_z by sedimentation in an analytical ultracentrifuge. The quantity a in the expression for the viscosity average molar mass varies from 0.5 to 0.8 and depends on the interaction between solvent and polymer in a dilute solution. In a typical distribution curve the average values are related to each other as follows $M_n < M_v < M_w < M_z$. Polydispersity of a sample is defined as M_w divided by M_n and gives an indication just how narrow a distribution is.

The most common technique for measuring molecular weight used in modern times is a variant of high pressure liquid chromatography (HPLC) known by the interchangeable terms of size exclusion chromatography (SEC) and gel permeation chromatography or (GPC). These techniques involve forcing a polymer solution through a matrix of cross-linked polymer particles at a pressure of up to several thousand psi. The limited accessibility of stationary phase pore volume for the polymer molecules results in shorter elution times for high molecular weight species. The use of low polydispersity standards allows the user to correlate retention time with molecular weight although the actual correlation is with the Hydrodynamic volume. If the relationship between molar mass and the hydrodynamic volume changes (i.e., the polymer is not exactly the same shape as the standard) then the calibration for mass is in error.

The most common detectors used for size exclusion chromatography include online methods similar to the bench methods used above. By far the most common is the differential refractive index detector which measures the change in refractive index of the solvent. This detector is concentration sensitive and very molecular weight insensitive so it is ideal for a single detector GPC system as it allows the generation of mass v's molecular weight curves. Less common but more accurate and reliable is a molecular weight sensitive detector using multi-angle laser light scattering. These detectors directly measure the molecular weight of the polymer and are most often used in conjunction with differental refractive index detectors. A further alternative is either low angle light scattering, which uses a single low angle to determine the molar mass or Right Angle Light Laser scattering in combination with a viscometer, although this last technique does not actually give an absolute measure of molar mass but one relative to the structural model used.

The molar mass distribution of a polymer sample depends on factors such as chemical kinetics and work-up procedure. Ideal step-growth polymerization gives a polymer with polydispersity of 2. Ideal living polymerization results in a polydispersity of 1. By dissolving a polymer an insoluble high molar mass fraction may be filtered off resulting in a large reduction in M_w and a small increase in M_n thus reducing polydispersity.

Weight Average Molecular Weight

The weight average molecular weight is a way of describing the molecular weight of a polymer. Polymer molecules, even if of the same type, come in different sizes (chain lengths, for linear polymers), so we have to take an average of some kind. For the weight average molecular weight, this is calculated by:

$$\overline{M}_w = \frac{\Sigma_i N_i M_i^2}{\Sigma_i N_i M_i}$$

where

N_i is the number of molecules of molecular weight M_i.

Intuitively, if the weight average molecular weight is w, and you pick a random monomer, then the polymer it belongs to will have a weight of w on average. The weight average molecular weight can be determined by light scattering, small angle neutron scattering (SANS), X-ray scattering, and sedimentation velocity.

An alternative measure of molecular weight for a polymer is the number average molecular weight; the ratio of the *weight average* to the *number average* is called the polydispersity index.

The *weight-average molecular weight*, M_w, is also related to the *fractional monomer conversion*, p, in step-growth polymerization as per Carothers' equation:

$$\overline{X}_w = \frac{1+p}{1-p} \quad \overline{M}_w = \frac{M_o(1+p)}{1-p}$$

where

M_o is the molecular weight of the repeating unit.

Number Average Molecular Weight

The number average molecular weight is a way of determining the molecular weight of a polymer. Polymer molecules, even ones of the same type, come in different sizes (chain lengths, for linear polymers), so the average molecular weight will depend on the method of averaging. The *number average* molecular weight is the ordinary arithmetic mean or average of the molecular weights of the individual macromolecules. It is determined by measuring the molecular weight of n polymer molecules, summing the weights, and dividing by n.

$$\overline{M}_n = \frac{\Sigma_i N_i M_i}{\Sigma_i N_i M_i}$$

The number average molecular weight of a polymer can be determined by gel permeation chromatography, viscometry (Mark-Houwink equation), and all colligative methods like vapor pressure osmometry or end-group titration.

An alternative measure of the molecular weight of a polymer is the weight average molecular weight. The ratio of the *weight average* to the *number average* is called the polydispersity index.

High Number-Average Molecular Weight Polymers may be obtained only with a high *fractional monomer conversion* in the case of step-growth polymerization, as per the Carothers' equation.

11.2 Measurement of Gas Solubility in Polymers

Solubility and diffusivity of gases in polymers are two important physical properties which determine the success of foaming a given polymer. These properties are also used to characterize the barrier property of the polymer in food packaging applications. Engineers at PPI have developed a unique method to measure these properties. Custom designed high pressure vessels are used for off-line solubility measurements. The gas solubility in a polymer is calculated by measuring the weight gain of the polymer samples after saturation time. Various thermodynamic models can be applied to correlate and predict the solubility data over wider range of pressure and temperature.

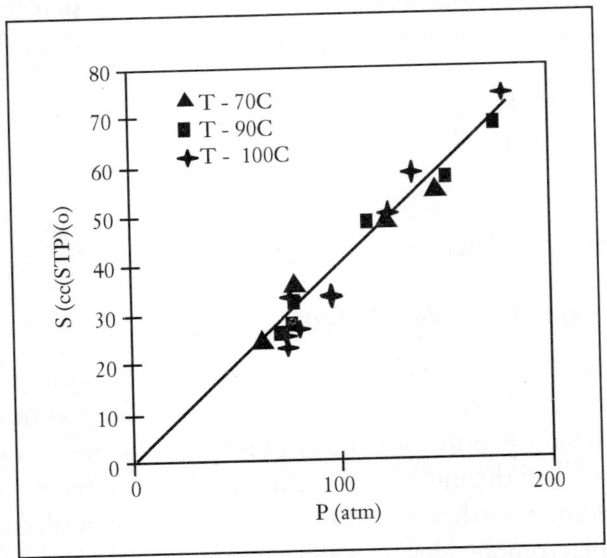

For example, the figure shows the solubility of CO_2 in polypropylene (PP) at three different temperatures. The line shows the Henry's law correlation for the data.

11.3 Electro Active Polymer Devices

Physicists and chemists have long sought to develop lightweight materials that grow or shrink significantly in length or volume when subjected to electric stimulation. Such substances could serve as the drivers for novel motion-generating devices (generally called actuators)—possible replacements for the ubiquitous electric motor, which is often too bulky and heavy for smaller scale applications. A new generation of electro-active polymer (EAP) materials displays sufficient physical response to electrical excitation to power new classes of actuators as well as innovative sensors and energy generators. Products based on this "artificial muscle" technology are just starting to hit the market.

Electro-active polymer technology could potentially replace common motion-generating mechanisms in positioning, valve control, pump and sensor applications, where designers are seeking quieter, power efficient devices to replace cumbersome conventional electric motors and drive trains. An EAP actuator is not only completely different from conventional electromechanical devices, but also separates itself from other high-tech approaches that are based on piezoelectric materials or shape-memory alloys by providing a significantly more power-dense package and, in many instances, a smaller footprint.

Shape-memory alloys contract with a thermal cycle, and piezoelectric technologies expand and contract with voltage at high frequencies. While both these technologies provide direct displacement, they are usually limited to a 1 per cent direct displacement. Electromagnetic solutions typically consist of a motor that rotates an output shaft, so there is no direct displacement from the motor itself. The output shaft connects to a "drive train," gear reducer transmission or other mechanical device that has several touching and moving parts, which create an "indirect" displacement.

According to a new market research study titled "Electro-active Polymer Actuators—Types, Applications, Developments, Industry Structure and Global Markets ", the global market for EAP actuators and sensors reached $15 million in 2007. This will increase to $247 million by 2012. North America has about 66 per cent market in 2007, followed by Europe at 21.3 per cent, Japan at 9.3 per cent, and rest of world at 3.3 per cent. The AAGR growth rate is expected to be 71.3 per cent to 91.8 per cent for the four major regions surveyed for the period 2007 to 2012.

The demand for EAP actuators and sensors in 2007 comes mostly from laboratories for research and development in new application areas. There is large scale EAP-related R&D activity in university and institutional labs in North America, resulting in initial large scale demand for EAP devices in 2007. The sensors (as used in smart fabrics and high strain applications) constitute about 16 per cent of the total EAP device market in 2007.

This new study segmented markets into four applications for electro-active polymer devices and products. These are medical devices, smart fabrics, digital mechtronics, and high strain sensing in construction. Manufacturers of electro-active polymers expect competition to persist and intensify in the future from a number of different sources. EAP devices are facing competition in a new rapidly evolving and highly competitive sector of the medical market. Increased competition could result in reduced prices and gross margins for EAP products and could require increased spending by research and development, sales and marketing, and customer support.

Medical devices will have the largest share in 2007, as much as 77.3 per cent, followed by smart fabrics with 13.3 per cent, digital mechtronics with 6.7 per cent, and high strain sensing in construction as the remaining 2.7 per cent of the market. While the medical devices will continue to maintain the lead in 2012, that sector will see the largest growth rate as well, as much as 92 per cent AAGR from 2007 to 2012.

According to the study, during 2007, there is low key activity in the manufacturing of EAP devices. Companies are catering to specific orders of low to medium volumes. Low penetration of EAP technology in the fragmented market is partly due to lack of standardization of product specifications among manufacturers. Adaptation of EAP technology during the period of replacement of bulky conventional actuators by OEMs will depend upon, besides cost, reliability and durability of the EAP devices.

Table 11.1: North American and global market for electro-active polymer actuators and sensors, 2007 and 2012.

	2007 ($ Millions)	*2007 (%)*	*2012 ($ Millions)*	*2012 (%)*	*AAGR (%) 2006-12*
North America	9.9	66	146	59.1	71.3
Global	15	100	247	100	75.1

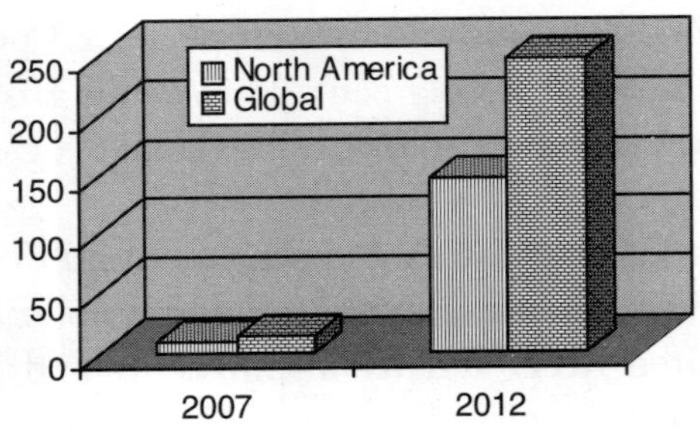

Fig. 11.4: North american and global market for electro-active polymer actuators and sensors, 2007 and 2012.

11.4 Fabrication and Study of Polymer Light Emitting Devices

Background

Polymer based light-emitting diodes (LED) were discovered in 1990 by Friend et al. The first polymer LEDs used poly(phenylene vinylene) (PPV) as the emitting layer. Since 1990 many different polymers have been shown to emit light under the application of an electric field (electroluminescence, EL). PPV and its derivatives are still the most commonly used materials, but polythiophenes, polypyridines, poly(pyridyl vinylenes), polyphenylenes and copolymers of these materials are now being used. A large effort in industry and academia to improve stability, lifetime, and efficiency has prompted the study of a vast number unique and novel configurations in polymer devices.

Recent Advances

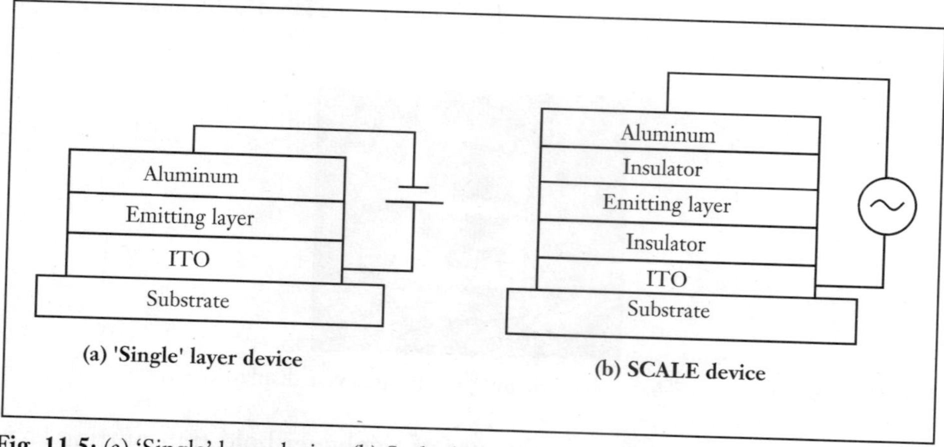

Fig. 11.5: (a) 'Single' layer device. (b) Scale device.

Our research efforts in the area of polymer light-emitting devices focuses on novel configurations using a variety of polymer and copolymer systems. Light-emitting devices consist of active/emitting layer(s) sandwiched between a cathode and an anode. Indium tin oxide (ITO) is typically used for the anode and aluminum or calcium as the cathode. Figure 11.5(a) shows the structure of a simple (single layer) device with only the electrodes and a single active layer. Simple 'single' layer devices typically work only under a forward DC bias. Figure 11.5(b) shows a novel configuration: a symmetrically configured alternating current light emitting (SCALE) device, which works under AC as well as forward and reverse DC bias.

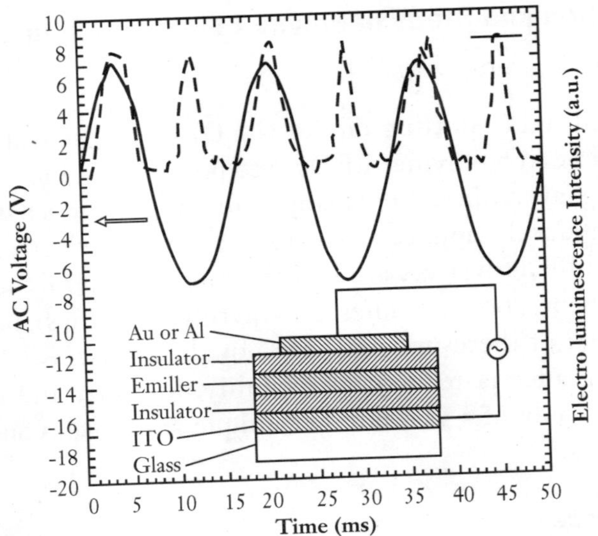

Fig. 11.6: Shows the response of a SCALE device driven at 60 Hz. The frequency of the light emission is 120 Hz.

Fig. 11.7: A picture of a 7-bar polymeric light emitting device display.

The device configuration makes use of a polymer hole-transporting layer and a polymer emitting layer. We continue to develop new device configurations as well as improving device performance.

11.5 Piezoelectric Ceramic, Polymer and Ceramic/Polymer Composite Devices

The piezoelectric device sector is actually not a single sector, but rather comprises a number of sectors with distinctly different characteristics. The sectors of most significance are:

- High production volume, piezoelectric device sector consisting of generic piezoelectric devices such as actuators, motors, sensors, accelerators, transducers for ultrasonic medical imaging and non-destructive testing acoustic devices, Lengevin actuators for ultrasonic welding and cleaning, ceramic resonators, and miscellaneous types of devices designed for special applications such as transformers, vibration and noise cancellation in structures limited to different grades of piezoelectric crystals, ceramics such as PZT, PVDF and composites;

- Sonars for military and civil use; and
- Niche applications such as energy harvesting, where piezoelectric devices such as generators offer a unique competitive advantage.

Piezo devices also include ultrasonic motors (USMs), which offer a high potential for miniaturization. These actuators produce no magnetic field since the excitation is quasi-electrostatic. Through their specific advantages compared to conventional electro-magnetic motors, USMs fill a gap in certain actuator applications. A key advantage of USMs over electromagnetic motors is their compactness, i.e. their high stall torque-mass ratio and high torque at low rotational speed, often making speed-reducing gears superfluous. Additionally, with no voltage applied, an inherent holding torque is present due to the frictional driving mechanism. It is also worthwhile to mention that their compactness and the high frequency electrical excitation make quick responses possible.

Study Goal and Objectives

This study focuses on key piezoelectric devices and provides data about the size and growth of the piezoelectric devices markets, company profiles and industry trends. The goal of this report is to provide a detailed and comprehensive multi-client study of the markets in North America, Europe, Japan, China, Korea and the rest of the world (ROW) for piezoelectric ceramic, polymer and ceramic/polymer composite devices, as well as potential business opportunities in the future.

A primary objective of this report is thorough coverage of underlying economic issues driving the piezoelectric ceramic and polymer devices business, as well as assessments of new, advanced piezoelectric devices that are being developed. Another important objective is to provide realistic market data and forecasts for piezoelectric devices. This study provides the most thorough and up-to-date assessment that can be found anywhere on the subject. The study also provides extensive quantification of the many important facets of market development in piezoelectric devices in the world. This, in turn, contributes to a determination of the kinds of strategic responses companies may adopt in order to compete in these dynamic markets.

Users of piezoelectric devices in developed markets must contend with twin pressures—to innovate and, at the same time, to reduce costs. New applications for piezoelectric devices have been proposed in recent years. This study condenses all of these business-related issues and opportunities.

Reasons for Doing the Study

The piezoelectric devices market is a diversified but attractive and still-growing multi-billion dollar market characterized by very high production volumes of a diversified range of piezoelectric devices that must be both extremely reliable and low in cost. Growth in the piezoelectric devices market continues to be driven by increasing demands in camera phones for autofocus mechanisms, piezo-transformers, energy harvesting devices, data storage, semiconductors, microelectronics production, precision mechanics, life science and medical technology, optics, photonics, nano-metrology, robots, toys, HVAC control systems, hand held consumer electronic devices, automotives sensors, ultrasonic transducers for medical imaging and non-destructive testing and vibration related applications, structure health monitoring, ultrasonic welding and cleaning, ceramic resonators for mobile phones and devices used for information and communication technologies. Since sonars for military and civil uses and other applications constitute an established market, we have included this segment in this report.

The diversified piezoelectric ceramic device business is complex and fast moving, with manufacturers increasingly adopting a truly global view of the market. Around the world, consumers are demanding a high power density as well as an extremely long cycle life. In this challenging market, manufacturers have attempted to achieve growth through company mergers and acquisitions and by implementing global strategies. Traditional piezoelectric devices have a broad customer base, and new applications such as ceramic resonators in mobile phones, transformers in notebooks, and energy harvesting are new areas that have now entered the mainstream and are showing significant sales volumes. With this background of new emerging technologies and applications, iRAP felt a need to conduct a detailed study and update technology developments and markets. The report identifies and evaluates piezoelectric ceramic devices and technologies which show potential growth.

Contributions of the Study

This study provides the most complete accounting of the current market and future growth in piezoelectric devices in North America, Europe, Japan, China and the rest of the world. It provides the most thorough and up-to-date assessment that can be found anywhere on the subject. The study also provides extensive quantification of the many important facets of market developments in emerging markets, such as China, for piezoelectric devices. This quantification, in turn, contributes to the determination of what kind of strategic response suppliers may adopt in order to compete in these dynamic markets. Audiences for this study include directors of technology, marketing executives, business unit managers and other decision makers in the piezoelectric devices companies, as well as in companies peripheral to this business.

Scope and Format

The market data contained in this report quantify opportunities for piezoelectric ceramic and polymer materials and devices. In addition to product types, this report also covers the merits and future prospects of the piezoelectric ceramic and polymer devices business, including corporate strategies, information technologies, and the means for providing these highly advanced product and service offerings. This report also covers in detail the economic and technological issues regarded by many as critical to the industry's current state of change. It provides a review of the piezoelectric ceramic devices industry and its structure, and of the many companies involved in providing these products. The competitive positions of the main players in the piezoelectric devices market and the strategic options they face are also discussed, along with such competitive factors as marketing, distribution and operations.

The study will benefit the existing manufacturers of piezoelectric actuators, ultrasonic motors, sensors, transducers, transformers, resonators and micro-energy harvesting devices that seek to expand revenues and market opportunities by expanding and diversify in vast applications of devices based on piezo materials like PZT, PVDF and composites of PZT and PVDF, which are positioned to become a preferred solution for many applications. Also, this study will benefit users of piezoelectric-operated actuators and motors that deal with actuators where electromagnetic field generation is an issue and operational performance parameters and space are important considerations, such as in autofocus lens mechanisms of camera phones, nano-metrology, precision linear/rotary drives, drug delivery systems, antenna array deployment, and other fields.

New applications are emerging for piezoelectric devices, which include actuators, ultrasonic motors, sensor arrays for structural health monitoring, transformers and micro energy harvesting devices which are an alternative to batteries in microwatt devices. Other new applications include high resolution ultrasonic medical imaging, computer disk drives, and accelerometers in mobile phones and notebooks.

Unlike other piezo devices, commercialization of piezoelectric-operated actuators and motors is likely to proceed in those markets where the specific advantages of high torque, high precision and lack of magnetic interference are particularly useful. When the costs can be lowered to competitive levels, and remaining technical problems such as frictional wear can be solved, piezoelectric motors may also become candidates in areas such as automotive accessories, where very high volume markets are possible.

The global market for traditional piezoelectric devices is quite mature in applications. However, the global market for new piezoelectric devices will see a robust two-digit growth rate in next five years.

Summary Table A shows the market in current US dollars for the four broad product types of piezoelectric devices for the years 2007 and 2012. The table also shows the market share for each segment. The two summary figures depict market shares for 2007 and 2012 respectively.

Major findings of this report are:

- The global market for the existing eleven generic types of piezoelectric devices now equals US $10.6 billion and is expected to reach $19.5 billion by 2012.
- Among the nine markets, information technology/robots (31.7%), is the clear leader, followed by semiconductor manufacturing and precision machines (18.6%), sonar (12.5%), and bio/medical (11.1%). Other sectors that complete the total market are ecology and energy harvesting (7%), accelerators and sensors (5.8%), non-destructive testing (5.7%) and miscellaneous (gas igniters, piezo printing heads, telecommunication devices) (4.5%), acoustic devices and resonators(3.1%)
- New devices such as piezoelectric generators will see the highest growth rate, estimated to be 51.5 per cent annually. This category is followed by ceramic resonators (27.5%); miscellaneous applications (high voltage devices-gas igniters, piezoelectric elements in laser mirror alignments, acousto-optic modulators, piezoelectric drivers and piezoelectric amplifiers (19.8%); accelerators and sensors (14%); ultrasonic motors (13.4%); and transformers(13.6%). Ultrasonic motors will see 14.6 per cent growth over five years.
- Traditional devices also will see growth, including acoustic devices (13.6%), actuators for computer disk drives (11.6%); Langevin actuators for welding and cleaning (14.6%); sonars (6.75%), transducers (14%); and gas igniters, piezo printing heads, diesel injectors, and piezo amplifiers (19.8%).
- In 2007, Japan has the highest market share of 26 per cent, followed by Europe with 24 per cent, china with 18 per cent, North America with 14 per cent, Korea with 8 per cent and the rest of the world with 10 per cent.

By 2012, china will occupy second position ahead of Europe, with a 22 per cent share of the global market.

11.6 Laser Fabrication of Polymer Micro-Fluidic Devices and Systems

Micro-fluidic devices and systems have many applications in clinical diagnostics, drug delivery, and micro-electronics cooling. So far, these micro-fluidic devices and systems are primarily fabricated using the well-established photo-lithography processes. In this work, laser-based techniques for fabricating polymer micro-fluidic devices and systems are studied. Polymer has many advantages as a micro-fluidic base material due to its low cost and various material properties suitable for micro-fluidic applications. In this work, basic characteristics of laser ablation for polymers are investigated. An ultraviolet laser is used for producing micro-structures on polymers. The ablation characteristics such as ablation depth, wall angle, and topography of the ablated surfaces are investigated. Two different laser fabrication techniques, direct laser writing and mask patterning, are proposed for machining various micro-structures. In order to assemble polymer micro-structures, special bonding processes such as silicon-on-glass bonding and laser bonding are developed and analyzed. The silicon-on-glass bonding provides a very strong and thin bonding layer. Laser bonding can be used to bond a transparent polymer with an opaque polymer and by controlling a laser power, precise local bonding can be achieved. A number of polymer devices such as micro-valves and diffuser micro-pumps are fabricated using the laser techniques. The polymer micro-check valve shows a great potential in its applications. A polymer diffuser micro-pump shows a flow rate up to 50 mm3/min at a frequency of 180 Hz. Finally, a micro-stamping process for rapid replication and mass production of micro-devices and systems is developed. New glass-ceramic stamps are developed for this process and the characteristics of these stamps are investigated. The fabricated glass stamps are then used in a hot-stamping process to replicate micro-structures. Many micro-structures such as micro-channels, micro-capillaries, and micro-diffusers are produced with this process. Overall, it has been found that the laser-based techniques are powerful and flexible, and are able to produce many structures that are difficult to make using photo-lithography.

Bibliography

"ATSDR Study Finds Dioxin Levels in Calcasieu Parish Residents Similar to National Levels," available at: http://www.atsdr.cdc.gov/NEWS/calcasieula031506.html; "ATSDR Study Finds Dioxin Levels Among Lafayette Parish Residents Similar to National Levels," available at: http://www.atsdr.cdc.gov/NEWS/lafayettela031606.html; ATSDR Report: Serum Dioxin Levels In Residents Of Calcasieu Parish, Louisiana, October 2005, Publication Number PB2006-100561, available from the National Technical Information Services, Springfield, Virginia, phone: 1-800-553-6847/1-703-605-6244

"Building quality with Air Products trimerisation catalysts" (pdf) (2003).

"Business Gives Styrofoam a Rare Redemption.", Stockton Record (21 September 2007).

"Calcasieu Cancer Rates Similar to State/National Averages." News Release, State of Louisiana Dept. of Health and Hospitals. January 17, 2002

"*Highly Efficient Triplet Chain Isomerization of Dewar Benzenes: Adiabatic Rate Constants from Cage Kinetics*." Merkel, P. B.; Roh, Y.; Dinnocenzo, J. P.; Robello, D. R.; Farid, S. J. Phys. Chem. A 2007, 111(7), 1188-1199.

"Indy-Car Upgrades Planned For 2008, '09", AutoWeek.com (2007-09-07).

"Jeffcat Amine Catalysts for the Polyurethane Industry" (pdf) (2006).

"New Twist on Green: 2008 Ford Mustang Seats Will Be Soy-Based Foam", Edmunds inside line (July 12, 2007).

"*Polymers for Photothermography: Controlled Thermal Release of Color Photographic Developer from Substituted Polystyrene Derivatives*." Robello, D. R.; Levy, D. H.; Reynolds, J. H.; Southby, D. T.; Twist, S. L. Macromolecules 2006, 39(17), 5686-5695.

"*Quantum Amplified Isomerization: A New Concept for Polymeric Optical Materials.*" Gillmore, J. G.; Neiser, J. D.; McManus, K. A.; Roh, Y.; Dombrowski, G. W.; Brown, T. G.; Dinnocenzo, J. P.; Farid, S.; Robello, D. R. Macromolecules 2005, 38, 7684-7694.

"*Refractive index imaging via a chemically amplified process in a solid polymeric medium.*" Robello, D. R.; Farid, S. Y.; Dinnocenzo, J. P.; Brown, T. G. SPIE Proceedings, 2006, 6117(Organic Photonic Materials and Devices VIII), 61170F/1-61170F/8.

"*Spontaneous Formation of an Exfoliated Polystyrene-Clay Nanocomposite Using a Star-Shaped Polymer.*" Robello, D. R.; Yamaguchi, N.; Blanton, T.; Barnes, C. J. Am. Chem. Soc. 2004, 126, 8118.

"*Synthesis and Characterization of Star Polymers Made from Simple, Multifunctional Initiators.*" Robello, D. R.; André, A.; McCovick, T. A.; Kraus, A.; Mourey, T. H., Macromolecules 2002, 35, 9334.

"Tattoo to monitor diabetes", BBC News (1 September 2002).

"Technical data sheet from Dow Chemical".

"The Socio-Economic Impact of Polyurethanes in the United States from the American Chemistry Council". The Polyurethanes Recycle and Recovery Council (PURRC), a committee of the Center for the Polyurethanes Industry (February 2004).

A Guide To Glycols, The Dow Chemical Company, 1992

A.J. Epstein and Y.Z. Wang, *Interface Control of Polymer and Oligomer-Based Light Emitting Devices*, **Semiconductive Polymers, ACS Symposium Series 735**, edited by Y. Wei and B. Hsieh (Oxford University Press, New York, 1999), Chapter 8,119-133 (1999).

Ashby, Michael and Jones, David. Engineering Materials. p. 191-195. Oxford: Butterworth-Heinermann, 1996. Ed. 2.

Askeland, Donald R.; Pradeep P. Phulé (2005). *The Science & Engineering of Materials*, 5th edition, Thomson-Engineering.

Beychok, M.R., *A data base of dioxin and furan emissions from municipal refuse incinerators*, Atmospheric Environment, Elsevier B.V., January 1987

Blackwell, J.; M.R. Nagarajan and T.B. Hoitink (1981). "The Structure of the Hard Segments in MDI/diol/PTMA Polyurethane Elastomers". Washington, D.C.: American Chemical Society. 0097-6156/81/0172-0179.

Blackwell, John; Kenncorwin H. Gardner (1979). "Structure of the hard segments in polyurethane elastomers". IPC Business Press. 0032-3861/79/010013-05.

Blair, G. Ron; Bob Dawe,Jim McEvoy, Roy Pask, Marcela Rusan de Priamus, Carol Wright (2007). "The Effect of Visible Light on the Variability of Flexible Foam Compression Sets"., Orlando, Florida: Center for the Polyurethane Industry.

Bornehag *et al.* (2004). "*The Association Between Asthma and Allergic Symptoms in Children and Phthalates in House Dust: A Nested Case-Control Study*". *Environmental Health Perspectives* **112** (14): 1393–1397.

Brandrup, J.; Immergut, E.H.; Grulke, E.A.; *eds* Polymer Handbook 4th Ed. New York: Wiley-Interscience, 1999.

Budinski, Kenneth G., and Michael K. Budinski. *Engineering Materials: Properties and Selection*, 7th Ed, 2002.

Business Wire (Nov 2005). "CHW Switches to PVC/DEHP-Free Products to Improve Patient Safety and Protect the Environment". *Business Wire.*

Callister, Jr., William D. (2000). *Materials Science and Engineering - An Introduction*, 5th edition, John Wiley and Sons.

Clayden, J., Greeves, N. *et al.* (2000), p1450-1466.

Clayden, J., Greeves, N. *et al.* (2000). "*Organic chemistry*" Oxford.

Costner, Pat, (2005), "Estimating Releases and Prioritizing Sources in the Context of the Stockholm Convention", International POPs Elimination Network, Mexico.

Creech and Johnson (Mar 1974). "*Angiosarcoma of liver in the manufacture of polyvinyl chloride.*". *Journal of occupational medicine: official publication of the Industrial Medical Association.* **16** (3): 150–1.

D Hosler, SL Burkett and MJ Tarkanian *Science* **284**, 1988 (1999).

D. E. Corpet, G. Parnaud, M. Delverdier, G. Peiffer and S. Tache (2000). "Consistent and Fast Inhibition of Colon Carcinogenesis by Polyethylene Glycol in Mice and Rats Given Various Carcinogens". *Cancer Res* **60** (12): 3160–3164. PMID 10866305.

D. Tanner, J. A. Fitzgerald, B. R. Phillips (1989). "The Kevlar Story - an Advanced Materials Case Study". *Angewandte Chemie International Edition in English* **28** (5): 649–654.

David Parker, Jan Bussink, Hendrik T. van de Grampel, Gary W. Wheatley, Ernst-Ulrich Dorf, Edgar Ostlinning, Klaus Reinking, "Polymers, High-Temperature" in Ullmann's Encyclopedia of Industrial Chemistry 2002, Wiley-VCH: Weinheim.

Di Palma JA *et al.* Am J Gastroenterol 2007 Sep 102:1964

Doetze J. Sikkema (2002). "Manmade fibers one hundred years: Polymers and polymer design". *J Appl Polym Sci, John Wiley & Sons, Inc.* (83): 484–488.

E. E. Magat (1980). "Fibers from Extended Chain Aromatic Polyamides, New Fibers and Their Composites". *Philosophical Transactions of the Royal Society of London Series A* **294** (1411): 463–472.

E. L. Nix and I. M. Ward, (1986) Ferroelectrics 67, 137.

E. L. Nix and I. M. Ward, (1986). Ferroelectrics 67, 137.

Eberhart, Mark (2003). *Why Things Break: Understanding the World by the Way It Comes Apart*. Harmony.

Eckhardt, Angela (November, 1998). "Paper Waste: Why Portland's Ban on Polystyrene Foam Products Has Been a Costly Failure". Cascade Policy Institute.

Ellis, D.A.; Mabury, S.A.; Martin, J.W.; Muir, D.C.G. (2001). "Thermolysis of fluoropolymers as a potential source of halogenated organic acids in the environment". *Nature* **412** (6844): 321–324.

Epidemiologic Notes and Reports Angiosarcoma of the Liver Among Polyvinyl Chloride Workers – Kentucky, Centers for Disease Control and Prevention Web site. 1997. Available at: http://www.cdc.gov/mmwr/preview/mmwrhtml/00046136.htm

Feske, Bert (October 2004). "The Use of Saytex RB-9130/9170 Low Viscosity Brominated Flame Retardant Polyols in HFC-245fa and High Water Formulations"., Las Vegas, NV: Alliance for the Polyurethane Industry Technical Conference.

FOMREZ Specialty Tin Catalysts for Polyurethane Applications (leaflet insert), Crompton Corporation, 2001-01

FOMREZ Specialty Tin Catalysts for Polyurethane Applications, Crompton Corporation, 2001-01

For example, in the online catalog[1] of Scientific Polymer Products, Inc., poly(ethylene glycol) molecular weights run up to about 20,000, while those of poly(ethylene oxide) have 6 or 7 digits.

G. Bittner el. al. (2005). "Melatonin enhances the in vitro and in vivo repair of severed rat sciatic axons". *Neouroscience Letters* **376**: 98–101..

Galloway, F.M. *et al* (1992) "Surface parameters from small-scale experiments used for measuring HCl transport and decay in fire atmospheres", *Fire Mater.*, 15:181-189

Gaskell, David R. (1995). *Introduction to the Thermodynamics of Materials*, 4th edition, Taylor and Francis Publishing.

Gordon, James Edward (1984). *The New Science of Strong Materials or Why You Don't Fall Through the Floor*, eissue edition, Princeton University Press.

Grillo, D.J.; Housel, T.L. (1992). "Physical Properties of Polyurethanes from Polyesters and Other Polyols". *Polyurethanes '92 Conference Proceedings*, New Orleans, LA: The Society of the Plastics Industry, Inc..

Gum, Wilson; Riese, Wolfram; Ulrich, Henri (1992). *Reaction Polymers*. New York: Oxford University Press.

Gum, Wilson; Riese, Wolfram; Ulrich, Henri (1992). *Reaction Polymers*. New York: Oxford University Press.

H. Yamakawa, "Helical Wormlike Chains in Polymer Solution", (Springer Verlag, Berlin, 1997).

Harper C.A., Handbook of plastic and elastomers, McGraw-Hill, New York, 1975, pp. 1-3,1-62, 2-42, 3-1

Harrington, Ron; Hock, Kathy (1991). *Flexible Polyurethane Foams*. Midland: The Dow Chemical Company.

History of Nylon US Patent 2,130,523 'Linear polyamides suitable for spinning into strong pliable fibers', U.S. Patent 2,130,947 'Diamine dicarboxylic acid salt' and U.S. Patent 2,130,948 'Synthetic fibers', all issued 20 September 1938

Howdeshell, KL; Peterman PH, Judy BM, Taylor JA, Orazio CE, Ruhlen RL, Vom Saal FS, Welshons WV (Jul 2003). "Bisphenol A is released from used polycarbonate animal cages into water at room temperature". *Environmental Health Perspectives* **111** (9): 1180–7.

Hunt, PA; Kara E. Koehler, Martha Susiarjo, Craig A. Hodges, Arlene Ilagan, Robert C. Voigt, Sally Thomas, Brian F. Thomas and Terry J. Hassold (April 2003). "Bisphenol A Exposure Causes Meiotic Aneuploidy in the Female Mouse". *Current Biology* **13** (7): 546–553.

International Fiber Journal (2007). "Teijin Launches Fourth Production Expansion in Six Years". *Fiber Journal* (February): 20.

International Union of Pure and Applied Chemistry, *et al.* (2000) "*IUPAC Gold Book*".

Introduction to Polymers 1987 R.J. Young Chapman & Hall.

IUPAC. "Glossary of Basic Terms in Polymer Science". Pure Appl. Chem. 1996, 68, 2287-2311.

IUPAC. "Glossary of Basic Terms in Polymer Science". Pure Appl. Chem. 1996, 68, 2287-2311

IUPAC. "Nomenclature of Regular Single-Strand Organic Polymers". Pure Appl. Chem. 1976, 48, 373-385.

J. Chiefari, Y. K. Chong, F. Ercole, J. Krstina, J. Jeffery, T. P. T. Le, R. T. A. Mayadunne, G. F. Meijs, C. L. Moad, G. Moad, E. Rizzardo, S. H. Thang, *Macromolecules* **1998**, *31(16)*, 5559–5562.

J. KAHOVEC, R. B. FOX and K. HATADA; "Nomenclature of regular single-strand organic polymers (IUPAC Recommendations 2002)"; Pure and Applied Chemistry; IUPAC; 2002; 74 (10): pp. 1921–1956.

Jim Morris, "In Strictest Confidence. The chemical industry's secrets," Houston Chronicle. Part One: "Toxic Secrecy," June 28, 1998, pgs. 1A, 24A-27A; Part Two: "High-Level Crime," June 29, 1998, pgs. 1,A, 8A, 9A; and Part Three: "Bane on the Bayou," July 26, 1998, pgs. 1A, 16A.]

JWS Hearle. *High-performance fibres*. Woodhead Publishing Ltd., Abington, UK - The Textile Institute.

K. Tashiro, M. Kobayashi, H. Tadokoro, and E. Fukada, (1980). Macromolecules 13, 691

K. Tashiro, M. Kobayashi, H. Tadokoro, and E. Fukada, (1980). Macromolecules 13, 691

Kadolph, Sara J. Anna L. Langford. Textiles, Ninth Edition. Pearson Education, Inc 2002. Upper Saddle River, NJ.

Katami, Takeo, *et al.* (2002) "Formation of PCDDs, PCDFs, and Coplanar PCBs from Polyvinyl Chloride during Combustion in an Incinerator" Environ. Sci. Technol., 36, 1320-1324. and Wagner, J., Green, A. 1993. Correlation of chlorinated organic compound emissions from incineration with chlorinated organic input. Chemosphere 26 (11): 2039-2054. and Thornton, Joe (2002) "Environmental Impacts of polyvinyl Chloride Building Materials", Healthy Building Network, Washington, DC.

Kaushiva, Byran D. (August 15, 1999). "*Structure-Property Relationships of Flexible Polyurethane Foams*". PhD Thesis. Virginia Polytechnic Institute.

Kawai, H., (1969). *Jpn. J. Appl. Phys*, 8, p975.

Kawai, H., (1969). *Jpn. J. Appl. Phys*, 8, p975.

Kenji, O., Hiroji, O., Keiko, K., (1996). J. Appl. Phys. 81, 2760.

Kenji, O., Hiroji, O., Keiko, K., (1996). J. Appl. Phys. 81, 2760.

Kleinert, Hagen, *Path Integrals in Quantum Mechanics, Statistics, Polymer Physics, and Financial Markets*, 4th edition, World Scientific (Singapore, 2004).

Koehler, KE; Robert C. Voigt, Sally Thomas, Bruce Lamb, Cheryl Urban, Terry Hassold, and Patricia A. Hunt (April 2003). "When disaster strikes: rethinking caging materials". *Lab Animal* **32** (4): 24–27.

Kuraray *via* K-Online.de (2007-06-22). "Kuraray Expands VECTRAN Superfiber Manufacturing".

Kyung-Jun Chu. Eur. Polym. J. Vol. 34, No. 3/4, pp. 577-580, 1998

L. Vollbracht and T.J. Veerman, US Patent 4308374 (1976).

Lars-Åke Kvarning, Bengt Ohrelius (1998), *The Vasa - The Royal Ship*, pp. 133-141

Lee Bowman (4 December 2004). "Study on dogs yields hope in human paralysis treatment", seattlepi.com.

Lewis, Peter Rhys, Reynolds, K and Gagg, C, *Forensic Materials Engineering: Case Studies*, CRC Press (2003).

Meyers and Chawla. Mechanical Behavior of Materials. pg. 41. Prentice Hall, Inc. 1999.

Michael Rubinstein and Ralph H. Colby, "Polymer Physics", Oxford University Press, 2003.

Modern Plastic Mid-October Encyclopedia Issue, Polyimide, thermoset, page 146.

Musselman, S.G.; Santosusso, T.M. and Sperling, L.H. (1998). "Structure Versus Performance Properties of Cast Elastomers". *Polyurethanes '98 Conference Proceedings*, Dallas, TX: The Society of the Plastics Industry, Inc..

National Emission Standards for Hazardous Air Pollutants (NESHAP) for Vinyl Chloride Subpart F, OMB Control Number 2060-0071, EPA ICR Number 0186.09 (Federal Register: September 25, 2001 (Volume 66, Number 186))

National Renewable Energy Laboratory, "Polyvinyl Chloride Plastics in Municipal Solid Waste Combustion," NREL/TP-430- 5518, Golden CO, April 1993

Newman, C.R.; Forciniti, D. (2001). "Modeling the Ultraviolet Photodegradation of Rigid Polyurethane Foams". *Ind. Eng. Chem. Res.* **40**: 3336–3352.

Niemeyer, Timothy; Patel, Munjal and Geiger, Eric (September, 2006). "A Further Examination of Soy-Based Polyols in Polyurethane Systems"., Salt Lake City, UT: Alliance for the Polyurethane Industry Technical Conference.

Oertel, Gunter (1985). *Polyurethane Handbook*. New York: Macmillen Publishing Co., Inc.

Oertel, Gunter (1985). *Polyurethane Handbook*. New York: Macmillen Publishing Co., Inc..

P. Mercandelli *et al.* / Journal of Organometallic Chemistry 692 (2007) 4784–4791

Painter, P and Coleman, M. "Fundamentals of Polymer Science". 1997, 96-100.

Pat Costner etal, "PVC: A Primary Contributor to the U.S. Dioxin Burden; Comments submitted to the U.S. EPA Dioxin Reassessment," (Washington, D.C. Greenpeace U.S.A., February 1995

Polyurethanes: A Class of Modern Versatile Materials Raymond B. Seymour George B. Kauffman J. Chem. Ed. 69, 909 **1992**

R. B. Borgens and D. Bohnert (2001). "Rapid recovery from spinal cord injury after subcutaneously administered polyethylene glycol". *Journal of Neuroscience Research* **66** (6): 1179–1186.

R. Kleinschmidt *et al.*rJournal of Molecular Catalysis A: Chemical 157(2000)83–90

R. Sun, Y. Wang, X. Zou, M. Fahlman, Q. Zheng, T. Kobayashi, T. Masuda, and A.J. Epstein, *Electroluminescence of Carbazole Substituted Polyacetylenes*, **Proceedings, SPIE Conference on Optical and Photonic Applications of Electroactive and Conducting Polymers**, San Diego, CA, 19-24 July 1998, Vol 3476, pp. 332-337 (1998).

R.G. Sun, Y.Z. Wang, D.K. Wang, Q.B.Zheng, E.M. Kyllo, T.L. Gustafson, and A.J. Epstein, *High Luminescent Efficiency In Light-Emitting Polymers Due To Effective Exciton Confinement*, **Applied Physics Letters 76**, 634-636 (2000).

R.G. Sun, Y.Z. Wang, Q.B. Zheng, H.J. Zhang, and A.J. Epstein, *1.54 µm Infrared Photoluminescence and Electroluminescence from an Erbium Organic Compound*, **Journal of Applied Physics 87**, 7589-7591 (2000).

R.J. Young and P.A. Lovell, Introduction to Polymers, 1991.

R.J. Young and P.A. Lovell, Introduction to Polymers, 1991.

Randall, David; Lee, Steve (2002). *The Polyurethanes Book*. New York: Wiley.

Rigo, H.G., Chandler, A. J., and Lanier, W.S., "The Relationship between Chlorine in Waste Streams and Dioxin Emissions from Waste Combustor Stacks," American Society of Mechanical Engineers Report CRTD, Vol 36, New York 1995

Ronald V. Joven. Manufacturing Kevlar panels by thermo-curing process. Los Andes University, 2007. Bogotá, Colombia.

Song *et al.* Macromol. Symp. 2004, 213, 173-185

Steiglitz, L., and Vogg, H., "Formation Decomposition of Polychlorodibenzodioxins and Furans in Municipal Waste," Report KFK4379, Laboratorium fur Isotopentechnik, Institut for Heize Chemi, Kerforschungszentrum Karlsruhe, Feb 1988.

Stepto, R. *et al.*, *Pure Appl. Chem.*, **75**, 1359 (2003).

T. L. Krause and G. D. Bittner (1990). "Rapid Morphological Fusion of Severed Myelinated Axons by Polyethylene Glycol". *PNAS* **87** (4): 1471–1475.

Tasaka, S., Miyata, S., (1981). *Ferroelectrics*, 32 (1), 17-23.

Tasaka, S., Miyata, S., (1981). *Ferroelectrics*, 32 (1), 17-23.

Textiles by Sara J. Kadolph.

This week 50 years ago in New Scientist, 28 April, 2007, p. 15

Thomas, Robert A. (March 8, 2005). "Where Might We Look for Environmental Heroes?". Center for Environmental Communications, Loyola University, New Orleans.

Tonya Johnson (21 April 2004). "Army Scientists, Engineers develop Liquid Body Armor".

Ulrich, Henri (1996). *Chemistry and Technology of Isocyanates*. New York: John Wiley & Sons, Inc.

US EPA, The Inventory of Sources and Environmental Releases of Dioxin-Like Compounds in the United States: The Year 2002 Update, May 2007.

Valentine, C; Craig, T.A.; Hager, S.L (1993). "Inhibition of the Discoloration of Polyurethane Foam Caused by Ultraviolet Light". *J. Cellular Plastics* **29**: 569–590.

Victor O. Sheftel (2000). *Indirect Food Additives and Polymers: Migration and Toxicology*. CRC, 1114-1116.

Vinogradov, A., Schwartz, M. (Ed.), (2002). "Piezoelectricity in Polymers", *Encyclopedia of Smart Materials, Volumes 1-2*, John Wiley & Sons, 780-792.

Vinogradov, A., Schwartz, M. (Ed.), (2002). "Piezoelectricity in Polymers", *Encyclopedia of Smart Materials, Volumes 1-2*, John Wiley & Sons, 780-792.

Vom Saal, FS; Hughes C (Aug 2005). "An extensive new literature concerning low-dose effects of bisphenol A shows the need for a new risk assessment". *Environmental Health Perspectives* **113** (8): 926–33. PMID 16079060.

Walker, Peter (Ed), (1993) *Chambers Dictionary of Materials Science and Technology*, Chambers Publishing.

Woods, George (1990). *The ICI Polyurethanes Book*. New York: John Wiley & Sons, Inc..

Y.Z. Wang and A.J. Epstein, *Interface Control of Light-emitting Devices Based on Pyridine-containing Conjugated Polymers*, **Accounts of Chemical Research 32**, 217-224 (1999).

Y.Z. Wang, D.D. Gebler, D.K. Fu, T.M. Swager, A.G. MacDiarmid, and A.J. Epstein, *Novel Light-Emitting Devices Based on Pyridine-Containing Conjugated Polymers*, **IEEE Transactions on Electron Devices 44 (8)**, 1263-1268 (1997).

Y.Z. Wang, D.D. Gebler, D.-K. Fu, T.M. Swager, and A.J. Epstein, *New Approach to Color Variable Light-Emitting Devices Based on Conjugated Polymers*, **Proceedings, International Society of Optical Engineers on Optical Probes of Conjugated Polymers**, vol. 3148, 27 July-1 August 1997, San Diego, CA,117-123 (1998).

Y.Z. Wang, D.D. Gebler, D.K. Fu, T.M. Swager, and A.J. Epstein, *Color Variable Bipolar/AC Light-Emitting Devices Based on Conjugated Polymers*, **Applied Physics Letters 70 (24)**, 3215-3217 (1997).

Y.Z. Wang, R.G. Sun, D.K. Wang, T.M. Swager and A.J. Epstein, *Polarity and Voltage Controlled Color Variable Light Emitting Devices Based on Conjugated Polymers*, **Applied Physics Letters 74**, 2593-2595 (1999).

Y.Z. Wang, R.G. Sun, F. Meghdadi, G. Leising, and A.J. Epstein, *Multicolor Multilayer Light Emitting Devices Based on Pyridine-Containing Conjugated Polymers and Para-Sexiphenyl Oligomer*, **Applied Physics Letters 74**, 3613-3615 (1999).

Y.Z. Wang, R.G. Sun, F. Meghdadi, G. Leising, T.M. Swager, and A.J. Epstein, *Color Variable Multilayer Light Emitting Devices Based on Conjugated Polymers and Oligomers*, **Synthetic Metals 102**, 889-892 (1999).

Year of Twaron. "Twaron - A history of innovation". *Twaron News* (June 2007): 10–11.

Zhang, Q. M., Bharti, V., Kavarnos, G., Schwartz, M. (Ed.), (2002). "Poly (Vinylidene Fluoride) (PVDF) and its Copolymers", *Encyclopedia of Smart Materials, Volumes 1-2*, John Wiley & Sons, 807-825.

Index